# Modern Mathematics

# Modern Mathematics

Edited by
Frank West

Willford Press
www.willfordpress.com

Published by Willford Press,
118-35 Queens Blvd., Suite 400,
Forest Hills, NY 11375, USA



ISBN: 978-1-68285-449-5

**Cataloging-in-Publication Data**

Modern mathematics / edited by Frank West.
p. cm.
Includes bibliographical references and index.
ISBN 978-1-68285-449-5
1. Mathematics. I. West, Frank.
QA7 .M63 2018
510--dc23

For information on all Willford Press publications
visit our website at www.willfordpress.com

Willford Press

# Contents

# Preface

Mathematics plays a central role in our life. It studies themes like space, quantity, structure, etc. Mathematics is applied in various fields like engineering, finance, medicine, natural science and social sciences. Mathematics can be divided into two distinct branches namely, pure mathematics and applied mathematics. This book provides significant information of this discipline to help develop a good understanding of mathematics and related fields. This book, with its detailed analyses and data, will prove immensely beneficial to professionals and students involved in this area at various levels.

The researches compiled throughout the book are authentic and of high quality, combining several disciplines and from very diverse regions from around the world. Drawing on the contributions of many researchers from diverse countries, the book's objective is to provide the readers with the latest achievements in the area of research. This book will surely be a source of knowledge to all interested and researching the field.

In the end, I would like to express my deep sense of gratitude to all the authors for meeting the set deadlines in completing and submitting their research chapters. I would also like to thank the publisher for the support offered to us throughout the course of the book. Finally, I extend my sincere thanks to my family for being a constant source of inspiration and encouragement.

**Editor**

# On a Non Logsymplectic Logarithmic Poisson Structure with Poisson Cohomology Isomorphic to the Associated Logarithmic Poisson Cohomology

Joseph Dongho[1]

[1] Department of Mathematics and Computer Science, University of Maroua, Cameroon

Correspondence: Joseph Dongho, Department of Mathematics and Computer Science, University of Maroua, Cameroon. E-mail: josephdongho@yahoo.fr

**Abstract**

The main purpose of this article is to show that there are non logsymplectic Poisson structures whose Poisson cohomology groups are isomorphic to corresponding logarithmic Poisson cohomology groups.

**Keywords:** log-symplectic structure, poisson cohomology, log-poisson cohomology, poisson structures

## 1. Introduction

Symplectic geometry was discoved in 1780 by Joseph Louis Lagrange when he considered the non constants variable and defined the bracket of two such elements. From symplectic manifold, Poisson defined his brackets as tool for classical dynamics. Charles Gustave Jacobi realized the importance of those bracket and elucidated their algebraic properties. Sophus Lie and others authors began the study of their geometry. Connection of poisson geometry with numbers of areas including harmonic analytic, mechanics of particles and continua; completely integrable systems, justify this recent development. It is interested to recall that number of proprieties and results in this theory was developed in the case of differential manifold. Too few authors have worked in the case of singular varieties. J. Huebschmann in (Huebschmann, J., 1990) study in 1990 Poisson algebra and apply its Lie-Rinehart cohomology in the study of their geometric quantization. A. Polishchuk in (Pichereau, A., 2006) study in 1997 the Poisson brackets in algebraic framework.

In 2002 Ryushi Goto (Goto, R., 2002), with the aim of generalizing the approach of the symplectic, Atiyah class to the construction of the invariants of knots, defined the logsymplectic manifold and study several examples. The notion of logsymplectic manifold is based on the theory of logarithmic differential forms extensively study in (Saito, K., 1980). Logsymplectic manifold is simply a complex manifold $X$ equipped with a symplectic form $\omega$ that has simple poles along a hypersurface $D \subset X$. In other words, Poisson structures defined on $X - D$ by any logsymplectic form $\omega$ extends to a Poisson bracket on all $X$ whose pfaffian in a reduced defining equation for $D$. Logsymplectic manifolds can arise when one attempts to compactify symplectic manifolds. Many modulis space in algebraic geometry and gauge theory come equipped with logsymplectic structure. Such Poisson structure can then play an important role in geometric quantization of many classical observable. According to I. Vaisman in (Vaisman, I., 1991), obstruction of quantization of such classical space is measure by Poisson cohomology. But it style very difficult to determine explicit form of Poisson cohomology as we can see in (Pichereau, A., 2006) and (Monnier, Ph., 2002).

In other to propose an alternative method in the computation of such Poisson cohomology, the first author introduce in (Dongho, J., 2012) the notion of logarithmic principal Poisson structure and prove that such Poisson structure induced a Lie-Rinehart structure on the module of logarithmic differential form along a finite generated ideal $\mathcal{I}$, from which he introduce the notion of logarithmic Poisson cohomology, and prove that such logarithmic Poisson cohomology are in general different to the associated Poisson cohomology. It was also prove in (Dongho, J., 2012) that when logarithmic Poisson structure are logsymplectic one, the two, Poisson cohomology and logarithmic Poisson cohomology are equivalent. The main objectif of this paper is to prove that there are non logsymplectic Poisson structure with isomorphic Poisson and logarithmic Poisson cohomology.

Recently in (Dongho, J. & Yotcha, S. R., 2016), the Differential Point of view of such cohomology has been study and and apply in the prequantization of such Poisson manifold.

More general theory of logarithmic Poisson cohomology and logarithmic Poisson algebra is in preparation in (Dongho, J., et al.).

The main results of this paper are:

**Proposition 1** *The Poisson cohomology groups of the Poisson algebra* $(\mathcal{A} = C[x,y], \{x,y\} = x^n), (n \in N^*)$ *are*

$$H_P^0 \cong C, \quad H_P^1 \cong C, \quad H_P^k \cong 0 \; \forall k \geq 2 \qquad when \quad n = 1$$

*and for all* $n \geq 2$,

$$\begin{aligned} H_P^0(x^n) &\cong C \\ H_P^1(x^n) &\cong C_{n-1}[x] \oplus C[y] \oplus xC[y] \oplus \cdots \oplus x^{n-2}C[y] \\ H_P^2(x^n) &\cong C[y] \oplus xC[y] \oplus \cdots \oplus x^{n-2}C[y] \\ H_P^k(x^n) &\cong 0 \; \forall k \geq 3 \end{aligned}$$

and

**Proposition 2** *The logarithmic Poisson cohomology groups of the Poisson structure defined by the logarithmic Poisson 2-form* $\pi = x^n \partial x \wedge \partial y$*, which is logarithmic along the ideal* $I = x^n C[x,y]$ *are*

$$\begin{aligned} H^n_{PS}(x^n) &\simeq 0 \qquad for \qquad n \geq 3 \\ H^2_{PS}(x^n) &\simeq \bigoplus_{i=0}^{n-2} x^i C[y] \\ H^1_{PS}(x^n) &\simeq C_{n-1}[x] \oplus \bigoplus_{i=0}^{n-2} x^i C[y] \end{aligned}$$

Those can be generalize to the case of any algebra with two generator over a non zero characteristic ring.

## 2. $A$-module of Differential Form

It follows that, $A = C[x,y]$, $\Omega_A$ is the $A$-module of differential form on $A$ and $Der_A$ the $A$-module on derivations of $A$. Then, $\Omega_A = \langle dx, dy\rangle_A$ and $Der_A = \langle \partial x, \partial y\rangle_A$.

For any $k \in N^*$, $Alt^k(\Omega_A, A)$ denote the $A$-module of $k$-multilinear skew symmetric form on $\Omega_A$. By convention, $Alt^0(\Omega_A, A) = A$ and then $Alt^k(\Omega_A, A) \cong \wedge^k Der_A$. Then,

$$\begin{aligned} Alt^1(\Omega_A, A) &\cong Der_A \cong A \times A \\ Alt^2(\Omega_A, A) &\cong Der_A \wedge Der_A = \langle \partial x \wedge \partial y\rangle_A \cong A \\ Alt^k(\Omega_A, A) &\cong 0 \quad \text{pour tout } k \geq 3 \end{aligned}$$

We deduce the following cochain complex

$$0 \longrightarrow A \xrightarrow{d^0} A \times A \xrightarrow{d^1} A \xrightarrow{d^2} 0$$

where $d^i$, $i = 0, 1, 2$ are associated Poisson differential and there are defined by

$$\begin{aligned} d^0\varphi(\alpha) &= H(\alpha)\varphi \quad \text{for} \quad \varphi \in A \\ d^1\varphi(\alpha_1, \alpha_2) &= H(\alpha_1)\varphi(\alpha_2) - H(\alpha_2)\varphi(\alpha_1) - \varphi([\alpha_1, \alpha_2]) \quad \text{for} \quad \varphi \in Alt^1(\Omega_A, A), \alpha_1, \alpha_2 \in \Omega_A \\ d^k &= 0 \quad \text{for every} \quad k \geq 2 \end{aligned}$$

$H : \Omega_A \longrightarrow Der_A$ is the Hamiltonian map defined by $H(da) = \{a, -\}$. It induce on $\Omega_A$ a bracket $[,]$ defined by $[da, db] = d\{a, b\}$. In particular, $(\Omega_A, [,], H)$ is a Lie-Rinehart-Poisson algebra. The cohomology groups are given by $H_P^k = \ker d^k / Im\, d^{k-1}$ $(k \in N)$. It therefore follows that

$$H_P^k \cong 0 \quad \forall k \geq 3$$

For a better understanding, we address the cases $n = 1$ and $n = 2$ and we end with a generalization.

## 3. The Case $n = 1$

### *3.1 Associated Poisson Differential*

Let $\varphi \in Alt^0(\Omega_A, A) = A$, $d^0\varphi \in Alt^1(\Omega_A, A) \cong Der_A$ that is $d^0\varphi = \varphi_1 \partial x + \varphi_2 \partial y$ with $\varphi_1, \varphi_2 \in A$. But $d^0\varphi(\alpha) = H(\alpha)\varphi$ for $\alpha \in \Omega_A$. Taking successively $\alpha = dx$ and $dy$ we obtain $\varphi_1 = H(dx)\varphi$ and $\varphi_2 = H(dy)\varphi$.

On the other hand, $H(dx) \in Der_A$ i.e; $H(dx) = \alpha_1\partial x + \alpha_2\partial y$, $\alpha_1, \alpha_2 \in A$. So, $H(dx)(x) = \alpha_1 = 0$ and $H(dy)(x) = \alpha_2 = \{x, y\} = x$. Therefore $H(dx) = x\partial y$. Similarly we can show that $H(dy) = -x\partial x$ therefore

$$d^0\varphi = x\frac{\partial\varphi}{\partial y}\partial x - x\frac{\partial\varphi}{\partial x}\partial y \approx (x\frac{\partial\varphi}{\partial y}, -x\frac{\partial\varphi}{\partial x})$$

For $\varphi = \varphi_1\partial x + \varphi_2\partial y \in Alt^1(\Omega_A, A)$, $d^1\varphi \in Alt^2(\Omega_A, A) \cong \wedge^2 Der_A$ i.e;

$d^1\varphi = \psi\partial x \wedge \partial y$, $\psi \in A$. Or $d^1\varphi(\alpha_1, \alpha_2) = H(\alpha_1)\varphi(\alpha_2) - H(\alpha_2)\varphi(\alpha_1) - \varphi([\alpha_1, \alpha_2])$ for $\varphi \in Alt^1(\Omega_A, A), \alpha_1, \alpha_2 \in \Omega_A$. In particular, for $\alpha_1 = dx$ and $\alpha_2 = dy$, we get $\psi = H(dx)\varphi(dy) - H(dy)\varphi(dx) - \varphi([dx, dy]) = x\frac{\partial\varphi_2}{\partial y} + x\frac{\partial\varphi_1}{\partial x} - \varphi_1$. Therefore,

$$d^1\varphi = (x\frac{\partial\varphi_1}{\partial x} + x\frac{\partial\varphi_2}{\partial y} - \varphi_1)\partial x \wedge \partial y$$

*3.2 Calculation of Cohomological Groups*

3.2.1 Expression of $H^0_P(x)$

By definition $H^0_P = \ker d^0 / Im\, d^{-1}$ with $d^{-1} : 0 \longrightarrow A$ i.e; $H^0 = \ker d^0$. Let $\varphi \in A$, $d^0\varphi = 0$ If and only if $x\frac{\partial\varphi}{\partial x} = x\frac{\partial\varphi}{\partial y} = 0$ That is $\varphi \in C$. Therefore,

$$H^0_P \cong C$$

3.2.2 Expression of $H^2_P = \ker d^2 / Im\, d^1$

Obviously, $\ker d^2 = A$ Let $\varphi \in A$, we can write

$$\varphi = -(-\varphi) + x\frac{\partial(-\varphi)}{\partial x} + x\frac{\partial}{\partial y}(\int \frac{\partial\varphi}{\partial x}dy) = d^1(-\varphi, \int \frac{\partial\varphi}{\partial x}dy)$$

$d^1$ is an epimorphism and we have

$$H^2_P \cong 0$$

3.2.3 Expression of $H^1_P(x)$

$(\varphi_1, \varphi_2) \in \ker d^1$ if and only if $\varphi_1 = x(\frac{\partial\varphi_2}{\partial y} + \frac{\partial\varphi_1}{\partial x})$. Thus, we can write $\varphi_1 = xu$, $u \in A$ therefore $\varphi_2 = -x\int \frac{\partial u}{\partial x}dy + b(x)$ with $b(x) \in C[x]$. It was therefore

$$\ker d^1 = \{(xu, -x\int \frac{\partial u}{\partial x}dy) + (0, b(x)); u \in A, b(x) \in C[x]\}$$

Let

$$\begin{array}{rcl} \beta : A & \longrightarrow & xA \times A \\ u & \longmapsto & (xu, -x\int \frac{\partial u}{\partial x}dy) \end{array}$$

$\beta$ is a monomorphism and we have

$$\ker d^1 = \beta(A) \oplus (0 \times C[x]) = \beta(A) \oplus (0 \times xC[x]) \oplus (0 \times C)$$

in other hand, $\beta(A) \oplus (0 \times xC[x]) \subseteq d^0(A)$ and $d^0(A) \cap (0 \times C) = 0$. Therefore, $\ker d^1 = \beta(A) \oplus (0 \times xC[x]) \oplus (0 \times C) \subseteq d^0(A) \oplus (0 \times C) \subseteq \ker d^1$. We deduce that $\ker d^1 = d^0(A) \oplus (0 \times C)$; that is

$$H^1_P \cong C$$

**4. The Case $n = 2$**

In this section, we recall and generalize the methods and results obtained in (Dongho, J., 2012). By using the reasoning above, the associated Poisson differential are:

$$d^0\varphi = x^2\frac{\partial\varphi}{\partial y}\frac{\partial}{\partial x} - x^2\frac{\partial\varphi}{\partial x}\frac{\partial}{\partial y}, \quad \varphi \in A$$

and

$$d^1(\varphi_1, \varphi_2) = x^2\frac{\partial\varphi_1}{\partial x} + x^2\frac{\partial\varphi_2}{\partial y} - 2x\varphi_1, \quad \varphi_1, \varphi_2 \in A$$

*4.1 Explicit Expression of Associated Cohomological Groups*

4.1.1 Expression of $H^0_P(x^2)$ and $H^1_P(x^2)$

By direct computation, we have:

$$H^0_P \cong C$$

By definition, we have $\ker d^2 = A$ and $Im\, d^1 \subseteq xA$. For $\varphi \in A$, we have

$$x\varphi = -2x(-\frac{1}{2}\varphi) + x^2\partial x(-\frac{1}{2}\varphi) + x^2\partial y\left(\int \partial x(\frac{1}{2}\varphi)dy\right) = d^1\left(-\frac{1}{2}\varphi, \int \partial x(\frac{1}{2}\varphi)dy\right).$$

That is $Im\, d^1 = xA$. In other hand $A = C[y] \oplus xA$. Therefore

$$H^2_P(x^2) \cong C[y]$$

4.1.2 Expression of $H^1_P(x^2)$

Let $\varphi_1, \varphi_2 \in A$; $(\varphi_1, \varphi_2)$ is a 1-cocycle if and only if $2\varphi_1 = x(\frac{\partial\varphi_1}{\partial x} + \frac{\partial\varphi_2}{\partial y})$. We deduce that $\varphi_1 = xu$, $u \in A$ and consequently $\varphi_2 = \int(1 - x\partial x)udy + b(x)$ with $b(x) \in C[x]$ i.e;

$$\ker d^1 = \{(xu, \int(1 - x\partial x)udy) + (0, b(x)), u \in A, b(x) \in C[x]\}$$

Let

$$\begin{aligned} \beta : A &\longrightarrow xA \times A \\ u &\longmapsto (xu, \int(1 - x\partial x)udy) \end{aligned}$$

$\beta$ is a monomorphism and we have $\ker d^1 = \beta(A) \oplus (0 \times C[x])$. Since $A = C[y] \oplus xA$, we obtain

$$\ker d^1 = \beta(C[y]) \oplus \beta(xA) \oplus (0 \times x^2C[x]) \oplus (0 \times C_1[x])$$

where $C_1[x] = \{a_0 + a_1x; a_0, a_1 \in C\}$ is the vector space of polynomials of degree less than or equal to 1. In other hand, we have $\beta(xA) \oplus (0 \times x^2C[x]) \subseteq d^0(A)$ et $d^0(A) \cap [(0 \times C_1[x]) \oplus \beta(C[y])] = 0$. Therefore

$$\ker d^1 \subseteq d^0(A) \oplus \beta(C[y]) \oplus (0 \times C_1[x]) \subseteq \ker d^1$$

i.e;

$$\ker d^1 = d^0(A) \oplus \beta(C[y]) \oplus (0 \times C_1[x])$$

then

$$H^1_P \cong C_1[x] \oplus C[y]$$

**5. Generalization** ($n \geq 2$)

At this stage, the calculation of differentials is no longer a secret. We therefore obtains

$$\begin{aligned} d^0\varphi &= (x^n\frac{\partial\varphi}{\partial y}, -x^n\frac{\partial\varphi}{\partial x}) \\ d^1(\varphi_1, \varphi_2) &= x^n\frac{\partial\varphi_1}{\partial x} + x^n\frac{\partial\varphi_2}{\partial y} - nx^{n-1}\varphi_1 \end{aligned}$$

with $\varphi, \varphi_1, \varphi_2 \in A$.

*5.1 Calculation of Associated Poisson Cohomological Groups*

In this section, we compute all Poisson cohomological groups associated to the above Poisson complex.

5.1.1 Calculation of $H^0_P$

For any $\varphi \in A$, we have $d^0\varphi = 0$ if and only if $\varphi = cte \in C$. We deduce that

$$H^0_P \cong C$$

5.1.2 Calculation of $H_P^2$

By definition, $\ker d^2 = A$ and $Im\ d^1 \subseteq x^{n-1}A$. For $u \in A$, we have

$$x^{n-1}u = -nx^{n-1}(-\frac{1}{n}u) + x^n\frac{\partial}{\partial x}(-\frac{1}{n}u) + x^n\frac{\partial}{\partial y}(\int \frac{\partial}{\partial x}(\frac{1}{n}u)dy)$$

We deduce that $Im\ d^1 = x^{n-1}A$. In addition,

$$A = C[y] \oplus xC[y] \oplus \cdots \oplus x^{n-2}C[y] \oplus x^{n-1}A.$$

Therefore

$$H_P^2 \cong C[y] \oplus xC[y] \oplus \cdots \oplus x^{n-2}C[y]$$

5.1.3 Calculation of $H_P^1$

Let $(\varphi_1, \varphi_2) \in A^2$, $d^1(\varphi_1, \varphi_2) = 0$ if and only if $n\varphi_1 = x(\frac{\partial\varphi_1}{\partial x} + \frac{\partial\varphi_2}{\partial y})$. Thus $\varphi_1 = xu$ with $u \in A$ and consequently $\varphi_2 = \int(n-1-x\frac{\partial}{\partial x})udy + b(x)$ with $b(x) \in C[x]$. Therefore,

$$\ker d^1 = \{(xu, \int (n-1-x\frac{\partial}{\partial x})udy) + (0, b(x)), u \in A, b(x) \in C[x]\}$$

Consider the application

$$\begin{array}{rcl} \beta : A & \longrightarrow & xA \times A \\ u & \longmapsto & (xu, \int (n-1-x\frac{\partial}{\partial x})udy) \end{array}$$

$\beta$ is a monomorphism and we have $\ker d^1 = \beta(A) \oplus (0 \times C[x])$. On the other hand, $A = C[y] \oplus xC[y] \oplus \cdots \oplus x^{n-2}C[y] \oplus x^{n-1}A$. Therefore

$$\ker d^1 = \beta(C[y]) \oplus \beta(xC[y]) \oplus \cdots \oplus \beta(x^{n-2}C[y]) \oplus \beta(x^{n-1}A) \oplus (0 \times x^nC[x]) \oplus (0 \times C_{n-1}[x])$$

with $C_{n-1}[x] = \{a_0 + a_1x + a_2x^2 + \cdots + a_{n-1}x^{n-1}; a_0, \cdots, a_{n-1} \in C\}$ denoting the vector space of polynomials of degree less than $n-1$. Let us show now that $\beta(x^{n-1}A) \oplus (0 \times x^nC[x]) \subseteq d^0(A)$. Let $u \in A$ and $b(x) \in C[x]$,

$$\begin{aligned} \beta(x^{n-1}u) + (0, x^nb(x)) &= \left(x^nu, \int (n-1-x\frac{\partial}{\partial x})(x^{n-1}u)dy + x^nb(x)\right) \\ &= \left(x^nu, -x^n \int \frac{\partial u}{\partial x}dy + x^nb(x)\right) \\ &= \left(x^nu, -x^n\frac{\partial}{\partial x}\left[\int\int \frac{\partial u}{\partial x}dydx - \int b(x)dx\right]\right) \\ &= \left(x^n\frac{\partial}{\partial y}\left[\int udy - \int b(x)dx\right], -x^n\frac{\partial}{\partial x}\left[\int udy - \int b(x)dx\right]\right) \\ &= d^0\left(\int udy - \int b(x)dx\right) \end{aligned}$$

By a simple computation, we have

$$d^0(A) \cap \left(\beta(C[y]) \oplus \beta(xC[y]) \oplus \cdots \oplus \beta(x^{n-2}C[y]) \oplus (0 \times C_{n-1}[x])\right) = 0$$

Furthermore, $\beta(C[y]), \beta(xC[y]), \cdots, \beta(x^{n-2}C[y]), (0 \times C_{n-1}[x]), d^0(A)$ are parts of $\ker d^1$. We deduce that

$$\ker d^1 \subseteq d^0(A) \oplus \beta(C[y]) \oplus \beta(xC[y]) \oplus \cdots \oplus \beta(x^{n-2}C[y]) \oplus (0 \times C_{n-1}[x]) \subseteq \ker d^1$$

and then,

$$\ker d^1 = d^0(A) \oplus \beta(C[y]) \oplus \beta(xC[y]) \oplus \cdots \oplus \beta(x^{n-2}C[y]) \oplus (0 \times C_{n-1}[x])$$

since $\beta$ is a monomorphism, we have

$$H_P^1 \cong C_{n-1}[x] \oplus C[y] \oplus xC[y] \oplus \cdots \oplus x^{n-2}C[y]$$

**Proposition 3** *The Poisson cohomology groups of the Poisson algebra* $(\mathcal{A} = C[x, y], \{x, y\} = x^n)$, $(n \in N^*)$ *are*

$$H_P^0 \cong C, \quad H_P^1 \cong C, \quad H_P^k \cong 0 \; \forall k \geq 2 \qquad when \quad n = 1$$

*and for all* $n \geq 2$,

$$\begin{aligned} H_P^0 &\cong C \\ H_P^1 &\cong C_{n-1}[x] \oplus C[y] \oplus xC[y] \oplus \cdots \oplus x^{n-2}C[y] \\ H_P^2 &\cong C[y] \oplus xC[y] \oplus \cdots \oplus x^{n-2}C[y] \\ H_P^k &\cong 0 \; \forall k \geq 3 \end{aligned}$$

## 6. Associated Logarithmic Poisson Cohomology

The Poisson 2-form remain $\pi = x^n \partial_x \wedge \partial_y$ and the module of 1-form logarithmic along $x^n A$ is $\Omega_A(LogI) = \frac{dx}{x}C[x, y] \oplus C[x, y]dy$ and the associated logarithmic Hamiltonian map is $H(\frac{dx}{x}) = x^{n-1}\partial_y, H(dy) = -x^n\partial_x$. This Hamiltonian map induced the following complex $0 \longrightarrow A \xrightarrow{\partial^0} A \otimes A \xrightarrow{\partial^1} A \longrightarrow 0$. Where $\partial^0 f = x^{n-1}(\partial y f, -x\partial x f)$ and $\partial^1(f_1, f_2) = x^{n-1}(\partial y f_2 + x\partial x f_1 - (n-1)f_1)$. It follow that The order zero logarithmic Poisson cohomology group is $H_{PS}^0 \simeq C$. In order to determine $H_{PS}^1$ and $H_{PS}^2$, $A$ is decomposed as follows: $A = C[y] \oplus xC[y] \oplus ... \oplus x^{n-2} \oplus x^{n-1}C[x, y]$. So; for all $g_0(y) + xg_1(y) + ... + x^{n-2}g(y) + x^{n-1}g_{n-1}(x, y) = g(x, y) \in A$, we have, $g \in \partial^1(A)$ if and only if $g_{n-1}(y) = \partial y f_2 + x\partial x f_1 - f_1$ and $g_i = 0$ for all $i \in \{0, ..., n-2, n\}$; for some $f_1, f_2 \in A$. Therefore, given $f_1 \in A$, there exist $a(x) \in C[x]$ such that $f_2 = \int(g_{n-1} + (n-1)f_1 - x\partial x f_1)dy + a(x)$. In particular, for $f_1 = g_{n-1}$ and $a(x) = 0$, we have $f_2 = \int(ng_{n-1} - x\partial x g_{n-1})dy$. Moreover, for all $0 \neq g_0(y) + xg_1(y) + ... + x^{n-2}g(y)$, the following equation

$$x^{n-1}(\partial y f_2 + x\partial x f_1 - (n-1)f_1) = g_0(y) + xg_1(y) + ... + x^{n-2}g(y).$$

haven't solution in $A \otimes A$. This implies that $A \simeq \partial^1(A \otimes A) \oplus \bigoplus_{i=0}^{n-2} x^i C[y]$. Therefore

$$H_{PS}^2 \simeq \bigoplus_{i=0}^{n-2} x^i C[y]$$

Let $(f_1, f_2) \in A \otimes A$. It is an element of $Ker(\partial^1)$ if and only if $\partial y f_2 = (n-1)f_1 - x\partial x f_1$. That is $f_2 = \int((n-1)f_1 - x\partial x f_1)dy + b(x)$. We define the following map $A \xrightarrow{\eta} A \otimes A$ by $\eta(u) = (u, \int((n-1)u - x\partial x u)dy)$. It is a monomorphism of $C$-modules and it follows from the above description of $Ker\partial^1$ that

$$\begin{aligned} Ker\partial^1 &\simeq \eta(A) \oplus C[x] \\ &\simeq \eta(\bigoplus_{i=0}^{n-2} x^i C[y]) \oplus \eta(x^{n-1}C[x, y]) \oplus x^n C[x] \oplus C_{n-1}[x] \end{aligned}$$

In addition, for all $g \in \eta(x^{n-1}C[x, y] \oplus (O_A \times x^n C[x]))$, there exist $u \in C[x, y]$ and $v \in C[x]$ such that $g = \eta(x^{n-1}u) + (0, x^n v(x)) = x^{n-1}(u, -x(\int \partial y u dy - v))$. This element is in $\partial^0(A)$ if and only if, there exist $a \in A$ such that $\partial^0(a) = g$. This imply that there exist $c(x) \in C[x]$ such that $a = \int u dy + c(x)$ and $\partial x(a) = \partial x u dy - v(x)$. But this imply that $c(x) = -\int v(x)dx$ and then $a = \int u dy - \int v(x)dx$ and $\eta(x^{n-1}C[x, y] \oplus (O_A \times x^n C[x])) \subset \partial^0(A)$. In other hand the following equation in $u$ have no solution in $C[x, y]$

$$\begin{cases} x^{n-1}\partial y u = \sum_{i=0}^{n-2} x^i g_i(y) \\ x^n \partial x u = \sum_{i=0}^{n-2} x^i \int g_i(y)dy + \sum_{i=0}^{n-1} a_i x^i \end{cases}$$

Therefore $Ker\partial^1 \simeq C_{n-1}[x] \oplus \eta(\bigoplus_{i=0}^{n-2} x^i C[y]) \oplus \partial^0(A)$ and then

$$H_{PS}^1(x^n) \simeq C_{n-1}[x] \oplus \eta(\bigoplus_{i=0}^{n-2} x^i C[y])$$

This complete the proof of the following proposition

**Proposition 3** *The logarithmic Poisson cohomology groups of the Poisson structure defined by the logarithmic Poisson 2-form $\pi = x^n \partial x \wedge \partial y$, which is logarithmic along the ideal $I = x^n C[x, y]$ are*

$$H^n_{PS} \simeq 0 \qquad for \qquad n \geq 0$$
$$H^2_{PS} \simeq \bigoplus_{i=0}^{n-2} x^i C[y]$$
$$H^1_{PS}(x^n) \simeq C_{n-1}[x] \oplus \eta(\bigoplus_{i=0}^{n-2} x^i C[y])$$

**Acknowledgement**

We will like to thanks Mister Shuntah Roland Yotcha that accepted to carry out the necessary English translation.

**References**

Braconnier, J. (1977). Algèbres de Poisson. *C. R. Acad. Sci. Paris Sér., A-B 284*(21). A1345-A1348.

Deligne, P. *Equations Diffrentielles à Points Singuliers Réguliers.* Lecture Notes in Mathematics. Berlin. Heidelberg. New York.

Dongho, J. (2012). Logarithmic Poisson cohomology: example of calculation and application to prequantization. *Annales de la facult?des sciences de Toulouse Mathématiques, 21*(4), 623-650. https://doi.org/10.5802/afst.1347

Dongho, J., & Yotcha, S. R. (2016). On logarithmic Prequantization of logarithmic Poisson manifolds. *Journal of Mathematical Sciences and Applications*, *4*(1), 4-13. http://dx.doi.org/10.12691/jmsa-4-1-2

Dongho, J., Yotcha, S. R., & Mbah, A. *On Theory of logarithmic Poisson Cohomology* in preparation.

Goto, R. (2002). Rozansky-Witten Invariants of log symplectic Manifolds. *Contemporary Mathematics, 309*, 69-84. https://doi.org/10.1090/conm/309/05342

Hochschild, G., Kostant, B., & Rosenberg, A. (1962). Differential Forms On Regular Affine Algebras. *Trans. Amer. Math. Soc., 102*, 383-408. https://doi.org/10.1090/S0002-9947-1962-0142598-8

Huebschmann, J. (1990). Poisson Cohomology and quantization, *J.Reine Angew. Math., 408*, 57-113.

Krasil'shchik, I. (1980). Hamiltonian cohomology of canonical algebras. *Dokl. Akad. Nauk SSSR, 251*(6), 1306-1309.

Lie, S. (1890). *Theorie der Transformations gruppen*, (Zweiter Abschnitt, unter Mitwirkung von Prof. Dr. Friederich Engel), Teubner, Leipzig.

Palais, R. (1961). The cohomology of Lie ring. *Proc. Symp. Pure Math., 3*, 130-137. https://doi.org/10.1090/pspum/003/0125867

Pichereau, A. (2006). Poisson (co)homology and isolated singularities. *J. Algebra, 299*(2), 747-777. https://doi.org/10.1016/j.jalgebra.2005.10.029

Rinehart, G. (1963). Differential forms on general commutative algebras. *Trans. Amer. Math. Soc., 108*(2), 195-222. https://doi.org/10.1090/S0002-9947-1963-0154906-3

Polishchuk, A. (1997). Algebraic geometry of poisson brackets, *Journal of Mathematical Sciences., 84*(5).

Monnier, Ph. (2002). Poisson cohomology in dimension two. *Israel J. Math., 129*, 189-207.

Saito, K. (1980). Theory of logarithmic differential forms and logarithmic vector fields, *Sec. IA, J.Fac.Sci. Univ. Tokyo., 27*, 265-291.

Vaisman, I. (1991). On the geometric quantization of Poisson manifolds. *J. Math., 32*, 3339-3345. https://doi.org/10.1063/1.529446

Vinogradov, A. M., & Krasil'shchik, I. S. (1975). What is Hamiltonian formalism? *Russian, Uspehi Mat. Nauk, 30*(1), 173-198.

Weinstein, A. (1983). The local structure of Poisson manifolds. *J. Differential Geometry, 18*, 523-557.

# 2

# Solving a Mixed Problem with Almost Regular Boundary Condition By the Contour Integral Method

S. T. Aleskerova[1]

[1] Institute of Mathematics and Mechanics of ANAS, Baku, Azerbaijan

Correspondence: S. T. Aleskerova, Azerbaijan State Pedagogical University, Azerbaijan, Department of Mathematical Analysis. E-mail: sabina.alesgerova75@gmail.com

**Abstract**

In the paper, a mixed problem for $\lambda$-complex parameter dependent, fourth order partial equation is considered. The estimations for the Green function for the eigenvalues of the considered problem outside of $\delta$ were obtained within the regular boundary condition. The solution of the mixed problem was given in the contour integral form.

**Keywords:** fundamental solutions, asymptotic, mixed problem, boundary conditions, sector, contour, eigenvalues, spectral problem

## 1. Introduction

The considered equation arises when considering heat and diffusion processes the boundary conditions are given in the general form. The paper deals with finding the solution of the mixed problem within second order almost regular boundary conditions. The solution of the problem is found by M.L. Rasulov's contour integral method. The studied mixed problem, mathematically given as follows. The following problem is considered

$$\frac{\partial^2 u(x,t)}{\partial t^2} - (p_1 + p_2\cdot)\frac{\partial^3 u(x,t)}{\partial t \partial x^2} + p_1 p_2 \frac{\partial^4 u(x,t)}{\partial x^4} =$$

$$= a(x)\frac{\partial u(x,t)}{\partial t} + b(x)\frac{\partial^2 u(x,t)}{\partial x^2}, 0 < x < 1,\ t > 0 \tag{1}$$

$$L_1 \equiv \sum_{k=1}^{4} \alpha_{1k}\frac{\partial^{(k-1)}u(0,t)}{\partial x^{k-1}} + \sum_{k=1}^{4} \beta_{1k}\frac{\partial^{(k-1)}u(1,t)}{\partial x^{k-1}} = 0$$

$$L_2 \equiv \sum_{k=1}^{4} \alpha_{2k}\frac{\partial^{(k-1)}u(0,t)}{\partial x^{k-1}} + \sum_{k=1}^{4} \beta_{2k}\frac{\partial^{(k-1)}u(1,t)}{\partial x^{k-1}} = 0$$

$$L_3 \equiv \sum_{k=1}^{4} \alpha_{3k}\frac{\partial^{(k-1)}u(0,t)}{\partial x^{k-1}} + \sum_{k=1}^{4} \beta_{3k}\frac{\partial^{(k-1)}u(1,t)}{\partial x^{k-1}} = 0$$

$$L_4 \equiv \sum_{k=1}^{4} \alpha_{4k}\frac{\partial^{(k-1)}u(0,t)}{\partial x^{k-1}} + \sum_{k=1}^{4} \beta_{4k}\frac{\partial^{(k-1)}u(1,t)}{\partial x^{k-1}} = 0 \tag{2}$$

$$u(x,0) = \varphi(x)$$

$$\frac{\partial u(x,0)}{\partial t} = \psi(x) \tag{3}$$

where $a(x), b(x), \varphi(x)$ are complex valued functions. $p_1$, $p_2$, $\alpha_{ij}$, $\beta_{ij}$ $(i,\ j = \overline{1,4})$ are complex numbers. The conditions $Re\, p_1 > 0$ and $Re\, p_2 > 0$ are satisfied.

The following spectral problem corresponding to mixed problem **??-??** is constructed

$$p_1 p_2 y^{IV} - (p_1 + p_2)\lambda^2 y'' + \lambda^4 y - b(x) y'' - a(x)\lambda^2 y = f(x, \lambda) \tag{4}$$

$$L_k(y) = 0 \tag{5}$$

where

$$f(x, \lambda) = \lambda^2 \varphi(x) + \psi(x) - a(x)\varphi(x) - (p_1 + p_2)\varphi''(x).$$

Partition the $\lambda$-complex plane into eight sectors

$$S_1 = \{\lambda | \lambda_2 > k_1 \lambda_1;\ \lambda_2 < k_4 \lambda_1\},$$

$$S_2 = \{\lambda | \lambda_2 > k_4 \lambda_1;\ \lambda_2 < k_3 \lambda_1\},$$

$$S_3 = \{\lambda | \lambda_2 > k_3 \lambda_1;\ \lambda_2 < k_2 \lambda_1\},$$

$$S_5 = \{\lambda | \lambda_2 > k_1 \lambda_1;\ \lambda_2 < k_4 \lambda_1\},$$

$$S_6 = \{\lambda | \lambda_2 > k_4 \lambda_1;\ \lambda_2 < k_3 \lambda_1\},$$

$$S_7 = \{\lambda | \lambda_2 > k_3 \lambda_1;\ \lambda_2 < k_2 \lambda_1\},$$

$$S_8 = \{\lambda | \lambda_2 > k_1 \lambda_1;\ \lambda_2 < k_2 \lambda_1\},$$

where

$$k_1 = ctg\psi_1; k_2 = \frac{|\omega_3| \cos\psi_3 - |\omega_1| \cos\psi_1}{|\omega_3| \sin\psi_3 - |\omega_1| \sin\psi_1};$$

$$k_3 = ctg\psi_3; k_4 = \frac{|\omega_1| \cos\psi_1 + |\omega_3| \cos\psi_3}{|\omega_1| \sin\psi_1 + |\omega_3| \sin\psi_3};$$

$$\omega_1 = |\omega_1| e^{\psi_1 i};\ \omega_2 = -\omega_1; \omega_3 = |\omega_3| e^{\psi_3 i}; \omega_4 = -\omega_3;$$

$$|\omega_1| = |p_1|^{-\frac{1}{2}};\ |\omega_3| = |p_3|^{-\frac{1}{2}};$$

$$\psi_k = -\frac{1}{2} arctg \frac{Im\, P_k}{Re\, P_k}, k = 1, 3,$$

$$0 < \psi_3 < \psi_1 < \frac{\pi}{4};\ |\omega_3| \sin\psi_3 - |\omega_1| \sin\psi_1 > 0.$$

In each sector of $S p\left(p = \overline{1, 8}\right)$ we find the asymptotics of fundamental solutions of equation 4 in the following way

$$\frac{d^k y_m(x, \lambda)}{dx^k} = (\lambda\omega_m)^k \left[1 + \frac{1}{\lambda} y^1_{mk}(x) + \frac{1}{\lambda^2} y^2_{mk}(x) + \frac{E_{mk}(x, \lambda)}{\lambda^3}\right] e^{\lambda\omega_m x},$$

where

$$y^1_{mk}(x) = \frac{1}{4q\omega_m^3 + 2p\omega_m}\left[\int_0^x a(\xi)\,d\xi + \omega_m^2\int_0^x b(\xi)\,d\xi\right],$$

$$y^2_{mk}(x) = \frac{1}{4q\omega_m^3 + 2p\omega_m}\left[\int_0^x \left(a(\xi)\,d\xi + \omega_m^2 b(\xi)\right)y^1(\xi)\,d\xi\right] -$$

$$-\frac{6q\omega_m^2 + p}{4q\omega_m^3 + 2p\omega_m}\frac{dy^1_{mk}(x)}{dx},$$

$$q = p_1p_2, p = -(p_1 + p_2).;$$

The functions $E_{mk}(x,\lambda)$ $\left(m = \overline{1,4};\ k = \overline{0,3}\right)$ are analytic functions and for large values of $|\lambda|$ they are bounded functions.

Let us give the following theorem connected with the asymptotics of eigen numbers of spectral problem (4)-(5) and the Green function.

**Theorem 1.** Assume that the coefficients of equation (4) and boundary conditions (5) satisfy the following conditions.

$$Re\,p_1 > 0, Re\,p_2 > 0, a(x), b(x) \in C^1[0, 1],$$

$$L(\alpha_3,\alpha_4,\beta_3,\beta_4) = 0, L(\alpha_3,\alpha_4,\beta_2,\beta_4) = 0, L(\alpha_2,\alpha_4,\beta_3,\beta_4) = 0$$

$$L(\alpha_2,\alpha_3,\alpha_4,\beta_4) \neq 0, L(\alpha_4,\beta_2,\beta_3,\beta_4) \neq 0.$$

Then the Green function of eigen-numbers of the spectral problem out of the vicinity of $\delta$ has the following estimation

$$|G(x,\xi,\lambda)| \le \frac{M}{|\lambda|^2}, \lambda \in S_p(p = \overline{1,8}), |\lambda \to +\infty|.$$

Thus the following formula is valid for the asymptotics of the eigen-numbers of spectral problem (4), (5)

$$\lambda_{k\nu} = -\frac{1}{\omega_k}\left\{\ln\left|\frac{2\pi\nu A_k}{\omega_k}\right| + i\left[2\pi\nu + \frac{\pi}{2}(2 - sgn\nu) + \arg A_k\right]\right\} +$$

$$+O\left(\frac{\ln|\nu|}{\nu}\right), k = 1, 2, 3, 4; (-1)^k\nu \to +\infty, \qquad (6)$$

where $A_k$ are complex numbers dependent on the coefficients of equation (4) and boundary conditions (5)

$$L\left(\gamma_n^1, \gamma_m^2, \gamma_p^3, \gamma_q^4\right) = \begin{vmatrix} \gamma^1_{1n} & \gamma^2_{1m} & \gamma^3_{1p} & \gamma^4_{1p} \\ \gamma^1_{2n} & \gamma^2_{2m} & \gamma^3_{2p} & \gamma^4_{2p} \\ \gamma^1_{3n} & \gamma^2_{3m} & \gamma^3_{3p} & \gamma^4_{3p} \\ \gamma^1_{4n} & \gamma^2_{4m} & \gamma^3_{4p} & \gamma^4_{4p} \end{vmatrix}.$$

In the paper the following theorem is proved.

**Theorem 2.** The coefficients of equation (1), boundary conditions (2) and initial conditions (3) satisfy the following conditions

$$Re\,p_1 > 0, Re\,p_2 > 0, a(x), b(x) \in C^2[0, 1],$$

$$\varphi(x) \in C^3[0, 1], \psi(x) \in C^2[0, 1],$$

$$\varphi(0)=\varphi(1)=\varphi'(0)=\varphi'(1)=\varphi''(0)=\varphi''(1)=0,$$

$$\psi(0)=\psi(1)=\psi'(0)=\psi'(1)=0,$$

$$L(\alpha_3,\alpha_4,\beta_3,\beta_4)=L(\alpha_3,\alpha_4,\beta_2,\beta_4)=L(\alpha_2,\alpha_4,\beta_3,\beta_4)=0$$

$$L(\alpha_2,\alpha_3,\alpha_4,\beta_4)\neq 0, L(\alpha_4,\beta_2,\beta_3,\beta_4)\neq 0.$$

then the solution of problem (1)-(3) will be found in the following contour integral form

$$U(x,t)=\frac{1}{\pi i}\int_{L_1}\lambda y(x,\lambda)e^{\lambda^2 t}d\lambda, \tag{7}$$

where the function $y(x,\lambda)$ is the solution of spectral problem (4)-(5).

$L_1$- contour is an infinitely extended unbounded curve and is given by the following formula

$$L_1=\left\{\lambda:\lambda=re^{i\varphi},\ |\varphi|\leq\frac{\pi}{4}+\delta,\ \lambda=Re^{\pm i\left(\frac{\pi}{4}+\delta\right)},\ R\geq r\right\}$$

$$r>0,\delta=\min\left\{\frac{\pi}{8}-\frac{\psi_1}{2},\frac{\pi}{8}-\frac{\psi_{12}}{2}\right\}.$$

**2. Proof**

$$\left|e^{\lambda^2 t}\right|=e^{tRe\lambda^2}=e^{t|\lambda|^2\cos 2\left(\frac{\pi}{4}+\delta\right)}=e^{-t|\lambda|^2\sin 2\delta}$$

As $0<\psi_k<\frac{\pi}{4}, k=1,2$ then $\sin 2\delta>0$.

This means that the function $e^{\lambda^2 t}$ as $t>0$, $\lambda\in L_1$, $|\lambda|\to+\infty$ decreases exponentially, i.e. the following integrals are regularly convergent

$$\int_{L_1}\lambda\frac{\partial^{p+q}}{\partial t^p\partial x^q}\left(y(x,\lambda)e^{\lambda^2 t}\right)d\lambda, p+q\in\{0,\ 1,\ 2,\ 3,\ 4\}.$$

Show that integral (7) satisfies equation (1)

$$\frac{\partial^2 u(x,t)}{\partial t^2}-(p_1+p_2)\frac{\partial^3 u(x,t)}{\partial t\partial x^2}+p_1p_2\frac{\partial^4 u(x,t)}{\partial x^4}-$$

$$-a(x)\frac{\partial u(x,t)}{\partial t}-b(x)\frac{\partial^2 u(x,t)}{\partial x^2}=\frac{1}{\pi i}\int_{L_1}\left[\lambda^4 y(x,\lambda)-\right.$$

$$\left.-(p_1+p_2)\lambda^2 y''(x,\lambda)+p_1p_2y^{IV}(x,\lambda)-\lambda^2 a(x)y(x,\lambda)-b(x)y''(x,\lambda)\right]$$

$$\lambda e^{\lambda^2 t}d\lambda=\frac{1}{\pi i}\left[\varphi(x)\int_{L_1}\lambda^3 e^{\lambda^2 t}d\lambda+(\psi(x)-a(x)\varphi(x)-\right.$$

$$\left.-(p_1+p_2)\varphi''(x))\int_{L_1}\lambda e^{\lambda^2 t}d\lambda\right]=0.$$

It is easy to show that the following expansion formula is true within the theorem conditions

$$\sum_{k=1}^{4}\sum_{n=1}^{\infty}\underset{\lambda=\lambda_{kn}}{res}\,\lambda^{s+2}\int_0^1\varphi(\xi)G(x,\xi,\lambda)d\xi=\begin{cases}\varphi(x), & s=1\\ 0, & s<1\end{cases}.$$

Using this formula, we can show that initial conditions (3) are satisfied. As the function $y(x, y)$ is the solution of spectral problem (4), (5) it is easy to verify that integral (7) satisfies boundary conditions (2). The theorem is proved.

### 3. Conclusion

The considered equation is higher order parabolic equation in the sense of Petrovsky. Boundary conditions (2) are called general form boundary conditions. At first a spectral problem corresponding to mixed problem (1)-(3) is constructed. For constructing asymptotics of fundamental solutions of the spectral problem. The $\lambda$-complex plane is divided into 8 sector. The asymptotics of fundamental solutions are found more exactly in each of these sectors. For the first order almost regular case, estimations for the Green function are found far from the $\delta$ vicinity of eigen numbers. Imposing smoothmen conditions and algebraic conditions on initial data, it was possible to find the solution of the problem in the form of the contour integral.

### References


Alesgerova, S. T., Akhmedov, S. Z., & Abbasova, A. K. (2015). Finding the asymptotics of eigen value of a fourth orde differential operator. Aktualniye problemy gumanitarnikh i yestestvennikh nauk. *Journal of Scientific publications,* 13-18, Moscow.

Alesgerova, S. T. (2015). *Finding the asymptotics of eigen numbers of fourth order differential operator within almost regular condition*, News of Pedagogical University, Section of Natural Sciences, *2*, 3-8.

Alesgerova, S. T., & Akhmedov, S. Z. (2015). *Finding the asymptotics of eigen numbers for fourth order equation corresponding to a spectral problem*, Proceeding of the scientific conference devoted to 85 years of Y.S. Mammadov, BSU, 57-60.

Akhmedov, S. Z., & Aleskerova, S. T. (2014). *On zeros of the characteristical determinant of a spectral problem, dependent on λ-complex parameter*, BDU-nun "Xeberleri", fiz. riy. el. ser., *4*, 36-45.

Akhmedov, S. Z., & Alesgerova, S. T. (2012). *Construction of asymptotics of fundamental solutions of equation*, BDU-nun "Xeberleri", fiz. riy. el. ser., *1*, 70-77.

Federyuk, M. V. (1983).*Asymptotic methods for linear ordinary differential equations*, Moscow, Nauka, Russian, 352.

Mamedov, Yu. A., & Nagiyeva, R. I. (2010). *Mathematical analysis of critical problems of electrodynamics*, Baku, Elm, Russian, 181.

Mamedov, Yu. A., & Akhmedov, S. Z. (2005). *Investigation of a characteristical determinant connected with the solution of spectral problem*, Vestnik BGU, ser. fiz. math. nauk, Russian, *2*, 5-12.

Mammadov, Y. A., & Alesgerova, T. A. (2012). *Estimation of Green function of a spectral problem with quasi-regular boundary condition*, Transactions Issue Mathematics, Series of physical-technical and mathematics science, Azerbaijan National Academy of Science, 113-119.

Rasulov, M. L. (1964). *Contour integral method*, Moscow, Nauka, Russian, *458*.

# 3

# Geometry of the 3D Pythagoras' Theorem

Luis Teia[1]

[1] Berlin, Germany

Correspondence: Dr. Luis Teia, Mehrower Allee 82, 12687 Berlin, Germany. E-mail: luistheya@gmail.com

**Abstract**

This paper explains step-by-step how to construct the 3D Pythagoras' theorem by geometric manipulation of the two dimensional version. In it is shown how $x + y = z$ (1D Pythagoras' theorem) transforms into $x^2 + y^2 = z^2$ (2D Pythagoras' theorem) via two steps: a 90-degree rotation, and a perpendicular extrusion. Similarly, the 2D Pythagoras' theorem transforms into 3D using the same steps. Octahedrons emerge naturally during this transformation process. Hence, each of the two dimensional elements has a direct three dimensional equivalent. Just like squares govern the 2D, octahedrons are the basic elements that govern the geometry of the 3D Pythagoras' theorem. As a conclusion, the geometry of the 3D Pythagoras' theorem is a natural evolution of the 1D and 2D. This interdimensional evolution begs the question – Is there a bigger theorem at play that encompasses all three?

**Keywords:** Geometry, Pythagoras, theorem, three dimensions

## 1. Introduction

For the past 2,500 years, the Pythagoras' theorem, arguably the most well known theorem in the world, ranging from the times of Babylon to today, has greatly helped mankind to evolve (Friberg 1981, Roy 2003, Strogaz 2015). Its useful right angles are everywhere, whether it is a building, a table, a graph with axes, the properties of light (Einstein, 1961) or the atomic structure of a crystal (Kelly, 2005). Its numerous proofs are universally applicable (Maor, 2007), but still the Pythagoras' theorem is exclusively bound to two dimensions. Since we live in a three dimensional world, the awareness of this gap in knowledge begs the question - How does the Pythagoras' theorem looks like in three dimensions?

## 2. Theory

### *2.1 The 1D and 2D Pythagoras' Theorem*

Let us start with the simple one dimensional version of the Pythagoras' theorem. The 1D Pythagoras' theorem is governed by line segments, where the geometric addition of two line segments gives a third, or $x + y = z$ (Figure 2a). This is by definition the mathematical process of summation. This can be transformed into the 2D Pythagoras' theorem via the following two steps:

> **Step 1.** Rotating line $x$ by 90 degrees about the middle point gives a right-angled triangle (Figure 2b).
> **Step 2.** Extruding all lines perpendicularly to their length gives squares. These squares result is the two dimensional version of the Pythagoras' theorem, in which the geometric addition of two squares gives a third, or $x^2 + y^2 = z^2$ (Figure 2c).

## 3. Hypothesis

It is assumed that the process used to convert the 1D Pythagoras' theorem into 2D (Figure 1) also applies from 2D to 3D. It is also assumed that the 3D Pythagoras' theorem has the same geometric elements as in 2D.

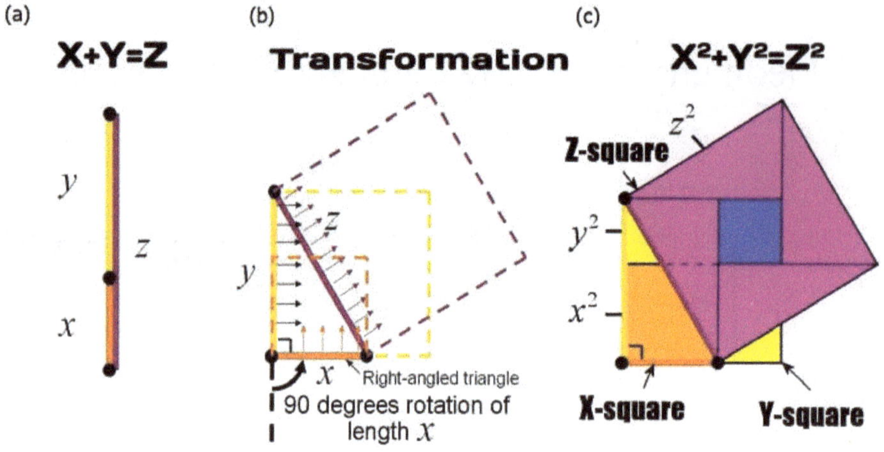


Figure 1. The geometry of (a) $x + y = z$, (b) the transformation, and (c) $x^2 + y^2 = z^2$.

## 4. Development

### *4.1 Location of the Axes*

The central square theory (Teia, 2015) is used to determine how, and where, the 2D Pythagoras' theorem is rotated when transforming into 3D. The theory states that the right side of the equation $x^2 + y^2 = z^2$ is composed geometrically of four right-angled triangles rotating around a central square $(y - x)^2$, which in turn when enclosed forms a new square about which other Pythagorean triples revolve (Figure 3a). This means that all triangles in the ternary tree of Pythagorean triples relate to each other via intermediate squares (Teia, 2016). Figure 3b shows the geometrical representation of the Pythagoras' family of triples constructed with the central square theory. For simplicity, from now on $x^2$, $y^2$, $z^2$ and $(y-x)^2$ are termed X-square, Y-square, Z-square and central square, respectively.

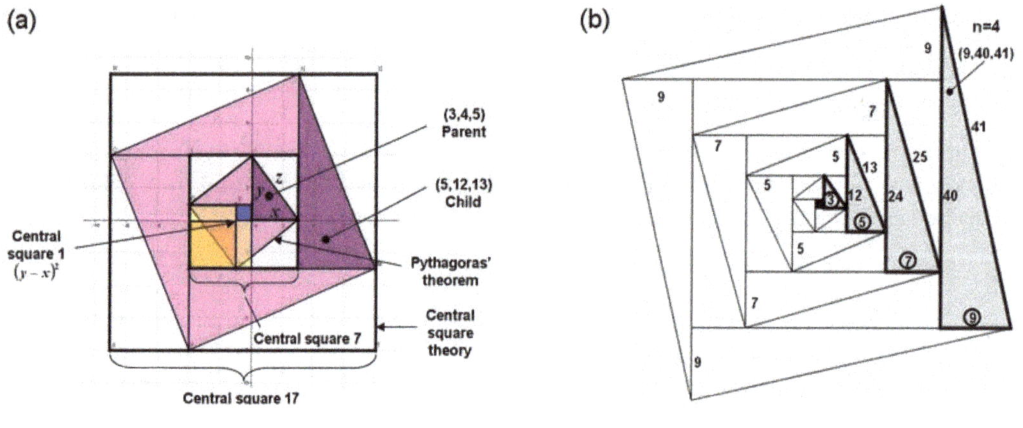


Figure 2. (a) The central square theory, and (b) the Pythagoras' family of triples (Teia, 2015).

More importantly, Figure 2b shows that the left side of the Pythagoras' theorem $x^2 + y^2 = z^2$ is governed by an axis of symmetry (redrawn and highlighted in Figure 3a). Similarly, the right side of the Pythagoras' theorem is composed of right-angled triangles revolving around an axis of rotation (redrawn and highlighted in Figure 3b). Ultimately, Figure 3c shows that central squares not only interconnect the left side of the Pythagoras' theorem (Figure 3a) with the right side (Figure 3b), but also they connect the axis of symmetry with the axis of rotation. That is, the squares are both symmetric about the axis of symmetry, as well as, revolve around the axis of rotation.

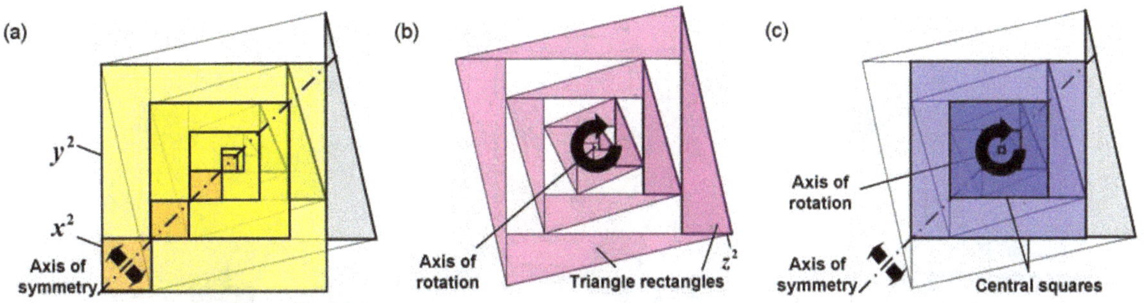


Figure 3. Definition of (a) axis of symmetry, (b) axis of rotation, and (c) their connection via central squares.

### *4.2 The 3D Pythagoras' Theorem*

The symmetry and rotation axes guide the two-step transformation (similar as defined previously in section 2.1) from $x^2 + y^2 = z^2$ to $x^3 + y^3 = z^3$. For simplicity, from now on $\sqrt{3}/2x^3$, $\sqrt{3}/2y^3$, $\sqrt{3}/2z^3$ and $\sqrt{3}/2(y-x)^3$ are termed X-, Y-, Z- and central octahedron, respectively. The steps are described as follows:

**Step 1.** Rotating the Z-square and central square by 90 degrees with respect to the X- and Y-square about the diagonal axis (Figure 4a) forms automatically a central octahedron (Figure 4b). Both the central square (2D) and central octahedron (3D) are a direct and inherent consequence of the process of transformation from 1D to 2D, and 2D to 3D, respectively. Hence, the geometric element that governs the 3D Pythagoras' theorem is an octahedron.

**Step 2.** Extruding all squares perpendicularly to their surfaces gives octahedrons. Finally, careful partition of the octahedrons gives the 3D Pythagoras' theorem shown in Figure 4c, which is equivalent to the 2D Pythagoras' theorem in Figure 4a.

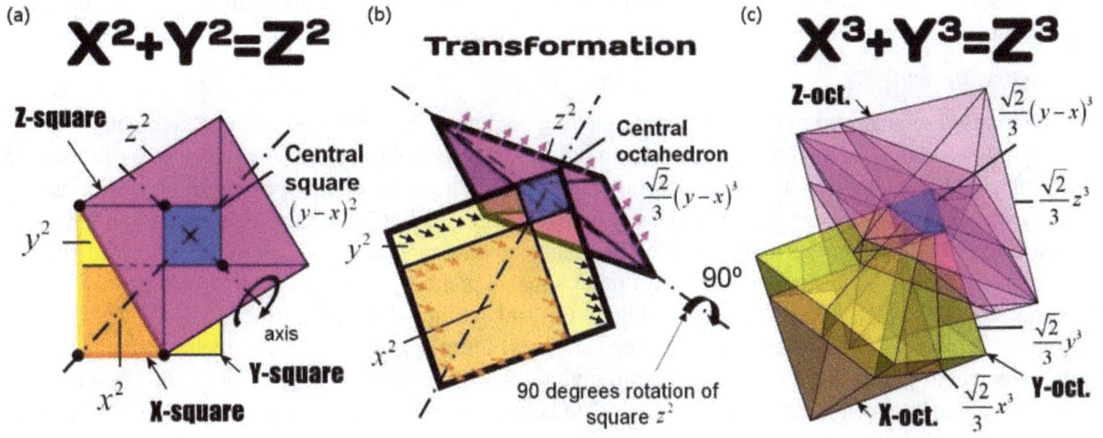


Figure 4. The geometry of (a) $x^2 + y^2 = z^2$, (b) the transformation, and (c) $x^3 + y^3 = z^3$.

The remainder of the paper focuses on explaining in more detail these two steps. First, the transformation of the X- and Y-squares into octahedrons is explained (section 4.3). Then, the formation of the 3D geometric equivalent to the symmetry rectangles is described (section 4.4). Similarly, the transformation of the Z-square into an octahedron is explained (section 4.5), and subsequently the formation of the 3D geometric equivalent to the revolving triangle is explained (section 4.6). Finally, the article concludes with the assembly of all 3D elements into the geometry of the 3D Pythagoras' theorem (shown above in Figure 4c). The details of the mathematical work involved are explained in Teia's book (2015).

### *4.3 Transformation of X- and Y-squares into Octahedrons*

The axis of symmetry shapes the geometry of the X- and Y-squares into its 3D equivalent. Let us start by looking at the left side of the Pythagoras' theorem. The left hand side of equation $x^2 + y^2 = z^2$ is geometrically composed of the X-square, and the central square, both superimposed on to the Y-square (Figure 5a). The area outside X-square and central square, but still inside the Y-square, forms the symmetry rectangles. These have a combined area $\sum A_s = 2x(y-x)$.

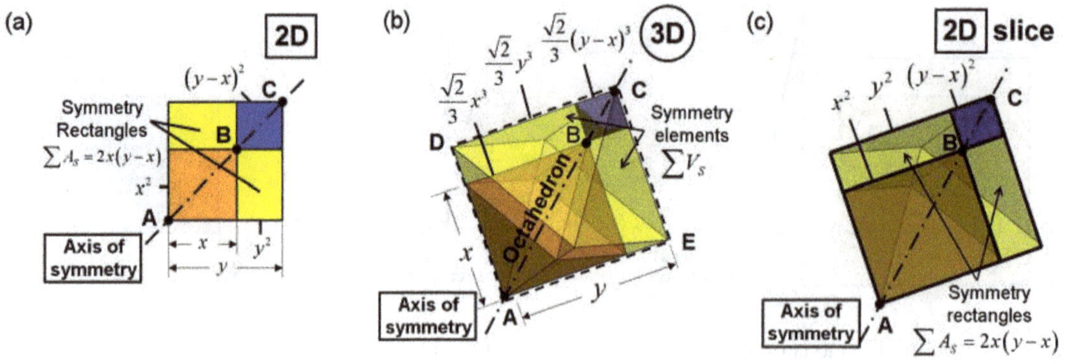


Figure 5. The geometrical transformation to 3D of $x^2$ and $y^2$.

The general rule to geometrically connect these elements is – they all share common vertices:

**Point A** connects the lower left corner of the X- and Y-square.
**Point B** connects the upper right corner of the X-square with lower left corner of the central square.
**Point C** connects the upper right corner of the Y-square with upper right corner of the central square.

It is assumed that when passing to three dimensions, the common points between elements (A, B and C in Figure 5a) remain the same. That is, the extreme corners of the three elements need to coincide with the axis of symmetry. Following the three-point ABC rule, the squares in Figure 5a are extruded around the axis of symmetry and form the octahedrons in Figure 5b. Taking a sectional view ADCE gives the two dimensional slice in Figure 5c, that is the same as Figure 5a. As one can imagine, the elements change size as point B translates along the axis of symmetry. In other words, as $x$ changes towards or away from $y$, point B slides along the axis of symmetry between points A and C. The left hand side of equation $x^3 + y^3 = z^3$ is therefore geometrically composed of X- and central octahedron overlapping onto the Y-octahedron. The volume outside the X- and central octahedron, but still inside Y-octahedron, forms the symmetry elements that possess a combined volume of $\sum V_S$. This is similar to the symmetry rectangles with a combined area $\sum A_S$, but in three dimensions.

*4.4 The Symmetry Element*

Just as the symmetry rectangles are symmetric about the diagonal axis, the three-dimensional symmetry element is also symmetric about the vertical axis. The symmetry element is found by splitting the Y-octahedron as follows. Splitting along a vertical plane gives half of the Y-octahedron (Figure 6a). The same element needs to revolve around the axis, hence splitting it again about the perpendicular plane results in a segment that is the same in all quadrants (Figure 6b). The element is further split along diagonal plane resulting in a segment whose shape is an eigth of the y-octahedron (Figure 6c). Since the symmetry rectangle does not overlap with other squares, neither does the symmetry element. Plotting a plane along the surface of the X-octahedron and cutting (Figure 6d) and another along the surface of the central octahedron and cutting again, results in the final shape of the revolving element (Figure 6f). As seen, 8 symmetry elements accommodate between the X-octahedron and the central octahedron, giving the end volume of the Y-octahedron. This is equivalent to the 2D case where two symmetry rectangles accommodate between the X-square and the central square, giving the area of the Y-square.

*4.5 Transformation of the Z-square into an Octahedron*

The axis of symmetry shapes the geometry of the Z-square into its 3D equivalent. The right hand side of equation $x^2 + y^2 = z^2$ is geometrically composed of four congruent right-angled triangles rotated around a central square, and hence also around an axis of rotation (Figure 7a). There are specific rules that define how the elements revolve around the central square. These are inferred from Figure 7a as:

**Rule 1.** The extended sides of the central square are in line with the vertices of the Z-square.
**Rule 2.** The elements revolving around the axis are identical, and have as shape right-angled triangles.

The three dimensional equivalent to Figure 7a is built by extruding the Z- and central squares perpendicularly along the axis of rotation forming two pyramids for each, which combined form two octahedrons (Figure 7a becomes Figure 7b). Taking a sectional view FGHI gives the two dimensional slice in Figure 7c, which is the same as Figure 7a. The central element is the central octahedron, and its midplane is the central square. The central octahedron was clocked

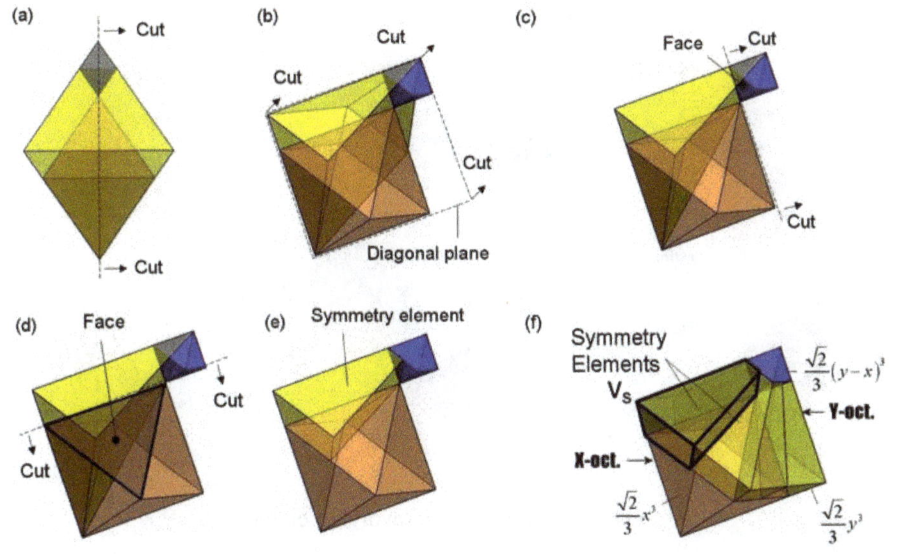


Figure 6. Formation of the symmetry element.

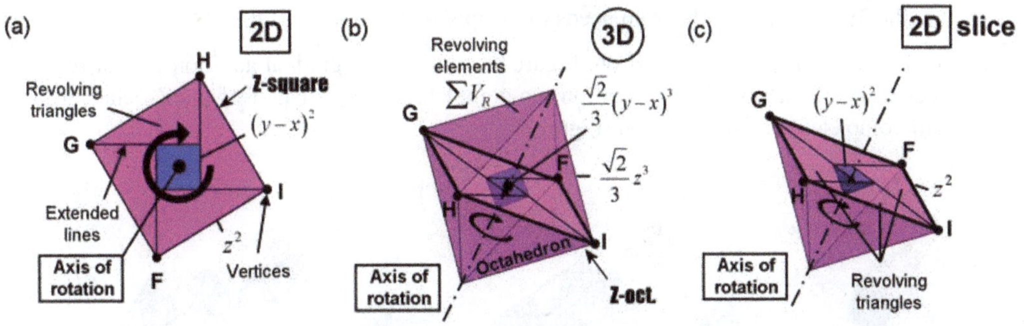


Figure 7. The geometrical transformation to 3D of $z^2$.

around the axis of rotation until its edges aligned with the corners FGHI of the Z-octahedron, satisfying Rule 1. The lines that fan out at midplane from the central octahedron indicate where the Z-octahedron needs to be split to form the revolving element, satisfying Rule 2. Even though the midplane section has the same "footprint" as 2D (i.e., a right-angled triangle), the 3D element that revolves around the central octahedron is not regular. The revolving element (not yet shown) is the 3D equivalent to the revolving triangle (Figure 7a). The volume outside the central octahedron, but still inside the Z-octahedron, forms the revolving elements that possess a combined volume of $\sum V_r$.

*4.6 The Revolving Element*

The revolving element is found by splitting the Z-octahedron as follows. Making an horizontal cut at the top and bottom edges of the central octahedron (Figure 8a) forms two pyramids and a central section (Figure 8b). Cutting at midplane and deleting the bottom allows us to simplify using symmetry. The element needs to follow the 2D rules, that is, its base is the 2D right-angled triangle. First, make a diagonal plane along one face of the central octahedron, and along one side of the triangle (Figure 8c). Second, another diagonal cut along another adjacent face and triangle side (Figure 8d).

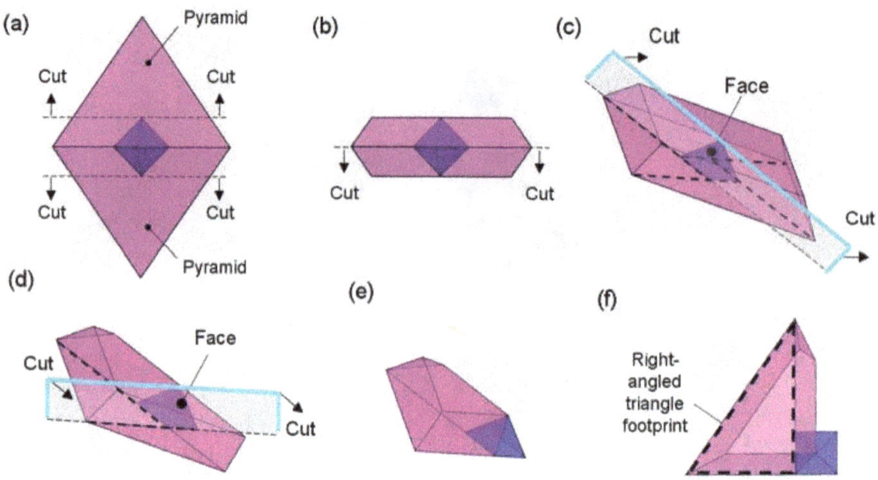


Figure 8. Formation of the revolving element.

The result is the shape of the rotating element in three dimensions (Figure 8e), which is the 3D equivalent to the right-angled triangle in 2D. Figure 8f shows it from a top view, where its compliance with the base right-angled triangle is observable. Note that the 3D Pythagoras' theorem needs to comply with the rules of the 2D Pythagoras' theorem.

Now to construct the new assembly of Z-octahedron, Figure 9a-d shows the gradual addition of four revolving elements around the central octahedron. Placing the pyramid on top completes the upper half of the Z-octahedron in Figure 9e. Adding the lower half completes the Z-octahedron (Figure 9f).

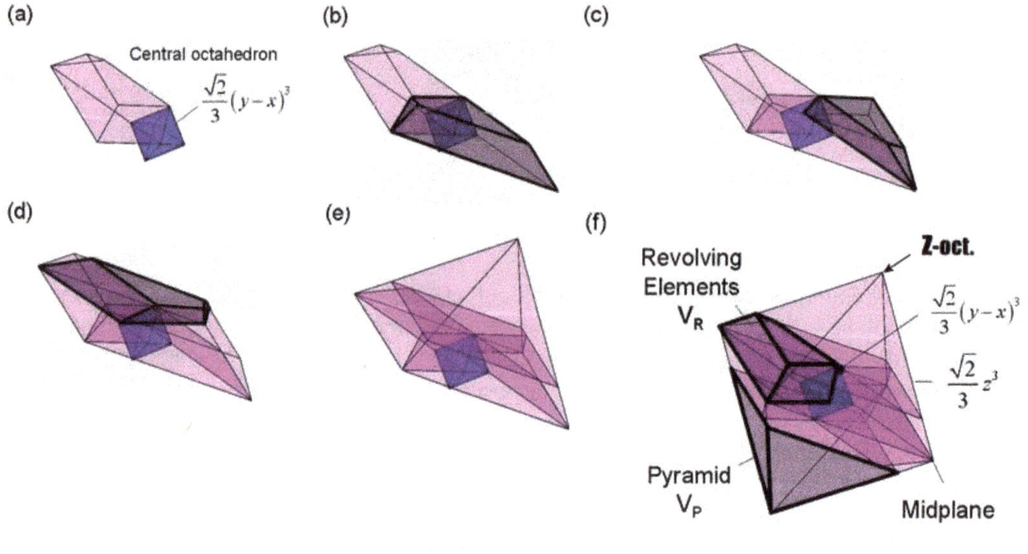


Figure 9. Assembly of the right hand side of $x^3 + y^3 = z^3$.

*4.7 Final Assembly*

The central octahedron connects the left hand side and right hand side. In three dimensions, the Y-octahedron contains symmetry elements, just like in two dimensions the Y-square contains symmetry rectangles. Similarly, in three dimensions, the Z-octahedron contains revolving elements, just like in two dimensions the Z-square contains revolving triangles. Merging the central octahedron on the left hand side (Figure 6f) to the central octahedron on the right hand side (Figure 9f) completes the geometry of $x^3 + y^3 = z^3$, shown in Figure 10 alongside with the geometric representation of the other dimensions, i.e. $x + y = z$ and $x^2 + y^2 = z^2$.

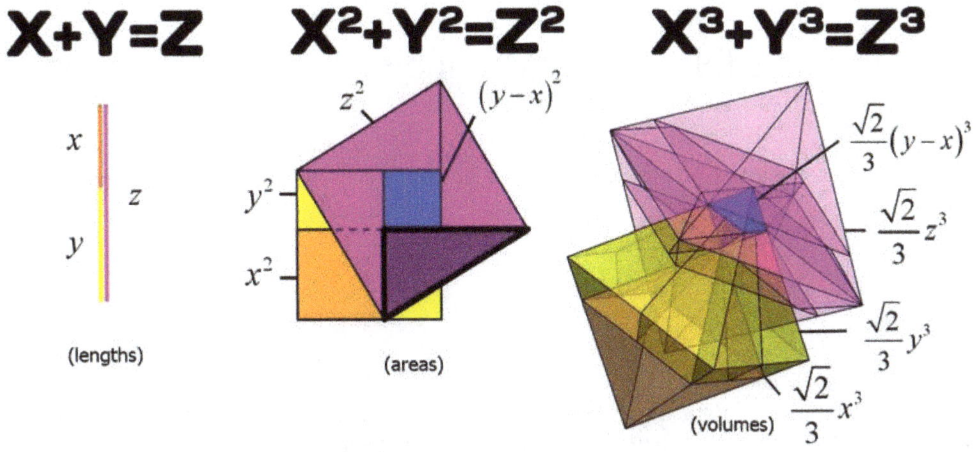


Figure 10. The 1D, 2D and 3D Pythagoras' theorem.

**5. Conclusion**

The three dimensional version of the Pythagoras' theorem is derived by geometric manipulation of the two dimensional version. Consequentially, they are interconnected. The 1D Pythagoras' theorem transforms to 2D in two steps: a 90-degree rotation, and a perpendicular extrusion. The 2D Pythagoras' theorem transforms to 3D using the same steps. Both the two and three dimensional version are governed by a symmetry and rotation axis, identified using the central square method. An octahedron emerges naturally during the process. Hence, just like squares in 2D, octahedrons are the basic elements that govern the geometry of the 3D Pythagoras' theorem. Each two dimensional element has a direct three dimensional equivalent. The geometry of the 3D Pythagoras' theorem is a natural evolution of the 1D and 2D. This evolution begs the question – Is there a bigger theorem at play that encompasses all three?

**References**

Einstein, A. (1961). *Relativity: The Special and the General Theory*, (Vol. 2). Crown Trade Paperbacks.

Friberg, J. (1981). Methods and Traditions of Babylonian Mathematics. *Historia Mathematica, 8*, 227–318. https:/doi.org/10.1016/0315-0860(81)90069-0

Kelly, B. S. and Splittgerber, A. G. (2005). The Pythagorean Theorem and the Solid State, *Journal of Chemical Education, 82*, 756

Maor, E. (2007). *The Pythagorean Theorem: A 4,000-year History*, Princeton University Press.

Roy, R. (2003). Babylonian Pythagoras' theorem, the Early History of Zero and a Polemic on the Study of the History of Science, *Resonance Journal, 8*, 30–40

Strogaz, S. (2015). *Einstein's Boyhood Proof of the Pythagorean Theorem*, The New Yorker, Retrieved from http://www.newyorker.com/tech/elements/einsteins-first-proof-pythagorean-theorem

Teia, L. (2015). Pythagoras' triples explained by central squares, *Australian Senior Mathematical Journal, 29*, 7–15

Teia, L. (2015). $X^3 + Y^3 = Z^3$*: The Proof*, self-published with Amazon.

Teia, L. (2016). Anatomy of the Pythagoras' tree, *Australian Senior Mathematical Journal, 30*, 38–47

# Heuristic Algorithms for Solving Multiobjective Transportation Problems

Ali Musaddak Delphi

Department of Mathematics, College of Basic education, University of Misan, Iraq. E-mail: alimathdelphi@yahoo.com

**Abstract**

In this paper, we proposed three heuristic algorithms to solve multiobjective transportation problems, the first heuristic algorithm used to minimize two objective functions (total flow time and total late work), the second one used to minimize two objective functions (total flow time and total tardiness) and the last one used to minimize three objective functions (total flow time, total late work and total tardiness), where these heuristic algorithms which is different from to another existing algorithms and providing the support to decision markers for handing time oriented problems.

**Keywords:** Transportation problems, Heuristic algorithms, Feasible solutions, Multiobjective problems.

## 1. Introduction

Decision making is the process of identifying and choosing alternatives based on the values and preferences of the decision maker. It is the process of sufficiently reducing uncertainly and doubt about alternatives to allow a reasonable choice to be made from among them. Decision making based solely on a single criterion appears insufficient as soon as the decision making-making process deals with the complex organization environment. So, one must acknowledge the presence of several criteria that lead to the development of multicriteria decision making. Optimization is a kind of the decision making, in which decision have to be taken to optimize one or more objectives under some prescribes set of circumstances. These problems may be single or multiobjective and are to be optimized (maximized or minimized) under a specified set of constraints. The constraints usually are in the form of inequalities or equalities. Such problems which often arise as a result of mathematical modeling of many real life situations are called optimization problems. Transportation problem is a special type of optimization problems which arise in many practical applications. In the beginning it was founded for determining the optimal shipping patter, so it is called transportation problems. The conventional and very well known transportation problem consists in transporting a certain product from each of $m$ origins $i = 1,\dots,m$ to any of $n$ destination $j = 1,\dots,n$. The origins are production facilities with respect capacities $a_1,\dots,a_m$ and the destination are warehouse with required levels of demand $b_1,\dots,b_n$. For the transport of a unit of the given product from the $i^{th}$ source to the $i^{th}$ destination a cost $c_{ij}$ is given for which, without loss of generally, we can assume $c_{ij} \geq 0$ for each $i$ and $j$. Hence, one must determine the amount $x_{ij}$ to be transported from all the origins $i = 1,\dots,m$ to all destination $j = 1,\dots,n$ in such a way that the total cost is minimized. This problem can be suitably modeled as a linear programming problem. Thus the conventional transportation problem can be mathematically expressed as:

Minimize $z = \sum_{i=1}^{m}\sum_{j=1}^{n} c_{ij}x_{ij}$

Subject to $\sum_{i=1}^{m} x_{ij} = b_j$ for all $j = 1,2,\dots,n$

$$\sum_{j=1}^{n} x_{ij} = a_j \quad \text{for all} \quad i = 1,2,\dots,m$$

$$x_{ij} \geq 0 \quad \text{for all} \quad i \text{ and } j$$

$$\sum_{i=1}^{m} a_i = \sum_{j=1}^{n} b_j \text{ ( Balanced condition)}$$

In real world cases transportation problem can be formulated as a multiobjective transportation problem because the complexity of the social and economic environment requires the explicit consideration of criteria other than the cost. Examples of additional concerns include: average delivery time of the commodities, reliability of transportation, accessibility to the users, product deterioration, among others. Until now, many researchers also have great interest in the multiobjective transportation problem, and a number of methods had been proposed for solving it like (Ammar & Youness, 2005) studied the multiobjective transportation problem, in which the cost factors, the sizes of supply and demand are all fuzzy numbers, (chakrobarty & chakrobarty, 2010) presenting method for minimizing the cost and time for transportation problem for this they using linear membership function, (Mahapatra and et al., 2010) presented multiobjective stochastic unbalanced transportation problem, they used fuzzy programming techniques and stochastic method for randomness of sources and destination parameters in inequality type constraints, (Huseen and et al., 2011) introduced the possibilistic multiobjective transportation problem is considered possibilistic parameters with known distribution into multiobjective programming frame work, recently (Sharma and et al., 2015) have used new proposed algorithm to minimize cost and time of goods which is supply from one source to another source. It is clear from the literature review of algorithms that a limited amount of work has been done in the area of multiobjective transportation problems. The majority of the research work has been focused in the area of transportation two objective problem. In this paper we will provide solution to multiobjective transportation problems using new heuristic algorithms.

## 2. Multiobjective Transportation Problem (MOTP)

Standard MOTP is to transfer goods from several origins to different destination subject to linear constraints such that all objective optimal. Suppose there are (m) origins of goods denoted by $O_1, O_2, ..., O_m$ having supplies $a_1, a_2, ..., a_m$ and (n) destinations denoted by $D_1, D_2, ..., D_n$ having demands $b_1, b_2, ..., b_m$. We assume total demand is equal to total supply. Mathematically $\sum_{i=1}^{m} a_i = \sum_{j=1}^{n} b_j$. For each of the objective, $c_{ij}^k$ be the cost or penalty of transferring one unit from $i^{th}$ origin to $j^{th}$ destination such that $z_k$ is minimum for $k = 1,2,...,K$. A available $x_{ij}$ represent the unknown quantity to be shipped from $i^{th}$ source to $j^{th}$ destination. Thus MOTP is as under :

Minimize $z_k = \sum_{i=1}^{m} \sum_{j=1}^{n} c_{ij}^k x_{ij}$ for all $k = 1,2,...,K$

Subject to $\sum_{i=1}^{m} x_{ij} = b_j$ for all $j = 1,2,...,n$

$\sum_{j=1}^{n} x_{ij} = a_j$ for all $i = 1,2,...,m$

$x_{ij} \geq 0$ for all $i$ and $j$

In the sense of multiobjective transportation problems, objective functions are usually conflicting each other in nature and the concepts of optimal solution gives place to the concept of Pareto optimal (efficient, nondominated, noninferior) solutions, for which the improvement of one objective function is attained only by sacrificing another objective function. Denoted the feasible set in decision space by $F$, for the multicriteria optimization we can get the following definitions for Pareto optimal solution and ideal point.

Definition (1)(Huseen and et al., 2011) : A feasible solution $x^* = [x_{ij}^*]$ is said to be a Pareto optimal solution if and only if there does not exist another $x \in F$ such that $z_k(x) \leq z_k(x^*)$ for every $k = 1,2,...,K$ and $z_p(x) \neq z_p(x^*)$ for some $p$. If $x^*$ is efficient solution, then $z(x^*)$ is called a non-dominated point. The set of all efficient solutions $x \in F$ is called efficient set and the set of all non-dominated points $z(x^*)$ is called non-dominated set or Pareto front.

Definition(2)(Huseen and et al.,2011): the ideal point is one that optimize each objective function simultaneously. It can then be defined as $z^* = [z_1^*, z_2^*, ..., z_k^*]$ where $z_k^*$ is the optimal value for the $k-th$ objective without considering other objectives.

## 3. Heuristic Proposed Algorithms

There are many algorithms that can be used for solving multiobjective transportation problems, which is to find the feasible solutions or at least approximation to it. The running time for the algorithm often increasing with the increase of the instance size. The purpose of any algorithm process is to find, for each instance a feasible solution called optimal, that minimize the objective function. This usual meaning of the optimum makes no sense in the multiobjective case because it doesn't exist, in most of the cases, a solution optimizing all objectives simultaneously. Hence we search for feasible solutions yielding the best compromise among objectives that constitutes a so called feasible solution set. These feasible solutions that cannot be improved in one objective without decreasing their performance in at least one of the others. It is clear that this feasible solutions set is difficult to find. Therefore, it could be preferable to have an approximation to that set in a reasonable amount of time. In this section, we shall try to find feasible solution for minimizing three objective transportation problem , where the first objective function is to minimize total flow time $F_{ij}$, the second objective function is to minimize total late work $V_{ij}$, where the late work for a task is the amount of processing of this task that is performed after its due date. and the last objective function is to minimize total tardiness $T_{ij}$, theses objective functions can be formulate into three transportation problems, the first transportation problem minimize two objective function total flow time $F_{ij}$ and total late work $V_{ij}$, and denoted by $(P_1)$. A mathematical model of MOTP with (2) objectives, (m) sources and (n) destinations can be written as: mathematically, the problem can be stated as

$$
\left.\begin{array}{lll}
\text{Minimize } z_1 = \sum_{i=1}^{m}\sum_{j=1}^{n} x_{ij}F_{ij} & & \\
z_2 = \sum_{i=1}^{m}\sum_{j=1}^{n} x_{ij}V_{ij} & & \\
\text{Subject to } \sum_{i=1}^{m} x_{ij} = b_j & \text{for all} & j = 1,2,\dots,n \\
\sum_{j=1}^{n} x_{ij} = a_j & \text{for all} & i = 1,2,\dots,m \\
x_{ij} \geq 0 & \text{for all} & i \text{ and } j
\end{array}\right\} \quad (P_1)
$$

The following heuristic algorithm (SCSVT) gives the feasible solution for problem $(P_1)$ .

Algorithm (1): Heuristic algorithm(SCTSE)

Step (0): check the given problem is balanced

Step (1): we check the number of rows and columns are equal or not. If number of row are not equal to number of columns and vice versa. The dummy row or dummy column must be added with zero flow time/tardiness/late work elements with zero demand/supply, so our matrix becomes square matrix.

Step(2): put $V_\circ = (\frac{\sum V_{ij}}{N}) - 1$, where $N$ the number of the distinct $V_{ij}$ cell in the matrix.

Step(3): find cell $(IJ)^* \in N$, such that $V_{(IJ)^*} \leq V_\circ$, note that for this $V_\circ$, $x_{ij} = o$ if $V_{ij}. > V_\circ$ for each $ij \in N$ and solve the problem with minimum total completion time objective function( $F_{ij}$ ), if $F_{(IJ)^*} = F_{iJ}$ choose the cell with smallest $V_{(IJ)^*}$, if no cells with $V_{(IJ)^*} \leq V_\circ$ go to step(4).

Step (4): stop.

Example(1): Consider the problem $(P_1)$ with the following data:

| | | DESTINATIONS | | | | SUPPLIES, |
|---|---|---|---|---|---|---|
| | i/j | B1 | B2 | B3 | B4 | Ai |
| SOURCES | A1 | 4<br>6 | 10<br>1 | 11<br>9 | 35<br>23 | 25 |
| | A2 | 38<br>5 | 25<br>20 | 10<br>6 | 49<br>11 | 15 |
| | A3 | 19<br>17 | 8<br>2 | 25<br>20 | 23<br>10 | 15 |
| | A4 | 11<br>4 | 12<br>8 | 15<br>3 | 13<br>7 | 30 |
| | DEMANDS, bi | 20 | 25 | 25 | 15 | 85 |

First, we calculate: $V_{\circ} = (\frac{\sum V_{ij}}{N}) - 1 \Rightarrow V_{\circ} = (\frac{1+2+3+4+5+6+7+8+9+10+11+17+20+23}{14}) - 1$ , and so

$V_{\circ} = (\frac{126}{14}) - 1 \Rightarrow V_{\circ} = 9 - 1 = 8.$ Now, we ignore all cells that have $V_{ij}. > 8$, and choose the remaining cells with lest completion time, as shown in the following table:

| | | DESTINATIONS | | | | SUPPLIES, |
|---|---|---|---|---|---|---|
| | i/j | B1 | B2 | B3 | B4 | ai |
| SOURCES | A1 | 15 4<br>6 | 10 10<br>1 | 11<br>9 | 35<br>23 | 25 |
| | A2 | 38<br>5 | 25<br>20 | 10 10<br>6 | 49<br>11 | 15 |
| | A3 | 19<br>17 | 15 8<br>2 | 25<br>20 | 23<br>10 | 15 |
| | A4 | 5 11<br>4 | 12<br>8 | 15 15<br>3 | 15 13<br>7 | 30 |
| | DEMANDS, bi | 20 | 25 | 25 | 15 | 85 |

then we have the following feasible solutions for each objective fuctions:

$$z_1 = \sum\sum x_{ij} F_{ij} = (15\times4) + (10\times10) + (10\times10) + (15\times8) + (5\times11) + (15\times15) + (15\times13) = 775$$

$$z_2 = \sum\sum x_{ij} V_{ij} = (15\times6) + (10\times1) + (10\times6) + (15\times2) + (5\times4) + (15\times3) + (15\times7) = 360.$$

the second transportation problem also minimize two objective function total flow time $F_{ij}$ and total tardiness $T_{ij}$, and denoted by $(P_2)$. A mathematical model of MOTP with (2) objectives, (m) sources and (n) destinations can be written as: mathematically, the problem can be stated as

$$\left.\begin{aligned} \text{Minimize } z_1 &= \sum_{i=1}^{m}\sum_{j=1}^{n} x_{ij} F_{ij} \\ z_2 &= \sum_{i=1}^{m}\sum_{j=1}^{n} x_{ij} T_{ij} \\ \text{Subject to } & \sum_{i=1}^{m} x_{ij} = b_j \quad \text{for all } j = 1,2,\ldots,n \\ & \sum_{j=1}^{n} x_{ij} = a_j \quad \text{for all } i = 1,2,\ldots,m \\ & x_{ij} \geq 0 \quad \text{for all } i \text{ and } j \end{aligned}\right\} \quad (P_2)$$

The following heuristic algorithm (SCSTV) gives the feasible solution for problem $(P_2)$ .

Algorithm (2): Heuristic algorithm(SCSTV)

Step (0): check the given problem is balanced

Step (1): we check the number of rows and columns are equal or not. If number of row are not equal to number of columns and vice versa. The dummy row or dummy column must be added with zero flow time/tardiness/late work elements with zero demand/supply, so our matrix becomes square matrix.

Step(2): put $T_{\circ} = (\frac{\sum T_{ij}}{N}) - 1$, where $N$ the number of the distinct $T_{ij}$ cell in the matrix.

Step(3): find cell $(IJ)^* \in N$, such that $T_{(IJ)^*} \leq T_{\circ}$, note that for this $T_{\circ}$, $x_{ij} = o$ if $T_{ij^*} > T_{\circ}$ for each $ij \in N$ and solve the problem with minimum total completion time objective function( $F_{ij}$ ), if $F_{(IJ)^*} = F_{iJ}$ choose the cell with smallest $T_{(IJ)^*}$, if no cells with $T_{(IJ)^*} \leq T_{\circ}$ go to step(4).

Step (4):stop.

Example(2): Consider the problem $(P_2)$ with the following data:

| | | DESTINATIONS | | | | SUPPLIES, |
|---|---|---|---|---|---|---|
| | i/j | B1 | B2 | B3 | B4 | ai |
| SOURCES | A1 | 50<br>17 | 41<br>5 | 52<br>16 | 42<br>6 | 30 |
| | A2 | 50<br>17 | 48<br>18 | 30<br>13 | 46<br>29 | 15 |
| | A3 | 70<br>14 | 65<br>25 | 54<br>23 | 44<br>15 | 15 |
| | A4 | 44<br>10 | 80<br>20 | 37<br>7 | 43<br>12 | 40 |
| | DEMANDS, bi | 20 | 15 | 35 | 30 | 100 |

First we calculate: $T_{\circ} = (\frac{\sum T_{ij}}{N}) - 1 \Rightarrow T_{\circ} = (\frac{5+6+7+10+12+13+14+15+16+17+18+20+23+25+29}{15}) - 1$ ,

and so $T_{\circ} = (\frac{230}{15}) - 1 \Rightarrow T_{\circ} = 15 - 1 = 14$. Now, we ignore all cells that have $T_{ij} > 14$, and choose the remaining cells with lest completion time, as shown in the following table:

| | | DESTINATIONS | | | | SUPPLIES, |
|---|---|---|---|---|---|---|
| | i/j | B1 | B2 | B3 | B4 | ai |
| SOURCES | A1 | 50<br>17 | 15 41<br>5 | 52<br>16 | 15 42<br>6 | 30 |
| | A2 | 50<br>17 | 48<br>18 | 15 30<br>13 | 46<br>29 | 15 |
| | A3 | 15 70<br>14 | 65<br>25 | 54<br>23 | 44<br>15 | 15 |
| | A4 | 5 44<br>10 | 80<br>20 | 20 37<br>7 | 15 43<br>12 | 40 |
| | DEMANDS, bi | 20 | 15 | 35 | 30 | 100 |

then we have the following feasible solutions for each objective functions:

$z_1 = \sum\sum x_{ij} F_{ij} = (15\times 41) + (15\times 42) + (15\times 30) + (15\times 70) + (5\times 44) + (20\times 37) + (15\times 43) = 4350$

$z_2 = \sum\sum x_{ij} T_{ij} = (15\times 5) + (15\times 6) + (15\times 13) + (15\times 14) + (5\times 10) + (20\times 7) + (15\times 12) = 940.$

the last transportation problem minimize three objective function total flow time $F_{ij}$ , total late work $V_{ij}$ and total tardiness $T_{ij}$, and denoted by $(P_3)$. A mathematical model of MOTP with (3) objectives, (m) sources and (n) destinations can be written as: mathematically, the problem can be stated as

$$\left.\begin{array}{lll} \text{Minimize} & z_1 = \sum_{i=1}^{m}\sum_{j=1}^{n} x_{ij} F_{ij} & \\ & z_2 = \sum_{i=1}^{m}\sum_{j=1}^{n} x_{ij} V_{ij} & \\ & z_3 = \sum_{i=1}^{m}\sum_{j=1}^{n} x_{ij} T_{ij} & \\ \text{Subject to} & \sum_{i=1}^{m} x_{ij} = b_j & \text{for all } j = 1,2,\ldots,n \\ & \sum_{j=1}^{n} x_{ij} = a_j & \text{for all } i = 1,2,\ldots,m \\ & x_{ij} \geq 0 & \text{for all } i \text{ and } j \end{array}\right\} \quad (P_3)$$

The following heuristic algorithm (SCSVST) gives the feasible solution for problem $(P_3)$.

Algorithm (3): Heuristic algorithm(SCSVST)

Step (0): check the given problem is balanced

Step (1): we check the number of rows and columns are equal or not. If number of row are not equal to number of columns and vice versa. The dummy row or dummy column must be added with zero flow time/tardiness/late work elements with zero demand/supply, so our matrix becomes square matrix.

Step(2): compute $V_\circ = (\frac{\sum V_{ij}}{N}) - 1$ and $T_\circ = (\frac{\sum T_{ij}}{N}) - 1$, where $N$ the number of the distinct $V_{ij}, T_{ij}$ cells in the matrix and put $\Omega = \min\{V_\circ, T_\circ\}$.

Step(3): if $\Omega = V_\circ$ , find cell $(IJ)^* \in N,$ such that $T_{(IJ)^*} \leq \Omega,$ note that for this $\Omega,$ $x_{ij} = o$ if $T_{ij}.> \Omega$ for each $ij \in N$ and solve the problem with minimum total completion time objective function($F_{ij}$), if $\Omega = T_\circ$ , find cell $(IJ)^* \in N,$ such that $V_{(IJ)^*} \leq \Omega,$ note that for this $\Omega,$ $x_{ij} = o$ if $V_{ij}.> \Omega$ for each $ij \in N$ and solve the problem with minimum total completion time objective function($F_{ij}$) , if $F_{(IJ)^*} = F_{iJ}$ choose the cell with smallest $T_{(IJ)^*}$ and if $T_{(IJ)^*} = T_{ij}$ choose the cell with smallest $V_{(IJ)^*}$, if no cells with $T_{(IJ)^*} \leq T_\circ$ go to step(4).

Step (4):stop.

Example(3): Consider the problem $(P_3)$ with the following data:

| | | DESTINATIONS | | | | SUPPLIES, ai |
|---|---|---|---|---|---|---|
| | i/j | B1 | B2 | B3 | B4 | |
| | A1 | 4 | 10 | 11 | 35 | 20 |
| | | 1 | 9 | 9 | 27 | |
| | | 6 | 1 | 1 | 20 | |
| SOURCES | A2 | 38 | 25 | 10 | 49 | 10 |
| | | 32 | 22 | 6 | 42 | |
| | | 13 | 15 | 6 | 11 | |
| | A3 | 19 | 8 | 25 | 35 | 15 |
| | | 17 | 2 | 22 | 28 | |
| | | 11 | 2 | 6 | 10 | |
| | A4 | 10 | 12 | 15 | 13 | 35 |
| | | 6 | 5 | 6 | 7 | |
| | | 6 | 2 | 3 | 2 | |
| | DEMANDS, bi | 25 | 25 | 15 | 15 | 80 |

First we calculate:

$$T_{\circ} = \left(\frac{\sum T_{ij}}{N}\right) - 1 \Rightarrow T_{\circ} = \left(\frac{1+2+5+6+7+9+17+22+27+28+32}{15}\right) - 1, \text{ and so}$$

$$T_{\circ} = \left(\frac{156}{11}\right) - 1 \Rightarrow T_{\circ} = 14 - 1 = 13.\ V_{\circ} = \left(\frac{\sum V_{ij}}{N}\right) - 1 \Rightarrow V_{\circ} = \left(\frac{1+2+3+6+10+11+13+15+20}{9}\right) - 1, \text{ and so}$$

$V_{\circ} = \left(\frac{81}{9}\right) - 1 \Rightarrow V_{\circ} = 9 - 1 = 8.$ Now, $\Omega = \min\{V_{\circ}, T_{\circ}\} = \min\{8,13\} = 8$, we ignore all cells that have $T_{ij}. > 8$, and choose the remaining cells with lest completion time, as shown in the following table:

| | | DESTINATIONS | | | | SUPPLIES, ai |
|---|---|---|---|---|---|---|
| | i/j | B1 | B2 | B3 | B4 | |
| | A1 | 20 4 | 10 | 11 | 35 | 20 |
| | | 1 | 9 | 9 | 27 | |
| | | 6 | 1 | 1 | 20 | |
| **SOURCES** | A2 | 38 | 25 | 10 10 | 49 | 10 |
| | | 32 | 22 | 6 | 42 | |
| | | 13 | 15 | 6 | 11 | |
| | A3 | 19 | 15 8 | 25 | 35 | 15 |
| | | 17 | 2 | 22 | 28 | |
| | | 11 | 2 | 6 | 10 | |
| | A4 | 5 10 | 10 12 | 5 15 | 15 13 | 35 |
| | | 6 | 5 | 6 | 7 | |
| | | 6 | 2 | 3 | 2 | |
| | DEMANDS, bi | 25 | 25 | 15 | 15 | 80 |

then we have the following feasible solutions for each objective functions:

$$z_1 = \sum\sum x_{ij} F_{ij} = (20\times 4) + (10\times 10) + (15\times 8) + (5\times 10) + (10\times 12) + (5\times 15) + (15\times 13) = 740$$

$$z_2 = \sum\sum x_{ij} T_{ij} = (20\times 4) + (10\times 6) + (15\times 2) + (5\times 6) + (10\times 5) + (5\times 6) + (15\times 7) = 385$$

$$z_2 = \sum\sum x_{ij} V_{ij} = (20\times 6) + (10\times 6) + (15\times 2) + (5\times 6) + (10\times 2) + (5\times 3) + (15\times 2) = 275.$$

## 4. Conclusion

In this paper, we conclude that the three heuristic algorithms found feasible solutions for the three problems optimality, these heuristic algorithms can be applied to different field such as military affair. It can be seen from the numerical examples that these heuristic algorithms provided in this paper is easier to compute and the results obtained are better.

## References

Ammar, E. E., & Youness, E. A. (2005). Study on multi-objective transportation problem wih fuzzy numbers. *Applies mathematics and computation, 166*, 241-253. http://dx.doi.org/10.1016/j.amc.2004.04.103

Chakraborty, A., & Chakraborty, M. (2010). Cost –time minimization in a transportation problem with fuzzy parameters: A case study. *Journal of Transpn. Sys. Engg. And IT., 10*(6), 53-63. http://dx.doi.org/10.1016/s1570-6672(09)60071-4

Huseen, M. L., Elbana, A. H., & Gaafer, A. G . (2011). Possibilistic multi-objective transportation problem. *International journal of optimization: theory, methods and applications, 3*(1), 1-20.

Mahapatra, D. R., Roy, S. K., & Biswal, M. P. (2010). Stochastic based multi-objective transportation problem involving normal randomness. *Advance modeling and optimization journal, 12*(2), 205-223.

Sharma, G., Abbas, S. H., & Gupta, V. K. (2015). Solving multi-objective transportation problem to reduce transportation cost and time. *Journal of advances in mathematics, 11*(1), 3908-3912.

# Approximation of a Second-order Elliptic Equation with Discontinuous and Highly Oscillating Coefficients by Finite Volume Methods

Bienvenu Ondami[1]

[1] Univesrité Marien Ngouabi, Brazzaville, Congo

Correspondence: Université Marien Ngouabi, Faculté des Sciences et Techniques, BP. 69, Brazzaville, Congo. E-mail: bondami@gmail.com

**Abstract**

In this paper we consider the numerical approximation of a class of second order elliptic boundary value problems with discontinuous and highly periodically oscillating coefficients. We apply both classical and modified finite volume methods for the approximate solution of this problem. Error estimates depending on $\varepsilon$ the parameter involved in the periodic homogenization are established. Numerical simulations for one-dimensional problem confirm the theorical results and also show that the modified scheme has a smaller constant of convergence than the classical scheme based on harmonic averaging for this class of equations.

**Keywords:** homogenization, elliptic equations, oscillating coefficients, finite volume method, finite difference

## 1. Introduction

There are many practical computational problems with highly oscillatory solutions e.g. computation of flow in heterogeneous porous media for petroleum and groundwater reservoir simulation (see, e.g., (Hornung, 1997) and the bibliographies therein). If a porous medium with a periodic structure is considered, with the size of the period is small enough compared to the size of the reservoir, and denoting their ratio by $\varepsilon$ $(0 < \varepsilon << 1)$ an asymptotic analysis, as $\varepsilon \longrightarrow 0$, is required.
In this paper we will consider problems that are described by a linear elliptic equation in divergence form with highly periodically oscillating coefficients. Especially we will consider the following model problem:

$$(P_\varepsilon) \quad \left\{ \begin{array}{rcll} -\text{div}\,(K^\varepsilon(x)\,\nabla u^\varepsilon) & = & f & \text{in } \Omega, \\ u_\varepsilon & = & 0 & \text{on } \Gamma. \end{array} \right.$$

$\Omega \subset \mathbb{R}^n$ $(n = 1, 2, 3)$ is a bounded polygonal convex domain with a periodic structure and smooth boundary $\Gamma$, $K^\varepsilon(x) = K(x/\varepsilon)$, $K$ is a symmetric and uniformly positive definite matrix in $\Omega$ which has jumps discontinuities across a given interface. The case of piecewise constant coefficient $K^\varepsilon$ is very important for the applications.
In porous medium flow, the problem $(P_\varepsilon)$ results from Darcy's law and continuity for a single phase, incompressible flow through a horizontal heterogeneous porous medium with periodic structure.

Using the homogenization tools (see, e.g., (Bakhvalov & Panasenko, 1989), ( Bensoussan, Lions & Papanicolaous, 1987), (Jikov, Kozlov & Oleinik, 1994) and (Sanchez-Palancia, 1980) ) original Problem $(P_\varepsilon)$ can be replaced by homogenized Problem, modeling some average quantity without the oscillations.

Whenever effective equations are applicable they are very useful for computational purposes. There are however many situations for which $\varepsilon$ is not sufficiently small so that the effective equations are not practical. In this cases the original equation has to be approximated directly.
The numerical approximation partial differential equations with highly oscillating coefficients has been a problem of interest for many years and many methods have been developed (see, e.g., ( Amaziane & Ondami, 1999), (Chen & Hou, 2002), (Versieux & Sarkis, 2008) and (Ondami, 2001, 2015) and the bibliographies therein).

The case where $K^\varepsilon$ has continuous coeffients is the most studied. The case of discontinuous coefficients has been addressed in some work as in (Bourgat, 1978) and (Bourgat & Dervieux, 1978) where Authors use a double-scale asymptotic expansion. In (Amaziane & Ondami, 1999) and (Ondami, 2001), the numerical approximation of the problem, in the case of discontinuous coefficients was done by finite elements methods. However no error estimate has been established and is still, to our knowledge, an opened issue.

Elliptic problems with discontinuous coefficients (often called interface problems) arise naturally in mathematical modeling processes in heat and mass transfer, diffusion in composite media flows in porous media etc.

In this paper the approximation will be done by finite volume methods (see, e.g., (Eymard, Gallouët & Herbin, 2000) and (Chernogorova, Ewing, Iliev & Lazarov, 2001)) and the study is limited to one-dimensional problem (1-D problem). This 1-D problem illustrates very clearly the dependence of numerical results to $\varepsilon$, the parameter of homogenization. Error estimates are established. The obtained results can be generalized in the two-dimensional and three-dimensional problems.

The paper is organized as follows. In section 2, a description of methods used is presented as well as the obtained error estimates. Section 3 is devoted to numerical simulations. Lastly, some concluding remarks are presented in section 4.

## 2. Methods

Our study will focus on the one-dimensional problem and we assume (without loss of generality) that $\Omega = ]0,1[$. In this case the problem $(P_\varepsilon)$ is written simply

$$\begin{cases} -\frac{d}{dx}\left(k^\varepsilon(x)\frac{du^\varepsilon}{dx}\right) = f \text{ in } (0,1), \\ u^\varepsilon(0) = u^\varepsilon(1) = 0. \end{cases} \tag{1}$$

where $k^\varepsilon(x) = k(x/\varepsilon) = k(y)$, with $y = x/\varepsilon$, $k$ is a discontinuous and periodic function of period 1, on $]0,1[$.

In all this paper we make the following assumptions

$(A1)$ $\alpha < k(y) \leqslant \beta$, a.e., in $]0,1[$, with some $\alpha, \beta \in \mathbb{R}_+^*$,

$(A2)$ $f \in L^2(]0,1[)$.

The assumptions $(A1)$ and $(A2)$ ensure the existence and uniqueness of the solution of the problem (1). From homogenization theory (see, e.g., (Sanchez-Palancia, 1980), (Bensoussan, Lions & Papanicolaous, 1987) and (Jikov, Kozlov & Oleinik, 1994)) follows

$$u_\varepsilon \rightharpoonup u \quad \text{in} \quad H_0^1(\Omega) \text{ (consists of functions in Sobolev space } H^1(\Omega) \text{ that vanish on 0 and 1) weakly,}$$

where $u$ (homogenized solution) satisfies the following homogenized problem,

$$\begin{cases} -\frac{d}{dx}\left(k^*\frac{du}{dx}\right) = f \text{ in } (0,1), \\ u(0) = u(1) = 0, \end{cases} \tag{2}$$

and the constant $k^*$ is the mean harmonic value of $k(y)$ on $(0,1)$.

This one-dimensional problem helps to clarify the eventual dependency of numerical results to the parameter $\varepsilon$. The error estimates obtained can be generalized in the two-dimensional and three-dimensional cases which uses, for instance, simplices or parallelepipedes mesh.

Two finite volume schemes will be compared: The classical scheme (see, e.g., (Eymard, Gallouët & Herbin, 2000)) and a modified scheme (see, (Ewing, Iliev & Lazarov, 2001)).

In order to compute a numerical approximation to the solution $u^\varepsilon$, let us define a mesh, denoted by $\mathcal{T}$, of the interval $(0,1)$ consisting of $N$ cells (or control volumes), denoted by $V_i$, $i = 1,...,N$, and $N$ points of $(0,1)$, denoted by $x_i$, $i = 1,...,N$, satisfying the following assumptions:

***Definition 1.*** *An admissible mesh of* $(0,1)$, *denoted by* $\mathcal{T}$, *is given by a family* $(V_i)_{i=1,...,N}$, *N a positive integer, such that* $V_i = \left(x_{i-\frac{1}{2}}, x_{i+\frac{1}{2}}\right)$, *and a family* $(x_i)_{i=0,...,N}$, *such that*

$$x_0 = x_{\frac{1}{2}} = 0 < x_1 < x_{\frac{3}{2}} < ... < x_{i-\frac{1}{2}} < x_i < x_{i+\frac{1}{2}} < ... < x_N < x_{N+\frac{1}{2}} = x_{N+1} = 1.$$

*One sets*

$$h_i = m(V_i) = x_{i+\frac{1}{2}} - x_{i-\frac{1}{2}},\ i = 1,...,N, \text{ and therefore } \sum_{i=1}^{N} h_i = 1,$$

$$h_i^- = x_i - x_{i-\frac{1}{2}},\ h_i^+ = x_{i+\frac{1}{2}} - x_i,\ i = 1,...,N,$$

$$h_{i+\frac{1}{2}} = x_{i+1} - x_i,\ i = 0,...,N,$$

$$size(\mathcal{T}) = h = max\{h_i,\ i = 1,...,N\}.$$

*2.1 Classical Finite Volume Scheme*

Let $\mathcal{T} = (V_i)_{i=1,...,N}$ be an admissible mesh, in the sense of *Definition 1* , such that the discontinuities of $k^\varepsilon$ coincide with the interfaces of the mesh.
Classical finite volume scheme consiste to integrate the first equation of the problem (1) over $V_i$ (see,e.g., (Eymard, Gallouët & Herbin, 2000)). So we have:

$$-\left(k^\varepsilon(x)\frac{du^\varepsilon(x)}{dx}\right)(x_{i+\frac{1}{2}}) + \left(k^\varepsilon(x)\frac{du^\varepsilon(x)}{dx}\right)(x_{i-\frac{1}{2}}) = \int_{V_i} f(x)dx, \quad i = 1,...,N. \tag{3}$$

Let

$$(u_i^\varepsilon)_{i=1,...,N} \text{ be the discrete unknows}$$

and let

$$k_i^\varepsilon = \frac{1}{h_i}\int_{V_i} k^\varepsilon(x)dx.$$

In order that the scheme be conservative, the discretization of the flux $-k^\varepsilon(x)\frac{du^\varepsilon(x)}{dx}$ at $x_{i+\frac{1}{2}}$ should have the same value on $V_i$ and $V_{i+1}$. To this purpose, one introduces the auxiliary unknown $u^\varepsilon_{i+\frac{1}{2}}$ (approximation of $u^\varepsilon$ at $x_{i+\frac{1}{2}}$). Since on $V_i$ and $V_{i+1}$, $k^\varepsilon$ is continuous, the approximation of $-k^\varepsilon(x)\frac{du^\varepsilon(x)}{dx}$ may be performe on each side of $x_{i+\frac{1}{2}}$ by using the finite difference principe:

$$H^\varepsilon_{i+\frac{1}{2}} = -k_i^\varepsilon \frac{u^\varepsilon_{i+\frac{1}{2}} - u_i^\varepsilon}{h_i^+} \quad \text{on } V_i, \quad i = 1,...,N,$$

$$H^\varepsilon_{i+\frac{1}{2}} = -k_{i+1}^\varepsilon \frac{u^\varepsilon_{i+1} - u^\varepsilon_{i+\frac{1}{2}}}{h_{i+1}^-} \quad \text{on } V_{i+1}, \quad i = 0,...,N-1,$$

with $u^\varepsilon_{1/2} = 0$ and $u^\varepsilon_{N+1/2} = 0$, for the boundary conditions. Requiring the two above approximation of $\left(k^\varepsilon\frac{du^\varepsilon}{dx}\right)\left(x_{i+\frac{1}{2}}\right)$ to be equal (conservativity of the flux) yields the value of $u^\varepsilon_{1+\frac{1}{2}}$ (for $i = 1,...,N-1$) :

$$u^\varepsilon_{1+\frac{1}{2}} = \frac{u^\varepsilon_{i+1}\frac{k^\varepsilon_{i+1}}{h^-_{i+1}} + u_i^\varepsilon\frac{k_i^\varepsilon}{h_i^+}}{\frac{k^\varepsilon_{i+1}}{h^-_{i+1}} + \frac{k_i^\varepsilon}{h_i^+}}$$

which, in turn, allows to give expression of the approximation $H^\varepsilon_{i+\frac{1}{2}}$ of $\left(k^\varepsilon\frac{du^\varepsilon}{dx}\right)\left(x_{i+\frac{1}{2}}\right)$:

$$H^\varepsilon_{i+\frac{1}{2}} = -\tau^\varepsilon_{i+\frac{1}{2}}\left(u^\varepsilon_{i+1} - u_i^\varepsilon\right), \quad i = 1,...,N-1, \tag{4}$$

$$H^\varepsilon_{\frac{1}{2}} = -\frac{k_1^\varepsilon}{h_1^-}u_1^\varepsilon, \tag{5}$$

$$H^\varepsilon_{N+\frac{1}{2}} = \frac{k_N^\varepsilon}{h_N^+}u_N^\varepsilon, \tag{6}$$

with

$$\tau^\varepsilon_{i+\frac{1}{2}} = \frac{k_i^\varepsilon k^\varepsilon_{i+1}}{h_i^+ k^\varepsilon_{i+1} + h^-_{i+1}k_i^\varepsilon}, \quad i = 1,...,N-1. \tag{7}$$

***Remark 1.*** If $h_i = h$, for all $i \in 1,...,N$, and $x_i$ is assumed to be center of $V_i$, then $h_i^- = h_i^+ = \frac{h}{2}$, so that

$$H^\varepsilon_{i+\frac{1}{2}} = -\frac{2k_i^\varepsilon k^\varepsilon_{i+1}}{k_i^\varepsilon + k^\varepsilon_{i+1}}\frac{\left(u^\varepsilon_{i+1} - u_i^\varepsilon\right)}{h},$$

and therefore the mean harmonic value of $k^\varepsilon$ is involved.

The numerical scheme for the approximation of Problem (1) is therefore,

$$H^\varepsilon_{i+\frac{1}{2}} - H^\varepsilon_{i-\frac{1}{2}} = h_i f_i, \quad \forall i \in 1,...,N. \tag{8}$$

with

$$f_i = \frac{1}{h_i}\int_{x_{i-\frac{1}{2}}}^{x_{i+\frac{1}{2}}} f(x)dx, \text{ for } i = 1, ..., N, \text{ and where } \left(H_{i+\frac{1}{2}}\right)_{i\in\{0,...,N\}} \text{ is defined by (4)-(6)}$$

Taking (7) and 4)-(6) into account, the scheme (8) yields a system of $N$ equations with $N$ unknowns $u_1^\varepsilon, ..., u_N^\varepsilon$.

***Remark 2.*** The fact that $k^\varepsilon$ is discontinuous, periodic (period $\varepsilon$) and with discontinuities that coincide with the interfaces of the mesh $\mathcal{T}$ leads to say that:

$\varepsilon = \frac{1}{n_p}$, where $n_p$ is a positive integer (periods number), and if $h_i = h = size(\mathcal{T})$, for all $i \in \{1, ..., N\}$ then $h = \frac{\varepsilon}{m}$, where $1 \leq m < N$.

***Note.*** Throughout the paper, we will denote by $c$ generic constants, even if they take different values at different places.

We now state the main result of this section.

**Theorem 1.** *Let $\mathcal{T} = (V_i)_{i=1,...,N}$ be an admissible mesh of $(0,1)$, in the sens of Definition 1, and uniform (i.e. $h_i = h$, $\forall i \in \{1, ...N\}$) such that*
*1) $x_i$ is the center of $(V_i)_{i=1,...,N}$, and the discontinuities of $k^\varepsilon$ coincide with the interfaces of the mesh,*
*2) $k^\varepsilon \in C^1\left(\overline{V_i}\right)$, and $f \in C^1\left(\overline{V_i}\right)$, for all $i = 1, ..., N$.*

*Let $e_i^\varepsilon = u^\varepsilon(x_i) - u_i^\varepsilon$, $e_i = u(x_i) - u_i^\varepsilon$ where $u^\varepsilon$ is the solution of Problem (1), $u$ is the homogenized solution and $u^{\varepsilon h} = (u_i^\varepsilon)_{i=1,...,N}$ is the solution of (4)-(8). Then there exists a constant $c$ independent of $\varepsilon$ and $h$ such that*

$$\left\|u^\varepsilon - u^{\varepsilon h}\right\|^2_{H^1(\Omega)} \equiv \sum_{i=1}^{N} \tau^\varepsilon_{i+\frac{1}{2}}\left(e^\varepsilon_{i+1} - e^\varepsilon_i\right)^2 \leqslant \frac{ch^2}{\varepsilon}, \tag{9}$$

where $\tau^\varepsilon_{i+\frac{1}{2}}$ is defined in (7), and

$$\left\|u^\varepsilon - u^{\varepsilon h}\right\|_{L^\infty(\Omega)} \equiv \max_{1\leq i\leq N} | e_i^\varepsilon | \leqslant c\,h, \tag{10}$$

$$\left\|u - u^{\varepsilon h}\right\|_{L^\infty(\Omega)} \equiv \max_{1\leq i\leq N} | e_i | \leqslant c\,h + c\varepsilon. \tag{11}$$

*Proof.*

The proof of the Theorem 1 is obtained by using the same gait as in (Eymard, Gallouët, & Herbin, 2000). Let

$$\overline{H}^\varepsilon_{i+\frac{1}{2}} = -\left(k^\varepsilon\frac{du^\varepsilon}{dx}\right)\left(x_{i+\frac{1}{2}}\right) \text{ and } H^{*,\varepsilon}_{i+\frac{1}{2}} = -\tau^\varepsilon_{i+\frac{1}{2}}\left(u^\varepsilon(x_{i+1}) - u^\varepsilon(x_i)\right), \text{ for } i = 0, ..., N, \text{ with } \tau^\varepsilon_{\frac{1}{2}} = \frac{k_1^\varepsilon}{h_1^-} \text{ and } \tau^\varepsilon_{N+\frac{1}{2}} = \frac{k_N^\varepsilon}{h_N^+}. \tag{12}$$

Let us first show there exists $c$ independent of $\varepsilon$ and $h$ such that

$$H^{*,\varepsilon}_{i+\frac{1}{2}} = \overline{H}^\varepsilon_{i+\frac{1}{2}} + T^\varepsilon_{i+\frac{1}{2}}, \quad \left|T^\varepsilon_{i+\frac{1}{2}}\right| \leq \frac{ch}{\varepsilon}, \; i = 0, ..., N. \tag{13}$$

In order to show this, let us introduce

$$H^{*,-,\varepsilon}_{i+\frac{1}{2}} = -k_i^\varepsilon\frac{u^\varepsilon(x_{i+\frac{1}{2}}) - u^\varepsilon(x_i)}{h_i^+} \text{ and } H^{*,+,\varepsilon}_{i+\frac{1}{2}} = -k^\varepsilon_{i+1}\frac{u^\varepsilon(x_{i+1}) - u^\varepsilon(x_{i+\frac{1}{2}})}{h^-_{i+1}} \tag{14}$$

Since $k^\varepsilon \in C^1\left(\overline{V_i}\right)$, one has $u^\varepsilon \in C^2\left(\overline{V_i}\right)$. By developing the first equation of (1) in $\overline{V_i}$ $(i = 1, ..., N)$, one obtains

$$-k^\varepsilon\frac{d^2u^\varepsilon}{d^2x} - \frac{1}{\varepsilon}\frac{dk\left(\frac{x}{\varepsilon}\right)}{dx}\frac{du^\varepsilon}{dx} = f \text{ in } \overline{V_i}, \quad i = 1, ..., N. \tag{15}$$

According to (Amaziane & Ondami, 1999) and (Ondami, 2001), one has

$$\|u^\varepsilon\|_{H^1(\Omega)} \leq c, \quad i = 1, ..., N.$$

Hence

$$\left\|\frac{du^{\varepsilon}}{dx}\right\|_{L^{\infty}(V_i)} \leq c, \quad i = 1, ..., N, \tag{16}$$

and

$$\left\|\frac{d^2u^{\varepsilon}}{d^2x}\right\|_{L^{\infty}(V_i)} \leq \frac{c}{\varepsilon}, \quad i = 1, ..., N. \tag{17}$$

By using (A1) and (14)-(17), one deduces that there exists $c$ independent of $\varepsilon$ and $h$ such that

$$H^{*,-,\varepsilon}_{i+\frac{1}{2}} = \overline{H}^{\varepsilon}_{i+\frac{1}{2}} + R^{-,\varepsilon}_{i+\frac{1}{2}}, \quad \text{where } \left|R^{-,\varepsilon}_{i+\frac{1}{2}}\right| \leqslant \frac{ch}{\varepsilon}, \quad i = 1, ..., N, \tag{18}$$

$$H^{*,+,\varepsilon}_{i+\frac{1}{2}} = \overline{H}^{\varepsilon}_{i+\frac{1}{2}} + R^{+,\varepsilon}_{i+\frac{1}{2}}, \quad \text{where } \left|R^{+,\varepsilon}_{i+\frac{1}{2}}\right| \leqslant \frac{ch}{\varepsilon}, \quad i = 0, ..., N-1. \tag{19}$$

This yields (13) for $i = 0$ and $i = N$.

The following equality:

$$\overline{H}^{\varepsilon}_{i+\frac{1}{2}} = H^{*,-,\varepsilon}_{i+\frac{1}{2}} - R^{-,\varepsilon}_{i+\frac{1}{2}} = H^{*,+,\varepsilon}_{i+\frac{1}{2}} - R^{+,\varepsilon}_{i+\frac{1}{2}}, \quad i = 1, ..., N-1 \tag{20}$$

yields that

$$u^{\varepsilon}\left(x_{i+\frac{1}{2}}\right) = \frac{\frac{k^{\varepsilon}_{i+1}}{h^{-}_{i+1}} u^{\varepsilon}(x_{i+1}) + \frac{k^{\varepsilon}_{i}}{h^{+}_{i}} u^{\varepsilon}(x_i)}{\frac{k^{\varepsilon}_{i}}{h^{+}_{i}} + \frac{k^{\varepsilon}_{i+1}}{h^{-}_{i+1}}} + S^{\varepsilon}_{i+\frac{1}{2}}, \quad i = 1, ..., N-1, \tag{21}$$

where

$$S^{\varepsilon}_{i+\frac{1}{2}} = \frac{R^{+,\varepsilon}_{i+\frac{1}{2}} - R^{-,\varepsilon}_{i+\frac{1}{2}}}{\frac{k^{\varepsilon}_{i}}{h^{+}_{i}} + \frac{k^{\varepsilon}_{i+1}}{h^{-}_{i+1}}}$$

So that

$$\left|S^{\varepsilon}_{i+\frac{1}{2}}\right| \leqslant \frac{1}{\alpha} \frac{h^{+}_{i} h^{-}_{i+1}}{h^{+}_{i} + h^{-}_{i+1}} \left|R^{+,\varepsilon}_{i+\frac{1}{2}} - R^{-,\varepsilon}_{i+\frac{1}{2}}\right|$$

Let us replace the expression (21) of $u^{\varepsilon}\left(x_{i+\frac{1}{2}}\right)$ in $H^{*,-,\varepsilon}_{i+\frac{1}{2}}$ defined by (14); this yields

$$H^{*,-,\varepsilon}_{i+\frac{1}{2}} = -\tau_{i+\frac{1}{2}} \left(u^{\varepsilon}(x_{i+1}) - u^{\varepsilon}(x_i)\right) - \frac{k^{\varepsilon}_{i}}{h^{+}_{i}} S^{\varepsilon}_{i+\frac{1}{2}}, \quad i = 1, ..., N-1. \tag{22}$$

Using (20), this implies that

$$H^{*,\varepsilon}_{i+\frac{1}{2}} = \overline{H}^{\varepsilon}_{i+\frac{1}{2}} + T^{\varepsilon}_{i+\frac{1}{2}}$$

where

$$\left|T^{\varepsilon}_{i+\frac{1}{2}}\right| \leqslant \left|R^{-,\varepsilon}_{i+\frac{1}{2}}\right| + \left|R^{+,\varepsilon}_{i+\frac{1}{2}} - R^{-,\varepsilon}_{i+\frac{1}{2}}\right| \frac{\beta}{2\alpha}. \tag{23}$$

Using (18) and (19), this last inequality yields that there exists c, independent of $\varepsilon$ and $h$ such that

$$\left|H^{*,\varepsilon}_{i+\frac{1}{2}} - \overline{H}^{\varepsilon}_{i+\frac{1}{2}}\right| = \left|T^{\varepsilon}_{i+\frac{1}{2}}\right| \leqslant \frac{ch}{\varepsilon}, \quad i = 1, ..., N-1.$$

Therefore (13) is proved.

Now, from (3) and (12), one has

$$\overline{H}^{\varepsilon}_{i+\frac{1}{2}} - \overline{H}^{\varepsilon}_{i-\frac{1}{2}} = h_i f_i, \quad \forall i \in \{1, ..., N\}. \tag{24}$$

Using (13) yields that

$$\overline{H}^{*,\varepsilon}_{i+\frac{1}{2}} - \overline{H}^{*,\varepsilon}_{i-\frac{1}{2}} = h_i f_i + T^{\varepsilon}_{i+\frac{1}{2}} - T^{\varepsilon}_{i-\frac{1}{2}}, \quad \forall i \in \{1, ..., N\}. \tag{25}$$

Let $e_i^\varepsilon = u^\varepsilon(x_i) - u_i^\varepsilon$ for $i = 1, ..., N$, and $e_0^\varepsilon = e_{N+1}^\varepsilon = 0$. Substracting (8) from (25) yields

$$-\tau^\varepsilon_{i+\frac{1}{2}}\left(e^\varepsilon_{i+1} - e^\varepsilon_i\right) + \tau^\varepsilon_{i-\frac{1}{2}}\left(e^\varepsilon_i - e^\varepsilon_{i-1}\right) = T^\varepsilon_{i+\frac{1}{2}} - T^\varepsilon_{i-\frac{1}{2}}, \quad \forall i \in \{1, ..., N\}.$$

Let us multiply this equation by $e_i^\varepsilon$, sum for $i = 1, ..., N$, reorder the summation. Therefore

$$\sum_{i=0}^{N} \tau^\varepsilon_{i+\frac{1}{2}}\left(e^\varepsilon_i - e^\varepsilon_{i-1}\right)^2 = \sum_{i=1}^{N} T^\varepsilon_{i+\frac{1}{2}}\left(e^\varepsilon_{i+1} - e^\varepsilon_i\right)$$

Thanks to (13), one has

$$\sum_{i=0}^{N} \tau^\varepsilon_{i+\frac{1}{2}}\left(e^\varepsilon_i - e^\varepsilon_{i-1}\right)^2 \leqslant \sum_{i=1}^{N} \frac{ch}{\varepsilon}\left|e^\varepsilon_{i+1} - e^\varepsilon_i\right|.$$

Denote by

$$A = \left(\sum_{i=0}^{N} \tau^\varepsilon_{i+\frac{1}{2}}\left(e^\varepsilon_{i+1} - e^\varepsilon_i\right)^2\right)^{\frac{1}{2}} \quad \text{and} \quad B = \left(\sum_{i=0}^{N} \frac{1}{\tau^\varepsilon_{i+\frac{1}{2}}}\right)^{\frac{1}{2}}$$

The Cauhy-Schwarz inequality yields

$$A^2 \leqslant \frac{ch}{\varepsilon} AB.$$

Now, since the mesh is uniform (i.e. $h_i = h, \forall i \in \{1, ..., N\}$), one has

$$\frac{1}{\tau^\varepsilon_{i+\frac{1}{2}}} \leqslant \frac{\beta}{\alpha^2}\left(h^-_{i+1} + h^+_i\right) = \frac{\beta}{\alpha^2} h.$$

Using Remark 2, one obtains

$$\frac{1}{\tau^\varepsilon_{i+\frac{1}{2}}} \leqslant \frac{\beta}{\alpha^2}\frac{\varepsilon}{m}, \quad \text{hence } B \leqslant c\varepsilon^{\frac{1}{2}}.$$

Therefore

$$A \leqslant \frac{ch}{\varepsilon^{\frac{1}{2}}} \text{ which yields Estimation (9).}$$

Remark that

$$\left|e^\varepsilon_i\right| \leqslant \sum_{j=1}^{N} \left|e^\varepsilon_j - e^\varepsilon_{j-1}\right|$$

Applying once again the Cauchy-Schwarz inequatity, one obtains

$$\left|e^\varepsilon_i\right| \leqslant AB, \quad \text{which yields Estimation (10).}$$

Theory from (Bensoussan, Lions & Papanicolaous, 1987) and (Jikov, Kozlov & Oleinik, 1994) on the estimate of the difference $u^\varepsilon - u$, with $u$ the homogenized solution implies

$$||u - u^\varepsilon||_{L^\infty(\Omega)} \leqslant c\varepsilon. \tag{26}$$

Using (26) and (10) we obtain (11). This completes the proof of Theorem 1. □

*2.2 Modified Finite Volume Scheme*

The modified finite volume approach (see, Ewing, Iliev & Lazarov, 2001) is to rewrite the problem (1) into its mixed form:

$$\begin{cases} q^\varepsilon = -k^\varepsilon(x)\frac{du^\varepsilon}{dx}, \ \frac{dq^\varepsilon}{dx} = f(x), \ 0 < x < 1 \\ u^\varepsilon(0) = u^\varepsilon(1) = 0. \end{cases} \tag{27}$$

$q^\varepsilon(x)$ is the flux dependent variable. Conditions for continuity of the fonction and the flux through interface points $\xi$ are added:

$$[q^\varepsilon] = [u^\varepsilon] = 0, \text{ for } x = \xi. \tag{28}$$

Here $[u^\varepsilon]$ deontes the difference of the right and left limits of $u^\varepsilon$ at the point of discontinuity. We introduce a standard uniform cell-centered grid $x_0 = 0$, $x_1 = \frac{h}{2}$, $x_i = x_{i-1} + h$, $i = 2, ..., N$, $x_{N+1} = 1$, where $h = \frac{1}{N}$. Note, that the endpoints $x_0 = 0$ and $x_{N+1} = 1$ are part of the grid, but they are at $\frac{h}{2}$ distance from their neighboring grid points.The internal grid points can be considered as centered around the volumes $V_i = \left(x_{i-\frac{1}{2}}, x_{i+\frac{1}{2}}\right)$ where $x_{i+\frac{1}{2}} = x_i + \frac{1}{2}h$, $x_{i-\frac{1}{2}} = x_i - \frac{1}{2}h$. The values of a funtion $f$ defined at the grid points $x_i$ are denoted by $f_i$. Non-uniform grids can be treated in a similar way. The finite volume method exploits the idea of writing the balance equation over the finite volume $V_i$, i.e. integrating the first equation of Problem (1) over each volume $V_i$.

$$q^\varepsilon_{i+\frac{1}{2}} - q^\varepsilon_{i-\frac{1}{2}} = hf_i,\ f_i = \frac{1}{h}\int_{x_{i-\frac{1}{2}}}^{x_{i+\frac{1}{2}}} f(x)dx,\ i = 1, 2, ..., N. \tag{29}$$

Next, we rewrite the flux equation in the form

$$-\frac{du^\varepsilon}{dx} = \frac{q^\varepsilon}{k^\varepsilon(x)}$$

and integrate this expression over the interval $(x_i, x_{i+1})$ :

$$\left(u^\varepsilon_{i+1} - u^\varepsilon_i\right) = -\int_{x_i}^{x_{i+1}} \frac{du^\varepsilon}{dx}dx = \int_{x_i}^{x_{i+1}} \frac{q^\varepsilon}{k^\varepsilon(x)}dx \tag{30}$$

One assumes that the flux is two times continuously differentiable on the interface, so it can be expanded around the point $x_{i+\frac{1}{2}}$ in th Taylor series

$$q^\varepsilon(x) = q_{i+\frac{1}{2}} + \left(x - x_{i+\frac{1}{2}}\right)\frac{dq_{i+\frac{1}{2}}}{dx} + \frac{\left(x - x_{i+\frac{1}{2}}\right)^2}{2}\frac{d^2q^\varepsilon(\eta)}{dx^2},\ \eta \in (x_i, x_{i+1}). \tag{31}$$

After replacing the first derivative of the flux at $x_{i+\frac{1}{2}}$ by a two-point backward difference one gets the following approximation of (30).

$$-\left(u^\varepsilon_{i+1} - u^\varepsilon_i\right) = q_{i+\frac{1}{2}}\int_{x_i}^{x_{i+1}} \frac{dx}{k^\varepsilon(x)} + \frac{q_{i+\frac{1}{2}} - q_{i-\frac{1}{2}}}{h}\int_{x_i}^{x_{i+1}} \frac{\left(x - x_{i+\frac{1}{2}}\right)}{k^\varepsilon(x)}dx + O(h^3). \tag{32}$$

Finally, by the same gait as in (Ewing, Iliev & Lazarov, 2001) we get the following finite difference approximation of the differential problem (27):

$$L_{\varepsilon h}u^{\varepsilon h}_i = f_i \text{ for } i = 1, ..., N. \tag{33}$$

where

$$L_{\varepsilon h}u^{\varepsilon h}_i \equiv \begin{cases} -\frac{4}{3}\frac{1}{h}\left(k^{\varepsilon h}_{\frac{3}{2}}\dfrac{u^{\varepsilon h}_2 - u^{\varepsilon h}_1}{h} - k^{\varepsilon h}_{\frac{1}{2}}\frac{2u^{\varepsilon h}_1}{h}\right) \text{ (for } i = 1), \\ -\left(1 + a^\varepsilon_{i+\frac{1}{2}} - a^\varepsilon_{i-\frac{1}{2}}\right)^{-1}\frac{1}{h}\left(k^{\varepsilon h}_{i+\frac{1}{2}}\dfrac{u^{\varepsilon h}_{i+1} - u^{\varepsilon h}_i}{h} - k^{\varepsilon h}_{i-\frac{1}{2}}\frac{u^{\varepsilon h}_i - u^{\varepsilon h}_{i-1}}{h}\right), & i \neq 1, N, \\ -\frac{4}{3}\frac{1}{h}\left(-k^{\varepsilon h}_{N+\frac{1}{2}}\dfrac{2u^{\varepsilon h}_N}{h} - k^{\varepsilon h}_{N-\frac{1}{2}}\frac{u^{\varepsilon h}_N - u^{\varepsilon h}_{N-1}}{h}\right) \text{ (for } i = N). \end{cases} \tag{34}$$

where

$$k^{\varepsilon h}_{i+\frac{1}{2}} = \left(\frac{1}{h}\int_{x_i}^{x_{i+1}} \frac{dx}{k^\varepsilon(x)}\right)^{-1}. \tag{35}$$

$$a^\varepsilon_{i+\frac{1}{2}} = k^{\varepsilon h}_{i+\frac{1}{2}}\frac{1}{h^2}\int_{x_i}^{x_{i+1}} \frac{\left(x - x_{i+\frac{1}{2}}\right)}{k^\varepsilon(x)}dx, \tag{36}$$

and $u^{\varepsilon h}_i$ denotes the approximation values of the exact solution. $k^{\varepsilon h}_{i+\frac{1}{2}}$ is the well known harmonic averaging of the coefficient $K^\varepsilon(x)$ over the cell $(x_i, x_{i+1})$, which has played a fundamental role in deriving accurate schemes for discontinuous coefficients (see e.g, (Samarskii, 1977) and (Samarskii & Andréev, 1978 )).

We now state the main result of this section.

**Theorem 2.** *Assume that the coefficient $k^\varepsilon(x)$ is a piecewise $C^1$–function and has a finite number of jump discontinuities, the grid is such that the discontinuities are at the points $x_{i+\frac{1}{2}}$, and the source term $f(x)$ is a $C^1$–function on $(0,1)$. Then the following estimate is valid:*

$$\left\|u^\varepsilon - u^{\varepsilon h}\right\|^2_{H^1(\Omega)} \equiv \sum_{i=1}^{N} k^{\varepsilon h}_{i+\frac{1}{2}} \left(e^\varepsilon_{i+1} - e^\varepsilon_i\right)^2 / h \leqslant \frac{ch^2}{\varepsilon}, \tag{37}$$

*where $k^{\varepsilon h}_{i+\frac{1}{2}}$ is given by (35), and c is a constant independent of $\varepsilon$ and h.*

*Proof.* The proof of Theorem 2 is the same as that of Theorem 1. □

**3. Numerical Results**

In this section, one presents numerical results, comparing the approximations described in this paper and an example of exact solution. More especially, we shall present numerical results obtained with following data of Problem (1):

$$k(y) = \begin{cases} k_1 & \text{if} \quad 0 < y < \frac{1}{2}, \\ k_2 & \text{if} \quad \frac{1}{2} < y < 1, \end{cases}$$

$k_1,\ k_2 \in \mathbb{R}^*_+$,

$$k^\varepsilon(x) = \begin{cases} k_1, & \text{if} \quad p\varepsilon < x < \left(p + \frac{1}{2}\right)\varepsilon, \\ k_2, & \text{if} \quad \left(p + \frac{1}{2}\right)\varepsilon < x < (p+1)\varepsilon, \end{cases}$$

$\varepsilon = \frac{1}{n_p}$, where $n_p$ is a positive integer; $0 < p < n_p - 1$ and the source function is $f = 1$.
Therefore the exact solution $u^\varepsilon$ is

$$u^\varepsilon(x) = \begin{cases} \frac{-x^2}{2k_1} + \frac{x}{2k_1} + \frac{(k_1-k_2)\varepsilon x}{4k_1(k_1+k_2)} - \frac{(k_1-k_2)p\varepsilon^2}{4k_1(k_1+k_2)} + \frac{(k_1-k_2)p\varepsilon}{4k_1k_2} - \frac{(k_1-k_2)p^2\varepsilon^2}{4k_1K_2}, & \text{if} \quad p\varepsilon < x < \left(p+\frac{1}{2}\right)\varepsilon, \\ \frac{-x^2}{2k_2} + \frac{x}{2k_2} + \frac{(k_1-k_2)\varepsilon x}{4k_2(k_1+k_2)} - \frac{(k_1-k_2)(p+1)\varepsilon^2}{4k_2(k_1+k_2)} - \frac{(k_1-k_2)(p+1)\varepsilon}{4k_1k_2} + \frac{(k_1-k_2)(p+1)^2\varepsilon^2}{4k_1k_2}, & \text{if} \quad \left(p+\frac{1}{2}\right)\varepsilon < x < (p+1)\varepsilon, \end{cases}$$

and the homogenized solution is

$$u(x) = \frac{(k_1+k_2)\,x(1-x)}{4k_1k_2}.$$

All the simulations presented have been done with uniform grids.

The first test problem involved simulation with $k_1 = 10^3$ and $k_2 = 1$.

Table 1. Error table with the classical finite volume method (CFVM) when $\varepsilon \to 0$.

| Volumes number=256 | $\varepsilon = 0.5$ | $\varepsilon = 0.25$ | $\varepsilon = 0.125$ | $\varepsilon = 0.0625$ |
|---|---|---|---|---|
| $\left\|u^\varepsilon - u^{\varepsilon h}\right\|_{L^\infty(\Omega)}$ | 1.907350e-06 | 1.907350e-06 | 1.907350e-06 | 1.907351e-06 |
| $\left\|u^\varepsilon - u^{\varepsilon h}\right\|_{L^2(\Omega)}$ | 1.348700e-06 | 1.348700e-06 | 1.348701e-06 | 1.348701e-06 |
| $\left\|u^\varepsilon - u^{\varepsilon h}\right\|_{H^1(\Omega)}$ | 1.543002e-03 | 2.069417e-03 | 2.843546e-03 | 3.961350e-03 |

Test Problem 1: This error table confirms the estimates of Theorem 1.

In the following graphics, Homog denotes the homogenized solution and Exact denotes the exact solution.

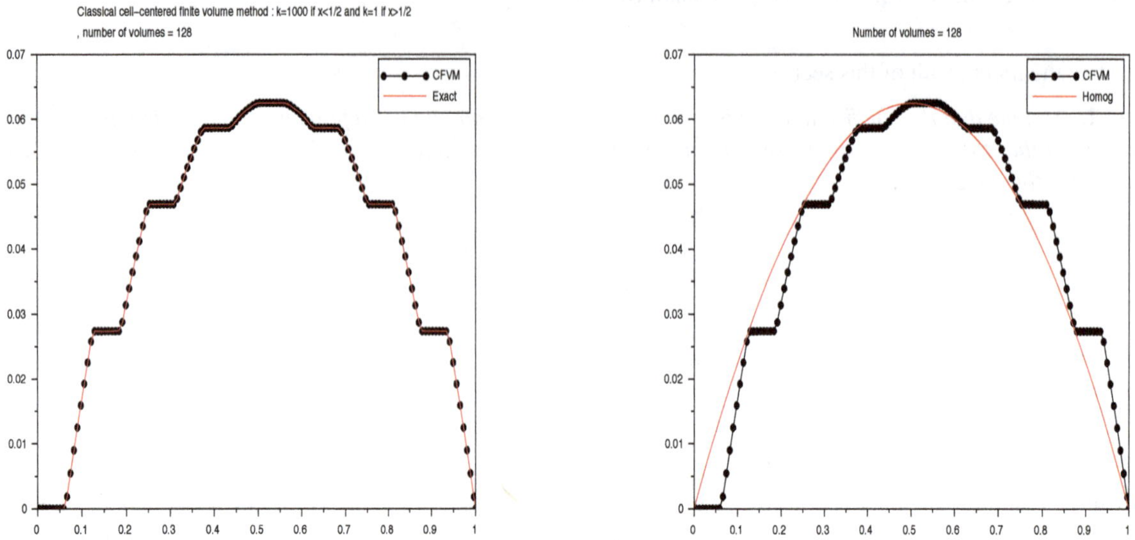


Figure 1. Test problem 1: $\varepsilon = 1/8$

Table 2. Error table with the modified finite volume method (MFVM) when $\varepsilon \to 0$.

| Volumes number=256 | $\varepsilon = 0.5$ | $\varepsilon = 0.25$ | $\varepsilon = 0.125$ | $\varepsilon = 0.0625$ |
|---|---|---|---|---|
| $\left\|u^{\varepsilon} - u^{\varepsilon h}\right\|_{L^{\infty}(\Omega)}$ | 1.897054e-06 | 1.897530e-06 | 1.897768e-06 | 1.897887e-06 |
| $\left\|u^{\varepsilon} - u^{\varepsilon h}\right\|_{L^{2}(\Omega)}$ | 9.119900e-07 | 1.001070e-06 | 1.049512e-06 | 1.074557e-06 |
| $\left\|u^{\varepsilon} - u^{\varepsilon h}\right\|_{H^{1}(\Omega)}$ | 1.193472e-03 | 1.823735e-03 | 2.670071e-03 | 3.838726e-03 |

Test Problem 1: This error table confirms the estimate of Theorem 2.

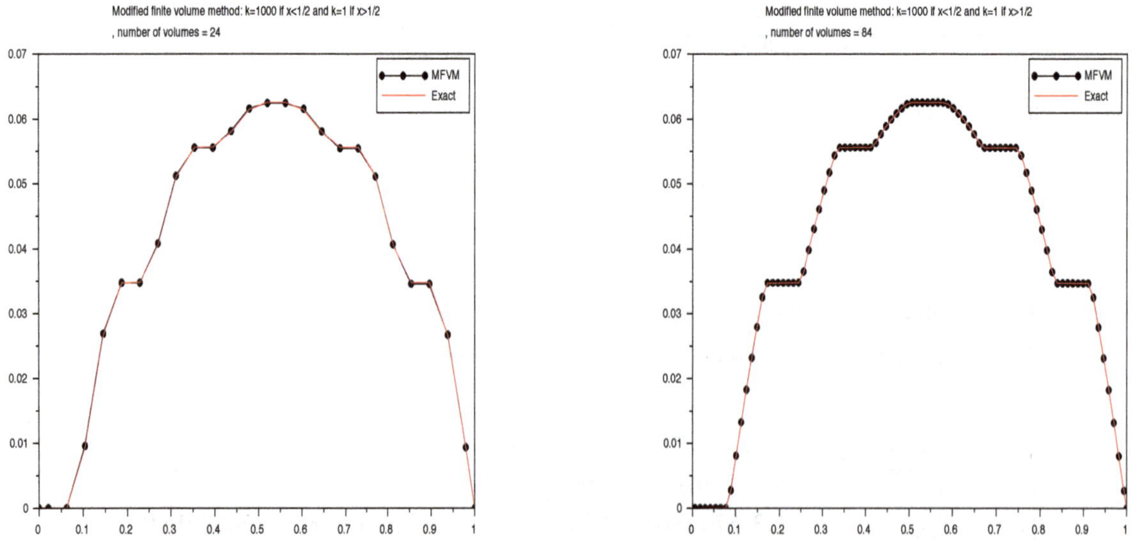


Figure 2. Test problem 1: $\varepsilon = 1/6$

The second test problem involved simulation with $k_1 = 1$ and $k_2 = 10^4$.

Table 3. Convergence test of the classical finite volume

| $\varepsilon$=0.1 | $h = 1/20$ | $h = 1/40$ | $h = 1/100$ | $h = 1/200$ |
|---|---|---|---|---|
| $\lVert u^{\varepsilon} - u^{\varepsilon h}\rVert_{L^\infty(\Omega)}$ | 3.125000e-04 | 7.812500e-05 | 1.250000e-05 | 3.125010e-06 |
| $\lVert u^{\varepsilon} - u^{\varepsilon h}\rVert_{L^2(\Omega)}$ | 2.209709e-04 | 5.524272e-05 | 8.838836e-06 | 2.209713e-06 |
| $\lVert u^{\varepsilon} - u^{\varepsilon h}\rVert_{H^1(\Omega)}$ | 4.049933e-02 | 2.024966e-02 | 8.099865e-03 | 4.049933e-03 |

Table 4. Convergence test of the modified finite volume method

| $\varepsilon$=0.1 | $h = 1/20$ | $h = 1/40$ | $h = 1/100$ | $h = 1/200$ |
|---|---|---|---|---|
| $\lVert u^{\varepsilon} - u^{\varepsilon h}\rVert_{L^\infty(\Omega)}$ | 2.968438e-04 | 7.616367e-05 | 1.237365e-05 | 3.109035e-06 |
| $\lVert u^{\varepsilon} - u^{\varepsilon h}\rVert_{L^2(\Omega)}$ | 1.736241e-04 | 4.342797e-05 | 6.949458e-06 | 1.737400e-06 |
| $\lVert u^{\varepsilon} - u^{\varepsilon h}\rVert_{H^1(\Omega)}$ | 3.847363e-02 | 1.924855e-02 | 7.702354e-03 | 3.851676e-03 |

The tables 3 and 4 demonstrate, as in (Ewing, Iliev & Lazarov, 2001) that the modified scheme has a smaller constant of convergence than the classical scheme based on harmonic averaging for this class of equations.

## 4. Concluding remarks

The purpose of this paper was to resolve, by finite volume methods, a class of second-order elliptic problems, with discontinuous and highly oscillating coefficients, in order to evaluate the effect of $\varepsilon$ the parameter involved in the periodic homogenization on the approximate solution. Study was limited to the academic Dirichlet problem and enabled to clarify the dependence of numerical experiments to $\varepsilon$. Error estimates were obtained. Numerical simulations confirm theorical results and also show that the modified finite volume scheme is much more accurate than the classical scheme in solving these problems.

The extension of these results to two-dimensional and three-dimensional problems is currently underway.

**Acknowledgements**

The author would like to thank the editor and anonymous reviewers for their positive and constructive comments.

**References**

Amaziane, B., & Ondami, B. (1999). Numerical approximations of a second order elliptic equation with highly oscillating coefficients. *Notes on numerical Fluid Mechanics, 70*, 1-11.

Bakhvalov, N., & Panasenko, G. (1989). *Homogenization: processes in periodic media*. Mathematics and its Applications, 36, Kluwer, Academic Publishers, London.

Bensoussan, A., Lions, J. L., & Papanicolaous, G. (1987). *Asymptotic analysis for periodic structures*. Studies in Mathematic and its Applications, 5, North-Holland, Amsterdan.

Bourgat, J. F. (1978). Numerical experiments of the homogenization method for operators with periodic coefficients. *Rapport de Recherche N° 277*, INRIA.

Bourgat, J. F., & Dervieux, A. (1978). Méthode d'homogénéisation des opérateurs à coefficients périodiques: étude des correcteurs provenant du développement asymptotique. *Rapport de Recherche N° 278*, INRIA.

Chernogorova, T., Ewing, R., Iliev, O., & Lazarov, R. (2000). On the Discretization of Interface Problems with Perfect and Imperfect Contact. *Lecture Notes in Physics, 552*, 93-103. http://dx.doi.org/10.1007/3-540-45467-5_7

Chen, Z., & Hou, T. Y. (2002). A mixed multiscale finite element method for elliptic problems with oscillating coefficients. *Mathematics of computation, 72(242)*, 541-576. http://dx.doi.org/10.1090/S0025-5718-02-01441-2

Ewing, R., Iliev, O. P., & Lazarov, R. D. (2001). A modified finite volume approximation of second-order elliptic equations with discontinuous coefficients. *SIAM J. Sci. Comp., 23*(4), 1334-1350. http://dx.doi.org/10.1137/S1064827599353877

Eymard, R., Gallouët, T., & Herbin, R. (2000). *The finite volume methods*. In Handbook of Numerical Analysis, Techniques of Scientific Computing, Part III, North-Holland, Amsterdam.

He, W. M., & Cui, J. Z., (2007). A new approximate method for second order elliptic problems with rapdly oscillating coefficients based on the method of multiscale asymptotic expansions. *J. Math. Anal.335, 657-668*. http://dx.doi.org/10.1016/j.jmaa.2007.01.085

Hornung, U. (1997). *Homogenization and porous media.* Interdisciplinary Applied Mathemlatics 6, Springer-Verlag, New York.

Ilive, O., & Rybak, I. (2008). On Numerical Upscaling for Flows in Heterogeneous Porous Media. *Comput. Methods Appl. Math., 8*(1), 60C76. http://dx.doi.org/10.2478/cmam-2008-0004

Jikov, V. V., Kozlov, S. M., & Oleinik, 0. A. (1994). *Homogenization of differentials operators and integral functionals.* Springer-Verlag, Berlin.

Ondami, B. (2001). *Sur quelques problèmes d'homogénéisation des écoulements en milieux poreux.* Thèse de Doctorat. Université de Pau et des Pays de l'Adour.

Ondami, B. (2015). The Effect of Numerical Integration on the Finite Element Approximation of a Second Order Elliptic Equation with Highly Oscillating Coefficients. *Journal of Interpolation And Approximation in Scientific computing, 2*, 128-136. http://dx.doi.org/10.5899/2015/jiasc-00084

Orive, R., & Zuazua, E. (2007). Finite difference approximation of homogenization problems for elliptic equations Multiscale. *Model. simul., 4*(1), 36-87. http://dx.doi.org/10.1137/040606314

Sanchez-Palancia, E. (1980). *Non-Homogeneous madia and vibration theory.* Lecture Notes in Physics, 127, Springer-Verlag, Berlin.

Samarskii, A. A. (1977). *Theory of difference schemes.* Nauka, Moscow.

Samarskii, A. A., & Andréev, V. B. (1978). *Méthodes aux Différences pour Équations Elliptiques.* MIR, Moscou.

Versieux, H., & Sarkis, M. (2008). Convergence analysis for the numerical boundary corrector for elliptic equations with rapidly oscillating coefficients. *SIAM J. Numer. Anal. 46*(2), 545-576. http://dx.doi.org/10.1137/060654773

# 6

# Bernoulli Algebra on Common Fractions and Generalized Oscillations

Alexander P. Buslaev[1,2] & Alexander G. Tatashev[2,1]

[1] Moscow Automobile and Road State Technical University, Moscow, Russia

[2] Moscow Technical University of Communications and Informatics, Moscow, Russia

Correspondence: Alexander P. Buslaev. E-mail: apal2006@yandex.ru

**Abstract**

Bernoulli algebra on set of proper common fractions with fixed denominator has been introduced and investigated. This algebra is one of most important components of a discrete dynamical system called logistic bipendulum.

**Keywords:** unary algebra, Bernoulli shift, number theory, dynamical system

## 1. Introduction

A dynamical system was introduced and investigated in (Kozlov, Buslaev, & Tatashev, 2015). This system is called a bipendulum. The bipendulum consists of two cells $V_0$ and $V_1$. A channel connects the cells. There are two particles $P_0$ and $P_1$. Each particle is in one of the cell at every instant. Two numbers $a^{(0)}$ and $a^{(1)}$ are given. These numbers belong to the segment $[0, 1]$. The binary representation of the number $a^{(i)}$

$$a^{(i)} = 0.a_1^{(i)}a_2^{(i)}\dots,\ i = 0, 1,$$

is called the plan of the particle $P_i$, $i = 0, 1$. If there are no delays, then the particle $P_i$ is located in the cell $a_T^{(i)}$ at the instant $T$, $i = 0, 1$; $T = 1, 2, \dots$ The plan implementation can be delayed. If one of the particles is located in the cell $V_0$ and, according to its plan, this particle has to come to the cell $V_1$ at next instant, and the other particle is in the cell $V_1$ and plans to come to the cell $V_0$, then only one particle moves in accordance with a given rule. The choice is realized in accordance with priorities (a dynamical system) or equiprobably (a Markov process). The average (in a certain sense) number of delays is interesting for investigations. Suppose that the plans of particles are periodical fractions

$$a^{(i)} = 0.a_1^{(i)}\dots a_l^{(i)}(a_{l+1},\dots,a_{l+m}),$$

i.e. proper common fractions. Shift onto a position to the left is equivalent to applying of the Bernoulli operation to a proper fraction. If we apply the Bernoulli shift to a fraction, we multiply the fraction by 2 and exclude the integer part. An algebra on proper fractions $k/N$, $k = 0, 1, \dots, N-1$, with respect to the Bernoulli shift has been introduced in (Kozlov, Buslaev, & Tatashev, 2015), and is called a Bernoulli algebra. The set of the algebra elements can be divided into disjoint subalgebras. It is found in (Kozlov, Buslaev, & Tatashev, 2015) that, if the plans are determined by elements of the same subalgebra, then the system comes to the state of synergy, i.e., there are no delays in the present instant and in the future. If plans of particles are given by elements of different orbits, then the tape velocities depend on orbits containing these elements. Bernoulli algebras are investigated in this paper.

The aim of this paper is to represent set of proper common fractions as Bernoulli algebras. These algebras are sets of logistic plans of bipendulums which was introduced earlier and are discrete dynamical systems with two positions and two pendulums. We discuss connection with Markov chains, graph theory, binary positional representational of numbers, and theory of functions. We have proved theorems about Bernoulli algebras.

## 2. Algebra of a Markov Process

In this section, we introduce a classification of elements of an algebra. This classification is similar to a known classification of Markov chain states.

Suppose algebra $G$ is a set of elements together with a unary operation $\omega$. We shall give definitions that allow to classify elements of the algebra. This classification is similar to the classification of Markov chain states (Gantmacher, 2004), (Borovkov, 1986). Suppose $x \in G$. The sequence $T(x) = \omega(x), \omega(\omega(x)), \dots$ is called the trajectory of an element $x$. We write $x \to y$ if $y \in T(x)$. Elements $x$ and $y$, $x \neq y$ are called *communicating* with each other if $y \in T(x)$ and $x \in T(y)$, i.e., $x \to y$ and $y \to x$. The element $x$ is called *inessential* if $x \notin T(y)$ for $\forall\, y \in T(x)$. The other elements are *essential*. The set of all essential elements can be divided into disjoint sets such that any two elements of the same set communicate with each other, and any two elements of different sets do not communicate with each other. These sets are called *classes of communicating essential elements* or *orbits.* An element $x$ is called *absorbing* if $\omega(x) = x$. If a class of communicating essential elements contains $d \geq 1$ elements, then this class is called *a periodical class of communicating essential elements* or *an orbit with period d.*

**3. Bernoulli Algebra on Common Fractions**

In Section 3, we give definition of Bernoulli algebra. We introduce the concept of connected components of algebras. This concept is similar to concept of connected components in graph theory.

Consider the set of proper common fractions with denominator $N$

$$G_N = \left\{ \frac{0}{N}, \frac{1}{N}, \frac{2}{N}, \ldots, \frac{N-1}{N} \right\}$$

with a unary operation, called the *Bernoulli shift,* (Schuster, 1984),

$$FB(x) = 2x - [2x],$$

where $[2x]$ is the integral part of $2x$.

Suppose $W_N$ is the algebra on the set $G_N$ with the operation $FB$, Figure 1. The number $i/N$ $(0 \leq i \leq N-1)$ generates a subalgebra. Denote by $W_N(i)$ the subalgebra generated by the element $i/N$, $i = 0, 1, \ldots, N-1$.

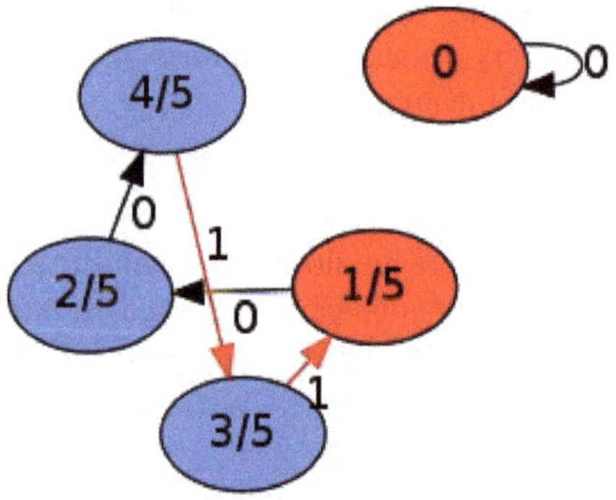


Figure 1. Algebra $W_5, 5 = 1[1] + 1[4]$

If $i/N$ and $j/N$ communicate with each other, then $W_N(j) = W_N(i)$, $i, j = 1, 2, \ldots, N-1$. If $j/N \notin W_N(i)$ and $i/N \notin W_N(j)$, then $W_N(i)$ and $W_N(j)$ are disjoint subalgebras. Identifying subalgebras $W_N(i)$ and $W_N(j)$ with each other for $W_N(i) = W_N(j)$, we divide the algebra $W_N$ into $k_N$ disjoint subalgebras

$$W_N(i_1), W_N(i_2), \ldots, W_N(i_{k_N}).$$

Each of these subalgebras contains a class of communicating essential elements and a set of inessential elements. This set of inessential elements can be empty. These subalgebras are called *connected components of the algebra.* We shall introduce the concept of the algebra $W_N$. graph. The graph of the algebra $W_N$ is directed graph (Harary, 1969) which contains $N$ vertices $v_0, v_2, \ldots, v_{N-1}$. The graph contains the arc $(v_i, v_j)$, where $v_i$ is the tail of the arc and $v_j$ is the head of the arc if and only if $FB(i/N) = j/N$. A connected component of the algebra $W_N$ graph corresponds to a connected component of this algebra.

**4. Binary Representation and Trajectory of the Element in the Bernoulli Algebra**

In Section 4, we consider binary representation and trajectories of elements in Bernoulli algebra. We give geometric interpretation of Bernoulli algebras in the form of graphs.

Consider a common fraction $k/N$, $N \geq 1$, $0 \leq k < N$. The following theorem allows to find the binary representation of this fraction

$$\frac{k}{N} = \sum_{i=1}^{\infty} a_i \cdot 2^{-i},$$

where $a_i$ is equal to 0 or 1, $i = 1, 2, \ldots$

**Theorem 1.** *Suppose*

$$b_0 = k/N,\ b_i = FB(b_{i-1}),\ i = 1, 2, \ldots$$

*Then* $a_i = 0$ *if* $b_i \geq b_{i-1}$, *and* $a_i = 1$ *if* $b_i < b_{i-1}$.

*Proof.* The fractional part of the number $2^{i-1}\frac{k}{N}$ is equal to that value if we apply $i-1$ times the operation $FB(\cdot)$, i.e., this value equals to $b_{i-1}$,

$$b_{i-1} = \sum_{j=i}^{\infty} a_j \cdot 2^{i-j-1},\ i = 1, 2, \ldots$$

This can be proved by induction on $i$. If $a_i = 0$, then $b_{i-1} < 0.5$ and, therefore, $b_i = 2b_{i-1} \geq b_{i-1}$. If $a_i = 1$, then $b_{i-1} > 0.5$, and, therefore, $b_i = 2b_{i-1} - 1 = b_{i-1} - (1 - b_{i-1}) < b_{i-1}$. Theorem 1 has been proved.

Suppose numbers $l$ $(l \geq 1)$ and $m$ $(m \geq 1)$ are such that $b_l = b_{l+m}$, and there are no numbers $i$ and $j$ that $i, j < l + m$ and $b_i = b_j$. Then $a_i = a_{i+m}$ for any $i \geq l$. Hence the binary representation of the common fraction $k/N$ contains an aperiodic part with length $l - 1$ and a repeating part with length $m$. We write

$$\frac{k}{N} = 0.a_1 \ldots a_{l-1}(a_l \ldots a_{l+m-1}).$$

We have formulated an approach to find the binary representation of a common fraction. This approach is similar to the approach of converting the decimal representation of a common fraction to the $p$-ary representation, (Broido & Ilyina, 2006). It is easy to reconstruct the trajectory of the element $k/N$ of the algebra $W_N$ if we know the binary representation of the number $k/N$, Figure 2.

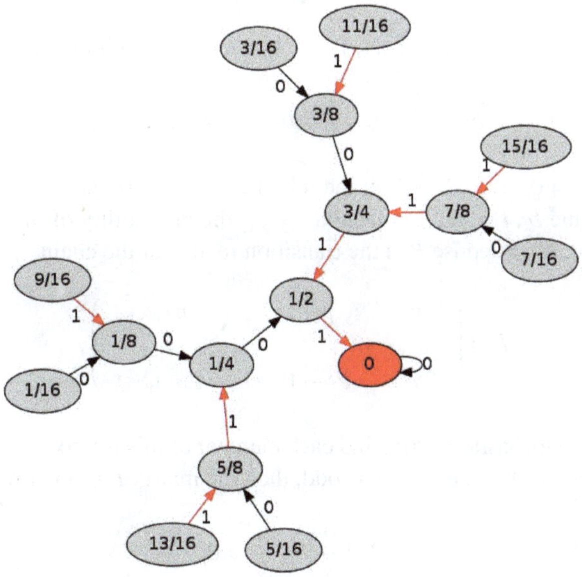


Figure 2. Algebra $W_{16}$, $16 = (8 \rightarrow 4 \rightarrow 2 \rightarrow 1 \rightarrow 1)$

## 5. Variation of the Binary Representation

In Section 5, we consider functional representation of plans. We give definition of variation which is one of most important characteristics of function. We also give definition of variation for binary representation of periodical fractions. This definition is equivalent to the general definition.

Suppose the binary representation of the number $k/N$ is

$$\frac{k}{N} = 0.a_1 \ldots a_l(a_{l+1} \ldots a_{l+m}).$$

The *variation* of this representation is defined as

$$V(k/N) = \frac{1}{m} \sum_{i=l+1}^{l+m} |a_{i+1} - a_i|,$$

where the addition, in indexes, is meant modulo $m$. The limit

$$Var(a) = \lim_{k\to\infty} \frac{1}{k} \sum_{i=1}^{k} |a_{i+1} - a_i|$$

is called the *variation of the binary representation* of a real number $a \in [0, 1)$

$$a = 0, a_1 a_2 \dots a_k \dots$$

if this limit exists. It is evident, that $V(k/N) = Var(k/N)$.

It is easy to give an example of a real number for that the variation of its binary representation is not defined. Suppose $a_1 = 1$, $a_j = 0$, $2^i \leq j < 2^i + 2^{i-1}$, $a_j = 1$, $2^i + 2^{i-1} \leq j < 2^i + 2^i$. Then the variation of the binary representation of the number $a$ is not defined. Some number theory problems, related to binary representation of numbers, are considered in (Uteshev, Cherkasov, Shaposhnikov, 2001).

**6. Markov Chain Interpretation of Algebra $W_N$**

In Section 6, we give Markov chain interpretation of Bernoulli algebras.

A Markov chain, (Gantmacher, 2004), (Borovkov, 1986), corresponds to the algebra $W_N$. The chain contains $N$ states $S_0, S_1, \dots, S_{N-1}$. The state $S_i$ corresponds to the element $i/N$ of the algebra $W_N$, $i = 0, 1, \dots, N-1$. The behavior of this chain is deterministic and depends only on the initial state. If at the instant $t-1$ the system is in the state $S_i$, and

$$FB\left(\frac{i}{N}\right) = \frac{j}{N},$$

then, at the instant $t$, the chain will be at the state $S_j$, $i, j = 0, 1, \dots, N-1$, with probability 1. Therefore, if

$$b_0 = \frac{k}{N}, \; b_i = FB(b_{i-1}), \; i = 1, 2, \dots,$$

and the chain is in the state $S_k$, $k = 0, 1, \dots, N-1$, at the initial instant $t = 0$, then, at the instant $t$, the change will be in the state corresponding to the value $b_t$, $t = 1, 2, \dots$ Denote by $p_{ij}$ the probability of the chain comes from the state $S_i$ to the state $S_j$ for a step, $i, j = 1, \dots, N$. Suppose $P$ is the transition matrix of the chain

$$P = \begin{pmatrix} p_{00} & p_{01} & \dots & p_{0,N-1} \\ \dots & \dots & \dots & \dots \\ p_{N-1,0} & p_{N-1,1} & \dots & p_{N-1,N-1} \end{pmatrix}.$$

On the one hand, the matrix $P$ is a stochastic matrix, and each element of this matrix is equal to 0 or 1. On the other hand, each row of the matrix $P$ contains exactly one 1. If $N$ is odd, then the matrix $P$ is a permutation matrix. Since

$$FB\left(\frac{0}{N}\right) = \frac{0}{N} = 0,$$

it follows that the state $S_0$ is absorbing, i.e., $p_{00} = 1$. If the state $S_k$ is inessential, and the chain comes from this state to the absorbing state $S_0$ for $l$ steps, then the binary representation of the number $k/N$ contains just $l$ positive digits

$$\frac{k}{N} = 0.a_1 \dots a_l = 0, a_1 \dots a_l(0).$$

If the initial state $S_k$ belongs to a periodic class of communicating states, and this class contains $d \geq 1$ states, then the repeating part of the number $k/N$ binary representation contains $d$ digits

$$\frac{k}{N} = 0.(a_1 \dots a_d).$$

If the initial state $S_k$ is inessential and the chain comes from this state after $l$ steps to a class of communicating states with period $d$, then the binary representation of the number $k/N$ is

$$\frac{k}{N} = 0.a_1 \dots a_l(a_{l+1} \dots a_{l+d}),$$

i.e., the aperiodic part of the binary representation contains $l$ digits, and the length of the repeating part equals $d$.

**7. Binary Representation of Common Fractions and Division of the Algebra $W_N$ into Classes**

In Section 7, we consider connection between Bernoulli algebras, binary representation, and Markov chains.

Consider examples. Suppose $N = 8$, Figure 3. There is an absorbing element $0/8 = 0.(0)$ and inessential elements

$$\frac{1}{8} = 0.001(0),\ \frac{2}{8} = \frac{1}{4} = 0.01(0),\ \frac{3}{8} = 0.011(0),$$

$$\frac{4}{8} = 0.1(0),\ \frac{5}{8} = 0.101(0)\ \frac{6}{8} = 0.11(0),\ \frac{7}{8} = 0.111(0).$$

Values of the variation functions are equal to

$$V\left(\frac{1}{8}\right) = 0,\ V\left(\frac{2}{8}\right) = 0,\ V\left(\frac{3}{8}\right) = 0, V\left(\frac{4}{8}\right) = 0,$$

$$V\left(\frac{5}{8}\right) = 0,\ V\left(\frac{6}{8}\right) = 0,\ V\left(\frac{7}{8}\right) = 0,$$

The transition matrix of the Markov chain of the algebra $W_8$ have the form

$$P = \begin{pmatrix} 1 & 0 & 0 & 0 & 0 & 0 & 0 & 0 \\ 0 & 0 & 1 & 0 & 0 & 0 & 0 & 0 \\ 0 & 0 & 0 & 0 & 1 & 0 & 0 & 0 \\ 0 & 0 & 0 & 0 & 0 & 0 & 1 & 0 \\ 1 & 0 & 0 & 0 & 0 & 0 & 0 & 0 \\ 0 & 0 & 1 & 0 & 0 & 0 & 0 & 0 \\ 0 & 0 & 0 & 0 & 1 & 0 & 0 & 0 \\ 0 & 0 & 0 & 0 & 0 & 0 & 1 & 0 \end{pmatrix}.$$

The state $S_0$ is absorbing. The other states are inessential, Figure 3.

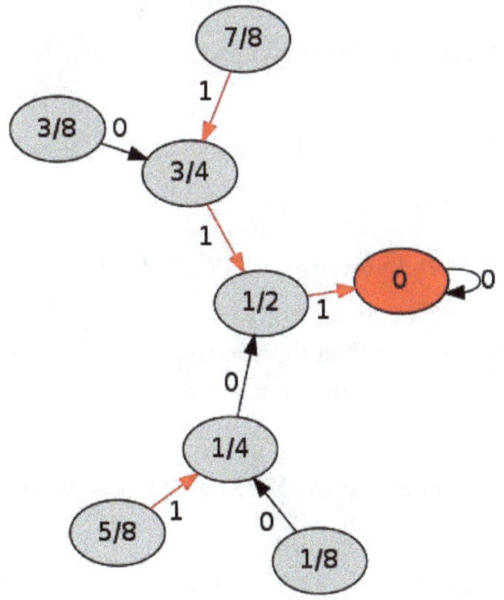


Figure 3. Algebra $W_8, 8 = (4 \to 2 \to 1 \to 1)$

**8. Algebras $W_N$ and $W_{2N}$**

In Section 8, we prove a theorem which allow to find the form of algebra $W_{2N}$ if the form of algebra $W_N$ is known.

Let us compare algebras $W_N$ and $W_{2N}$ with each other.

**Theorem 2.** *Algebras $W_N$ and $W_{2N}$ contain the same number of connected components and, for any $d$, the same number of orbits with period $d$. The algebra $W_{2N}$ contains a subalgebra isomorphic to the algebra $W_N$. This subalgebra consists* *of elements $2i/2N$, $i = 0, 1, \ldots, N-1$. The algebra $W_{2N}$ also contains N inessential elements $x = 2i + 1/2N$, $i = 0, 1, \ldots, N-1$, such that $FB(x) = y$ for an element y of the subalgebra isomorphic to algebra $W_N$.*

*Proof.* The subalgebra $W_{2N}$, with elements $2i/2N$, $i = 0, 1, \ldots, N-1$, is isomorphic to the algebra $W_N$ as the values of the elements $2i/2N$ of the algebra $W_{2N}$ and the element $i/N$ of the algebra $W_N$, $i = 0, 1, \ldots, N-1$, are the same. The elements $2i + 1/2N$, $i = 0, 1, \ldots, N-1$, are inessential as, applying the Bernoulli shift, we obtain an element which belongs to the subalgebra isomorphic to the algebra $W_N$. Theorem 2 has been proved.

**9. Structure of the Algebra $W_N$ for an Odd $N$**

In Section 9, we prove theorems about Bernoulli algebras in the case of odd values of N.

*9.1. Preliminary Definitions and Results*

Suppose the canonical representation of the number N, $N \geq 2$, has the form

$$N = p_1^{s_1} \ldots p_l^{s_l}, \tag{1}$$

where $2 < p_1 < \cdots < p_l$ are prime numbers, $s_1, \ldots, s_l$ are natural numbers. We write $N = \overline{p}^{\overline{s}}$, where $\overline{p} = (p_1, \ldots, p_l)$, $\overline{s} = (s_1, \ldots, s_l)$. Denote by $E(N)$ the Euler's function (Vinogradov, 1972),

$$E(N) = \left(p_1^{s_1} - p_1^{s_1-1}\right) \ldots \left(p_l^{s_l} - p_l^{s_l-1}\right).$$

**Euler's theorem on numbers, (Vinogradov, 1972).** *The value of $E(N)$ is equal to number of positive integers less than N that are coprime to N.*

This function can be also represented as

$$E(N) = N\left(1 - \frac{1}{p_1}\right)\left(1 - \frac{1}{p_2}\right) \ldots \left(1 - \frac{1}{p_l}\right).$$

**Euler's theorem on divisibility, (Vinogradov, 1972).** *Suppose k and N are coprime natural numbers. Then N is a divisor of $k^{E(N)} - 1$.*

Fermat's little theorem is a special case of Euler's theorem on divisibility.

**Fermat's little theorem, (Vinogradov, 1972).** *If k is prime, then k is a divisor of $N = 2^k - 2$.*

Denote by $LCM(a_1, \ldots, a_L)$ the least common multiple of numbers $a_1, \ldots, a_l$, and N has the form (1). *Generalized Euler's function* is the function $e(N)$, $e(1) = 1$,

$$e(N) = LCM(p_1^{s_1-1}(p_1 - 1), \ldots, p_l^{s_l-1}(p_l - 1)).$$

It is clear that any divisor of $e(N)$ is a divisor of $E(N)$.

**The generalized Euler's divisibility theorem, (Vinogradov, 1972).** *Let k and N be coprime positive integers. Then N is a divisor of $k^{E(N)} - 1$.*

According to this theorem, if k and N are coprime then there exists a positive number l such that N is a divisor of $k^l - 1$. *The smallest positive number l such that N is a divisor of $k^l - 1$ is called the order of k modulo N, (Vinogradov, 1972). Denote by $m(N, k)$ the order of k modulo N.*

**The generalized Euler's order theorem, (Vinogradov, 1972).** *Let k and N be coprime positive integers. Then $m(N, k)$ is a divisor of $e(N)$.*

Denote by $m(N)$ the order of 2 modulo N, i.e., $m(N) = m(N, 2)$. The number $m(N)$ is the smallest number m such that N, is a divisor of $2^m - 1$. The elements of the algebra $W_N$ can be divided into two classes. They are *the class of noncancelable fractions and the class cancelable fractions*

$$W_N = W_N^{copr[ime]} + W_N^{canc[elable]}.$$

Consider the set of all elements $k/N$ of the algebra $W_N$ such that $k/N$ is a noncancelable fraction. The number of these elements is equal to the number of positive integers less than N and coprime to N, i.e., the function $e(N)$ is equal to Euler's function. Suppose

$$\delta(k, E(N)) = \begin{cases} \frac{E(N)}{m(N)}, & k = m(N), \\ 0, & k \neq m(N). \end{cases} \tag{2}$$

In accordance with section 2 an orbit of period $d$ is a set of elements of the algebra $W_N$

$$\frac{k_1}{N}, \frac{k_2}{N}, \ldots, \frac{k_d}{N},$$

$0 \leq k_i \leq N-1$, $i = 1, \ldots, d$ such that

$$\frac{k_{i+1}}{N} = FB\left(\frac{k_i}{N}\right) \neq \frac{k_1}{N}, \; i = 1, \ldots, d-1,$$

$$\frac{k_1}{N} = FB\left(\frac{k_d}{N}\right).$$

Denote by $O(N,k)$ *the number of orbits with period* $k$ $(k \geq 2)$.

*9.2. Simple Case*

Suppose $N = p^s$, where $p \geq 3$ is prime, $s$ is a natural number;

$$E(N) = p^s - p^{s-1}.$$

**Theorem 3.** *The following formula is true*

$$O(N,k) = \sum_{r=1}^{s} \delta(k, \varphi(p^r)) = \sum_{r=1}^{s} \delta(m(p^r), E(p^r)).$$

The proof is based on Lemma 1.

**Lemma 1.** *Suppose* $r \leq s$, $k/p^r$ *is an element of the algebra* $W_{p^r}$ *such that it is noncancelable fraction. Then this element belongs to an orbit of period* $m(p^r)$. *The set of cancelable fractions of the algebra* $W_{p^r}$ *is a subalgebra isomorphic to the algebra* $W_{p^{r-1}}$.

*Proof.* Suppose the number $a$ satisfies the equation

$$2^m \cdot \frac{k}{N} = \frac{k}{N} + a.$$

The number $a$ is natural if and only if the number $N$ is the divisor of $2^m - 1$. We take into account that $k$ and $N$ are comprime, and $k < N$. Applying to a noncancelable element the Bernoulli shift operation $m(N)$ times, we obtain the same element. If the Bernoulli shift is applied to this element less than $m(N)$ times, then the same element cannot be obtained. It is obviously that the subalgebra of cancelable fractions of the algebra $W_{p^r}$ and the algebra $W_{p^{r-1}}$ are isomorphic to each other. Lemma 1 has been proved.

The graphs of algebras $W_{13}$, $W_{23}$, $W_{27}$ are shown in Figures 4 – 6.

*9.3. Common Case*

Let $N$ be as (1).

**Lemma 2.** *Any element* $k/N$ *of the algebra* $W_N$ *belongs to an orbit, and, if* $k/N \in W_N^{copr}$, *i.e., the element* $k/N$ *is noncancelable fraction, then the period of this orbit equals* $m(N)$.

*Proof.* Apply the Bernoulli shift operation to the element $k/N$ $i$ times. We obtain the same element if and only if the equation

$$2^i \cdot \frac{k}{N} = \frac{k}{N} + a, \tag{3}$$

contains natural number $a$. If $i = e(N)$ in (3), then the number $a$ is natural. Take into account, that $N$ is coprime to 2. Hence there exists an orbit of period not greater than $e(N)$, such that this orbit contains the element $k/N$. If the fraction $k/N$ is a nocancelable fraction, then $i = m(N,2) = m(N)$ is the smallest value of $i$ such that the number $a$ is a natural number. From this Lemma 2 follows.

**Lemma 3.** *Let* $N$ *be an odd natural number. Two elements* $a \in W_N^{copr}$, $b \in W_N^{canc}$ *of the algebra* $W_N$ *cannot belong the same orbit, i.e., if one of these elements is a noncancelable fraction, then the other element is a cancelable fraction.*

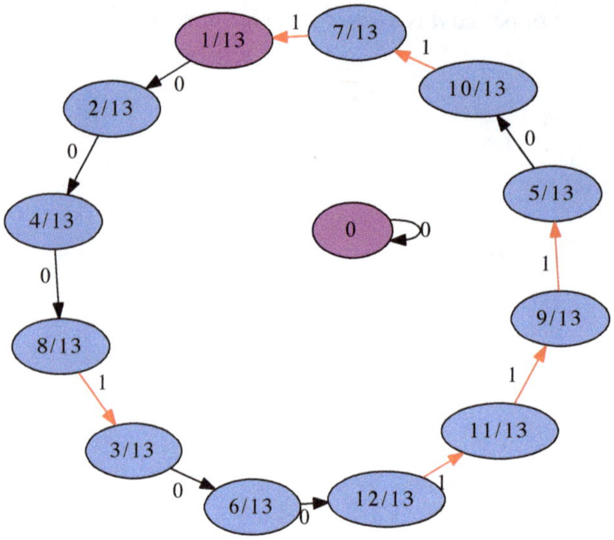


Figure 4. Algebra $W_{13}, 13 = 1[1] + 1[12]$

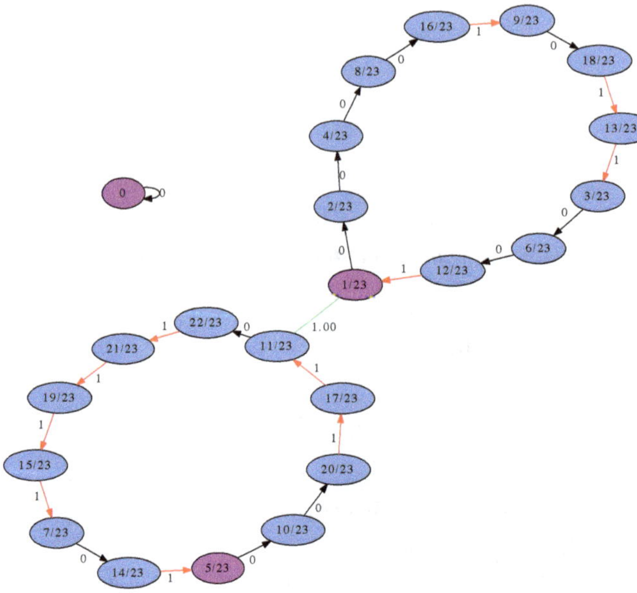

Figure 5. Algebra $W_{23}, 23 = 1[1] + 2[11]$

*Proof.* If an element is cancelable fraction, and we apply the Bernoulli shift operation to this element any times, we can obtain no noncancelable fraction.

Denote by $\delta_{d,N}$ the number of orbits with period $d$ in the subalgebra $W_N^{copr}$. In accordance with Lemmas 2 and 3,

$$\delta_{m(p^r),p^r} = \frac{E(p^r)}{m(p^r)}.$$

If $d \neq m(p^r)$, then $\delta_{d,p^r} = 0$.

Suppose $R = R(N)$ is the set of vectors $r = (r_1, \ldots, r_l)$ with integer nonnegative numbers, $0 \leq r_1 \leq s_1, \ldots, 0 \leq r_l \leq s_l$, and at least one of numbers $r_1, \ldots, r_l$ is positive.

**Theorem 4.** *Suppose $O(N,k)$ is the number of orbits with period $k$ ($k \geq 2$) in algebra $W_N$. Then*

$$O(N,k) = \sum_{\overline{r} \in R} \delta(k, \overline{p}^{\overline{r}}),$$

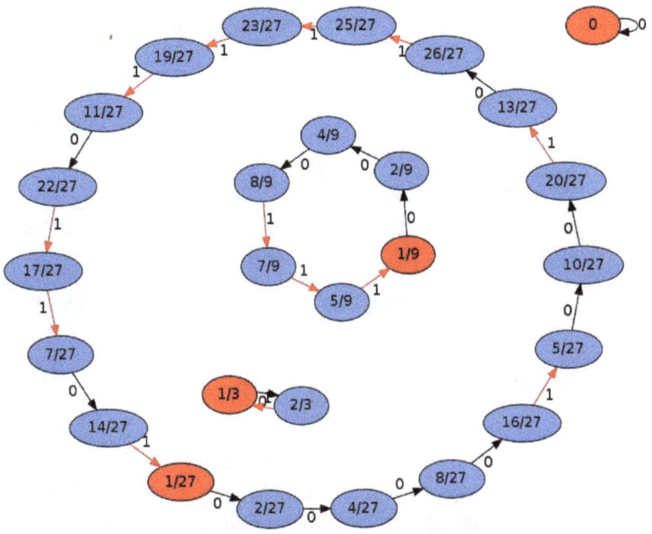


Figure 6. Algebra $W_{27}$, $27 = 1[1] + 1[2] + 1[6] + 1[18]$, (Kozlov, Buslaev, & Tatashev, 2015)

*where $\delta(k, p^r)$ is calculated in accordance with (2).*

*Proof.* If $r_1, r_2 \in R(N)$, then, for $N_1 = \overline{p}^{\overline{r}_1}$, $N_2 = \overline{p}^{\overline{r}_2}$, $N_1 \neq N_2$ we have

$$W_{N_1}^{copr} \cap W_{N_2}^{copr} = \emptyset.$$

Therefore any orbit is contained either in $W_{N_1}^{copr}$ or in $W_{N_2}^{copr}$. From this fact and Lemmas 2 and 3, Theorem 4 follows.

**10. Algebra $W_N$ for an even $N$**

In Section 10, we prove theorems about Bernoulli algebras in the case of even values of $N$.

Consider the case of an even $N$. Suppose $h$ is a natural number,

$$N = 2^h N_1,$$

where $N_1 = p_1^{s_1} \dots p_l^{s_l}$, $2 < p_1 < \dots < p_l$ are prime numbers.

**Lemma 4.** *The element $i/N$ ($i = 2^h k$) of the subalgebra $W_N$ belongs to an orbit with period $m$ if and only if the element $k/N$ of the algebra $W_{N/2^h}$ also belongs to an orbit with period $m$.*

*Proof.* The number $N/2^h$ is odd. The element $k/N$ of the algebra $W_{N/2^h}$ belongs to an orbit in accordance with Lemma 2. The period of this orbit is equal to $m(N_1)$, where $m(N_1)$ is the minimum $l$ such that in the equation

$$2^l \cdot \frac{k}{N_1} = \frac{k}{N_1} + a \tag{4}$$

the value $a$ is a natural number. We can rewrite (4) as

$$2^l \cdot \frac{2^h k}{2^h N_1} = \frac{2^h k}{2^h N_1} + a.$$

Hence, if we apply the Bernoulli shift operation to the element $\frac{i}{N}$ of the algebra $W_N$ $m(N_1)$ times, then we obtain the same element, and, if we apply the Bernoulli shift operation to the element $i/N$ less than $m(N_1)$ times, we do not obtain the same element. From this, Lemma 7 follows.

**Lemma 5.** *Suppose the element $\frac{i}{N}$ is such that i is no multiple of $2^h$. Then the element $i/N$ is inessential.*

*Proof.* If we apply the Bernoulli operation to the element $i/N$ $2^h$ times, then we obtain $j/N$, where $j$ is a multiple of $2^h$. We do not obtain the element $i/N$ again if we apply the Bernoulli operation any times. From this, the lemma follows.

**Theorem 5.** *Suppose h is a natural number,*

$$N = 2^h N_1,$$

*where* $N_1 = p_1^{s_1} \dots p_l^{s_l}$, $2 < p_1 < \dots < p_L$ *is prime numbers. Then algebras* $W_N$ *and* $W_{N/2^h}$ *contain the same number of orbits with each period.*

*Proof.* Theorem 5 follows from lemmas 4 and 5.

**Theorem 6.** *Suppose*

$$N = 2^h N_1,$$

*where* $N_1 = p_1^{s_1} \dots p_l^{s_l}$, $2 < p_1 < \dots < p_l$, *are prime numbers. Then the algebra* $W_N$ *contains* $N_1$ *essential elements. The other* $N - N_1$ *elements are nonessential elements.*

*Proof.* If the number of the Bernoulli shift algebra is odd, then, in accordance with Theorem 4, all elements of this algebra are essential. From this fact and Theorem 5, Theorem 6 follows.

**Theorem 7.** *Suppose h is a natural number,*

$$N = 2^h N_1,$$

*where* $N_1 = p_1^{s_1} \dots p_l^{s_l}$, $2 < p_1 < \dots < p_l$ *are prime numbers. Then the graph of the algebra* $W_N$ *contains, as its subgraphs,* $N_1$ *trees, and the depth, or, in other words, height, of each of these tree equals h. Each of* $N_1$ *essential elements of the algebra* $W_N$ *is the root of just one tree, and the degree of the root equals 1. The degree of other vertices, for any tree, is equal to 2 save the vertices of the level h. Any tree contains* $2^k$ *elements. One of these elements is the root of the tree, and there are* $2^{i-1}$ *elements that are vertices of the level i for any* $i = 1, \dots, h$.

*Proof.* We shall prove that the roots of trees are essential elements of the set $M_0$ containing elements

$$\frac{2^h i}{N}, \quad i = 0, 1, \dots, N_1 - 1,$$

and the vertices of the level 1 are elements of the set $M_1$

$$\frac{2^{h-1}(2i+1)}{N}, \quad i = 0, 1, \dots, N_1 - 1.$$

Indeed, suppose $i < \frac{N_1 - 1}{2}$. If we apply the Bernoulli shift operation to the element $\frac{2^{h-1}(2i+1)}{N}$, we obtain the element $\frac{2^h(2i+1)}{N}$. Suppose $i = \frac{N_1-1}{2}$, the number $N_1 - 1$ is a multiple of 2 as $N_1$ is even. If we apply the Bernoulli shift to the element $\frac{(2i+1)2^{h-1}}{N}$, we obtain the element $0/N$. Suppose $i > \frac{N_1-1}{2}$. Let us apply the Bernoulli shift to the element $\frac{2^{h-1}(2i+1)}{N}$. We have

$$2 \cdot \frac{2^{h-1}(2i+1)}{N} = 2 \cdot \frac{2\left(\frac{N_1}{2} + i + \frac{1}{2} - \frac{N_1}{2}\right)2^{h-1}}{N} =$$

$$= \frac{2^h \cdot (2i + 1 - N_1)}{N}.$$

Hence each element of the set $M_1$ corresponds to exactly one element of the set $M_0$, i.e., each element of the set $M_1$ corresponds to its successor belonging to the set $M_0$. Let $M_j$ be the set of elements

$$\frac{2^{h-j}(2i+1)}{N},$$

$j = 2, \dots, h$, $i = 0, 1, \dots, 2^{j-1}N_1 - 1$. It is easy to see that numerators of the elements of the set $M_j$ contain numbers that are multiples of $2^{h-j}$ but are not multiples of $2^{h-j+1}$. We shall prove that $M_j$ is the set of the vertices of the level $j$, $j = 1, 2, \dots, h$, and, if $j < h$, any vertex of the set $M_j$ is the vertex of degree 2. Assume that $i = i_0$ $(0 \le i_0 \le 2^{j-2}N_1)$. In the set $M_j$, there is a successor of the elements $\frac{2^{h-j}(2i_0+1)}{N}$ and $\frac{2^{h-j}(2i_0+2^{j-1}N_1+1)}{N}$. Namely, this successor is $\frac{2^{h-j+1}(2i_0+1)}{N}$, $j = 1, \dots, h$. There are no any other predecessors of the element $\frac{2^{h-j+1}(2i_0+1)}{N}$ of the set $M_{j-1}$. From this the theorem follows.

## 11. Algebras $W_N$ and $W_d$, Where $d$ is a Divisor of $N$

In Section 11, we prove theorems which allow to compare algebra $W_N$ with algebra $W_d$, where $d$ is a divisor of $N$.

Consider algebras $W_N$ and $W_d$, where $d$ is a divisor of the number $N$.

**Theorem 8.** *Let d be a divisor of the number N. If the element i/d of the algebra* $W_d$ *belongs to the orbit with period m, then the element* $\frac{N}{d}\frac{i}{N}$ *of the algebra* $W_N$ *belongs to an orbit with period m. If the elements i/d and j/d of the algebra* $W_d$ *belong to the same orbit, then elements* $\frac{N}{d}\frac{i}{N}$, $\frac{N}{d}\frac{j}{N}$ *of the algebra* $W_N$ *also belong to the same orbit. If the elements i/d* *and jd of the algebra $W_d$ belong to different orbits, then the elements $\frac{N}{d}\frac{i}{2N}$ and $\frac{N}{d}\frac{j}{2N}$ of the algebra $W_N$ also belong to the different orbits. If the element $i/d$ of the algebra $W_d$ is inessential, then the element $\frac{N}{d}\frac{i}{N}$ of the algebra $W_N$ is also inessential. Elements $\frac{N}{d}\frac{i}{N}$ and $\frac{N}{d}\frac{j}{N}$ of the algebra $W_N$ belong to the same components if and only if the elements $i/N$ and $j/N$ of the algebra $W_N$ also belong to the same connected component.*

*Proof.* The value of the element $\frac{N}{d}\frac{i}{N}$ of the algebra $W_N$ and the value of element $i/d$ of the algebra $W_d$ are the same for any $i = 1, 2, \ldots, d-1$. From this, Theorem 8 follows.

**Theorem 9.** *Let d be a divisor of the number N. Then the number of orbits with period m, the number of the connected components and the number of elements $W_N$ are not less than in the algebra $W_d$.*

Theorem 9 follows from Theorem 8.

## 12. Calculation of Numbers and Periods of Orbits of the Algebra $W_{225}$

In Section 12, we give an example. We consider algebra $W_{225}$. We calculate number of orbits of each period.

1) Divisors of the number $225 = (3,5)^{(2,2)}$, not equal to 1 and 225, are 3, 5, 9, 15, 25, 45, 75.

2) Values of the generalized Euler's function $e$ for divisors, not equal to 1 and 225, of the number 225 are $e(3) = 2$, $e(9) = 6$, $e(15) = 8$, $e(25) = 20$, $e(45) = 12$, $e(75) = 30$, $e(225) = 60$.

3) There exist no divisors, not equal to 1 and 2, of the number $e(3) = 2$, Therefore, $m(3) = e(3) = 2$.

4) There exist a divisor 2 of the number $e(5) = 4$. The number 5 is not a divisor of $2^2 - 1$. Therefore, $m(5) = e(5) = 4$.

5) There exist divisors 2 and 3 of the number $e(9) = 6$. The number 9 not a divisor of $2^2 - 1$ or $2^3 - 1$. $m(9) = e(9) = 6$.

6) There exist divisors 2 and 4 of the number $e(15) = 8$. The number 15 is not a divisor of $2^2 - 1$ but 15 is a divisor of $2^4 - 1$. $m(15) = 4$.

7) There are divisors 2, 4, 5, 10, not equal to 1 and 20, of the number $E(25) = 20$. The number 25 is not a divisor of $2^2 - 1$, $2^4 - 1$, $2^5 - 1$, or $2^{10} - 1$. Therefore, $m(25) = 20$.

8) There exist divisors 2, 3, 4, 6 of the number $e(45) = 12$. The number 45 is not a divisor of $2^2 - 1$, $2^3 - 1$, $2^4 - 1$, or $2^6 - 1$. Therefore, $m(45) = 12$.

9) There exist divisors 2, 4, 5, 10 of the number $e(75) = 20$. The number 75 is not a divisor of $2^2 - 1$, $2^4 - 1$, $2^5 - 1$, $2^8 - 1$, $2^{10} - 1$. Therefore, $m(75) = 20$.

10) Let us calculate the number $m(225)$. There exist divisors 2, 3, 4, 6, 10, 12, 15, 20, 30, 60 of the number $e(225) = 60$. The number 225 is not a divisor of $2^2 - 1$, $2^3 - 1$, $2^4 - 1$, $2^6 - 1$, $2^8 - 1$, $2^{10} - 1$, $2^{12} - 1$, $2^{15} - 1$, $2^{20} - 1$, or $2^{30} - 1$. Therefore, $m(225) = 60$.

11) Let us calculate the number

$$\delta\left(m\left(\overline{p}^{\overline{r}}\right), \overline{p}^{\overline{r}}\right)$$

for divisors of the number 225, not equal to 1, i.e., for any $\overline{r} \in R \cup \overline{s}$ :

$$m(3) = 2, \delta(2,3) = \delta(m(3),3) = \frac{E(3)}{m(3)} = \frac{2}{2} = 1,$$

$$m(5) = 4, \delta(4,5) = \delta(m(5),5) = \frac{E(4)}{m(4)} = \frac{4}{4} = 1,$$

$$m(9) = 6, \delta(6,9) = \delta(m(9),9) = \frac{E(9)}{m(9)} = \frac{6}{6} = 1,$$

$$m(15) = 4, \delta(4,15) = \delta(m(15),15) = \frac{E(15)}{m(15)} = \frac{8}{4} = 2,$$

$$m(25) = 20, \delta(20,25) = \delta(m(25),25) = \frac{E(25)}{m(25)} = \frac{20}{20} = 1,$$

$$\delta(12,45) = \delta(m(45),45) = \frac{E(45)}{m(45)} = \frac{24}{12} = 2,$$

$$\delta(20,75) = \delta(m(75),75) = \frac{E(75)}{m(75)} = \frac{40}{20} = 2,$$

$$\delta(60,225) = \delta(m(225),225) = \frac{E(120)}{m(60)} = 2.$$

12) Let us calculate the value $A(k,225)$ for any $k$ such that $\delta(k,225) > 0$ :

$$A(2,225) = \delta(2,3) = 1,$$

$$A(4,225) = \delta(4,5) + \delta(4,15) = 3,$$

$$A(6,225) = \delta(6,9) = 1,$$

$$A(12,225) = \delta(12,45) = 2,$$

$$A(20,225) = \delta(20,25) + \delta(20,75) = 3,$$

$$A(60,225) = \delta(225,60) = 2.$$

Thus the algebra $W_{225}$ contains an orbit with period 1; an orbit with period 2; 3 orbits with period 4; one orbit with period 6; 2 orbits with period 12; 3 orbits with period 20; 2 orbits with period 60. Since the number 225 is odd, the depth of each orbit equals 0.

## 13. Conclusion

We can use the concept of Bernoulli algebras in analysis of a dynamical system, (Kozlov, Buslaev, & Tatashev, 2015). The same mathematical subject, which is considered in the paper, can be interpreted as an algebra, a Markov process, or a graph. and related to the binary representation of numbers and a dynamical system, called a bipendulum. We can consider a more general system. There are $M$ particles $P_1, \dots, P_M$ and $K > 1$ vertices $V_1, \dots, V_K$. Behavior of the particle $P_i$ is determined by its plan $a_i$, $i = 1, \dots, M$. The plan of each particle is the $K$-ary representation of a number which belongs to the segment $[0, 1]$. The generalized Bernoulli algebra $W_N(K)$ is related to the dynamical system, $N \geq 1$. It is the algebra of proper fractions $\{\frac{0}{N}, \frac{1}{N}, \dots, \frac{N-1}{N}\}$ with respect to a single unary operation. If this operation is applied to a fraction, then the fraction is multiplied by $K$ and the integer part is excluded. If $K = 2$, then this algebra is a Bernoulli algebra, i.e., $W_N(2) = W_N$.

## References

Borovkov, A.A. (1986) *Probability theory (2nd ed.).* Moscow, Nauka. (In Russian.)

Broido, W.P., Ilyina, O.P. (2006). *Computer and systems architecture.* Saint Petersburg, Piter. (In Russian.)

Gantmacher, F.R. (2004). *Theory of matrices (5th ed.).* Moscow, Fizmatlit. (In Russian.)

Harary, F. (1969). *Graph theory.* Massachusetts, Addison-Wesley Publishing Company.

Kozlov, V.V., Buslaev, A.P., & Tatashev, A.G. (2015). On real-valued oscillations of a bipendulum. *Applied Mathematics Letters* (Vol. 46, pp. 44 – 49). http://dx.doi.org/10.1016/j.aml.2015.02.003

Schuster, H.G. (1984). *Deterministic chaos: An introduction.* Weinheim, Physik–Verlag.

Uteshev, A.Ju., Cherkasov, T.M., Shaposhnikov, A.A. (2001). *Digits and codes.* Saint Petersburg, Saint Petersburg University. (In Russian.)

Vinogradov, I.M. (1972). *Foundations of number theory.* Moscow, Nauka. (In Russian.)

# 7

# On Dynamical Systems for Transport Logistic and Communications

Alexander P. Buslaev[1,2] & Alexander G. Tatashev[2,1]

[1] Moscow Automobile and Road State Technical University, Moscow, Russia

[2] Moscow Technical University of Communications and Informatics, Moscow, Russia

Correspondence: Alexander P. Buslaev. E-mail: apal2006@yandex.ru

**Abstract**

In this paper a discrete dynamical system is considered . There is a dial with $N$ positions (vertices) and $M$ particles. Particles are located in vertices. Each particle moves, at every time unit, in accordance with its plan. The plan is logistics, given through a real number which belongs to the segment $[0, 1]$. The number is represented in positional numeral system with base $N$ equal to the number of vertices. A competition takes place if particles must move in opposite directions simultaneously. A rule of competition resolution is given. Systems characteristics are investigated for sets of rational and irrational plans. Some algebraic constructions are introduced for this purpose. Probabilistic analogues (random walks) are also considered.

**Keywords:** discrete dynamical systems; Markov process; number theory; ergodic theory

## 1. Introduction

A dynamical system, named bipendulum, was introduced and investigated in (Kozlov, Buslaev, & Tatashev, 2015c). This system is the simplest version of the transport-logistic problem. The geometric interpretation of this system is the following. Each particle is located in one of two vertices. A channel connects the vertices. Particles must move or not move at each discrete time instant in accordance with plans. Delays are due to the condition that the particles cannot move towards each other. A competition takes place if two particles attempt to move in opposite directions. A rule of competitions resolution is given. The implementation of plans of losing competitions particles is delayed.

Plans of particles are given by binary representation of numbers which belong to the segment $[0, 1]$. A rational number determines a plan, if the representation of this number is a periodic positional fraction or a common fraction. In the first of these cases, we assume that the particle plan is written on the tape, and the particle reads digits of the plan successively. The transition to implementation of the next digit of the plan is equivalent to the following. Digits of plans are shifted to the left onto a position, and the digit to the left of the point is excluded. Therefore this transition is equivalent to multiplication of a number by 2 and exclusion of the integral part. This operation is called the Bernoulli shift (Schuster, 1984). Let us consider a set of plans defined by proper common fractions with the same denominator $N$. An algebra $W_N$ of elements of this set with respect to the operation of Bernoulli shift is introduced in (Kozlov, Buslaev, & Tatashev, 2015c), (Buslaev, & Tatashev, 2016). The algebra $W_N$, depending on $N$, is represented uniquely as a sum of connected components which are subalgebras $W_{N_i}$, where $N_i$ are divisors of $N$. In this case, dynamics of particles can be interpreted as motion of particles on the subalgebra graph, where weights of arcs generate logistics. The concept of particle tape velocity was introduced. Thus this algebra is an important component of the dynamical system called bipendulum.

A system with any number of vertices and particles is considered in this paper. The system is called real-valued pendulum. The plan of each particle is given by a number which belongs to the segment $[0, 1]$. This number is represented in the numeral positional system with the base equals to the number of vertices. If particles try to move towards each other, then delays take place as in the case of bipendulum. Transition to the next digit of the plan is equivalent to applying of the generalized Bernoulli shift operation to the number which determines the plan. The generalized Bernoulli shift is defined as follows. The number is multiplied by $N$ and the integral part is excluded. The base of positional system is equal to the number of vertices. Thus algebra of proper fractions generate real-valued $N$–pendulum as Bernoulli algebra is the abstract basis for bipendulum.

Some discrete dynamical systems with symmetric periodic structures were introduced and investigated in (Kozlov, Buslaev, Tatashev & Yashina ,2014) (Kozlov, Buslaev & Tatashev, 2015b), (Kozlov, Buslaev & Tatashev, 2015a).

## 2. Formulation of Problem

In Section 2, the problem of formulation is given in terms of dynamical systems.

### *2.1. Particles Dislocation*

Suppose there are $N$ vertices $V_0, \dots, V_{N-1}$ and $M$ particles $P_1, \dots, P_M$, Figure 1. Each particle is in one of $N$ vertices at every moment of time. There are logistic plans of particles dislocations. Each plan is a real number $a^{(j)}$ belonging to the segment $[0, 1]$. The plan of a particle determines a sequence of indexes of vertices in which the particle must be located at successive moments of time. This sequence is given by $N$-ary representation of the number $a^{(i)}$. The vertices are connected by channels. If two particles try to move towards each other, then a competition takes place. In this case only particles, trying to move in one of two directions, realize the attempt. This direction is chosen in accordance with a given rule. Implementation of plans of losing competition particles is delayed. The main essence of this system is total planning of particles dislocation.

The system state can be described for one of special cases with a binary matrix. The rows of the matrix correspond to particles. Each row contains a single "one". The index of the column, containing this "one", is equal to the index of the vertex, containing the particle at present time. This index is equal to one of numbers $0, 1, 2, \dots, N-1$, Figure 1.

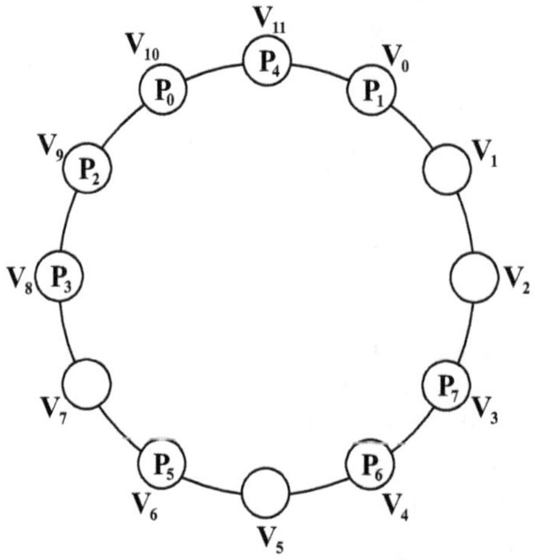


Figure 1. Round table and dislocation of particles

### *2.2. Plan Logistics*

A real number $a^{(i)}$ is given. This number is called the plan of the particle $P_i$ , $i = 1, 2, \dots, M$. This number is represented in $N$-ary system

$$a^{(i)} = 0.a_1^{(i)} a_2^{(i)} \dots a_k^{(i)} \dots, \tag{1}$$

where each digit value is equal to one of the numbers $0, 1, \dots, N-1$. We assume that the number $a^{(i)}$ is recorded on the tape of the particle $P_i$, $i = 1, \dots, M$. Each particle reads a digit recorded on its tape at every discrete time moment $T = 1, 2, \dots$. This digit determines the index of the vertex such that the particle tries to occupy this vertex, Figure 2.

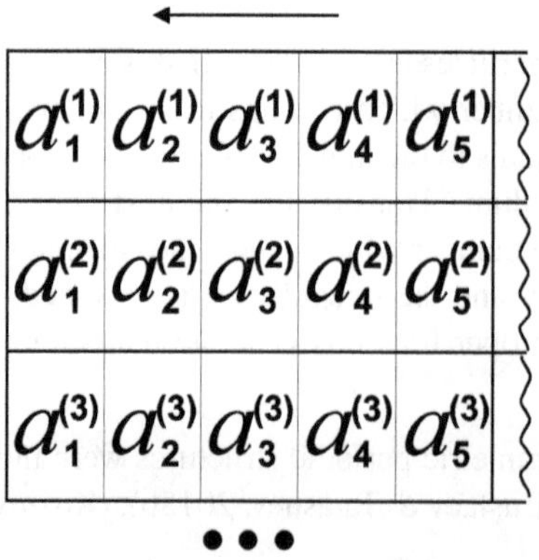


Figure 2. Turing tapes and plan logistics

### 2.3. Random Walk Logistics

Random walk is considered as an alternative for the logistics plan. In the case of random walk, the next planned dislocation of particle $P_i(T+1)$, $i = 1, \ldots, M$, will be determined, when $P_i(T)$ has been realized. Each digit of the number $a^{(i)}$ is played before the particle reads this digit, and the value of the digit is equal to $j$ with probability $p_{i,j}$, $j = 0, \ldots, N-1$; the numbers $p_{i,j}$ are given, $0 < p_{i,j} < 1$, $p_{i,0} + p_{i,1} + \cdots + p_{i,N-1} = 1$, $1 \le i \le M$. The finally configuration is formed if possible competitions have been taken into account.

### 2.4. Rules of Activity and Competition for Plan Logistics

The particle $P_i$ must be located in the vertex $V_{a_{jT}}$ in accordance with the plan of this particle, at the moment $T$, $i = 1, \ldots, M$, $T = 1, 2, \ldots$ If the particle $P_i$ is in the vertex $V_{a_T^{(i)}}$ at the moment $T$, then at the moment $T+1$, $T = 0, 1, 2, \ldots$, this particle will be in the vertex $V_{a_{T+1}^{(i)}}$ except the case of competition. *A competition takes place if*, at present moment of time, there are particles which try to come from the vertex $V_k$ to the vertex $V_l$, and particles which try to come from the vertex $V_l$ to the vertex $V_k$, $k \ne l$, $0 \le k, l \le N-1$. If $s_{k,l}$ particles try to move from the vertex $V_k$ to the vertex $V_l$ and $s_{l,k}$ particles try to move from the vertex $V_l$ to the vertex $V_k$, then all particles, trying to move from the vertex $V_k$ to the vertex $V_l$, win the competition with probability, Figure 3,

$$\frac{s_{k,l}}{s_{k,l} + s_{l,k}}. \tag{2}$$

If a competition takes place at the moment of time $T$, then all competing particles, winning the competition, will be, at the moment of time $T+1$, in the vertex, determined by the plan, and losing particles do not move. The tape of winning particle will read next digit at the moment of time $T+1$. The losing particle will read at the moment of time $T+1$ the digit corresponding the moment of time $T$. Initial conditions are given. These conditions determine the location of particles at the moment of time $T = 0$.

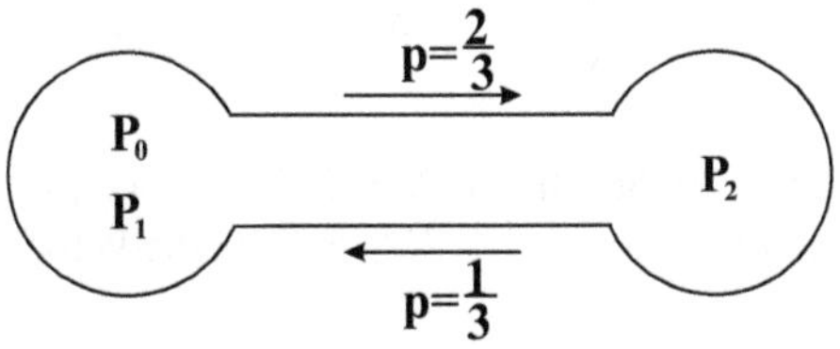


Figure 3. Competition resolution

### 2.5. Quantitative and Qualitative Characteristics

Denote by $D_i(T)$ the number of transitions of the particle $P_i$ tape during the time interval $[0; T]$, $i = 1, \ldots, M$; $T = 1, 2, \ldots$; $H(t)$ is the number of competitions during the time interval $(0; T]$; $H_i(t)$ is the number of lost competitions of the particle $P_i$ during time interval $(0; T]$, $T > 0$,

$$H_1(T) + H_2(T) + \cdots + H_M(T) = H(T).$$

The limit

$$w_i = \lim_{T\to\infty} \frac{D_i(T)}{T}, \; i = 1, \ldots, M, \tag{3}$$

is called *the velocity of the particle $P_i$ tape*, $i = 1, \ldots, M$, if this limit exists.

The limit

$$h = \lim_{T\to\infty} \frac{H(T)}{T}, \; i = 1, \ldots, M, \tag{4}$$

is called *the intensity of competitions* if this limit exists.

The limit

$$h_i = \lim_{T\to\infty} \frac{H_i(T)}{T}, \; i = 1, \ldots, M, \tag{5}$$

is called the intensity of lost particle $P_j$ competitions, if this limit exists.

The limits (3)–(5) depend on the process realization. These limits can exist or not exist depending on the realization.

The system is in *the state of synergy* after a moment of time $T_{syn}$ if, after the moment of time $T_{syn}$, no competitions take place.

### 3. Rational Logistic Plans

In Section 3, we introduce concepts related to real-valued pendulum, and prove theorems about particles tapes velocities.

If the number $a^{(i)}$ is rational, then this number can be represented as a periodic fraction

$$a^{(i)} = a_1^{(i)}a_2^{(i)}\dots a_{l_i}^{(i)}\left(a_{l_i+1}^{(i)}a_{l_i+2}^{(i)}\dots a_{l_i+m_i}^{(i)}\right),$$

where $l_j$ is the length of the aperiodic part of the number $a^{(i)}$ representation, and $m_i$ is the length of the repeating part of the representation, $i = 1,\dots,M$.

*3.1. Extreme Rational Bipendulum*

Consider the following example. Suppose $N = M = 2$; $a^{(1)}$ and $a^{(2)}$ are the numbers 1/3 and 1/5 respectively. This numbers are represented in the binary system as periodic fractions

$$a^{(1)} = \frac{1}{3} = 0.(01),\ a^{(2)} = 0.(0011).$$

**Lemma 1, (Kozlov, Buslaev, & Tatashev, 2015c).** *Suppose*

$$a^{(1)} = 0.(01),\ a^{(2)} = 0.(0011).$$

*Then limits (3), (4) and (5) exist with probability 1, and*

$$h = \frac{2}{5},\ h_1 = h_2 = \frac{1}{5},\ w_1 = w_2 = \frac{4}{5}.$$

It is proved in (Kozlov, Buslaev, & Tatashev, 2015c) that, if $N = 2$, $M = 2$, i.e., in the case of bipendulum, the velocity of particles is not less than $\frac{4}{5}$. We shall prove the following. If $N \geq 3$, then the velocity of particles can be equal to $\frac{3}{4}$.

**Lemma 2.** *Suppose* $N = 3$, $M = 2$.

$$a^{(1)} = 0.(012),\ a^{(2)} = 0.(021).$$

*Then* $h = \frac{1}{2}$, $h_1 = h_2 = \frac{1}{4}$, $w_1 = w_2 = \frac{3}{4}$.

*Proof.* The system is a Markov chain (Feller, 1970), (Kemeny, & Snell, 1976), (Borovkov, 1986) with states $(i_1, i_2)$, $0 \leq i_1, i_2 \leq 2$, where $i_j$ is a digit such that the particle $P_j$, $j = 1, 2$, reads this digit. There are 9 states

$$S_0 = (0,0),\ S_1 = (0,1),\ S_2 = (0,2),\ S_3 = (1,0),\ S_4 = (1,1),$$

$$S_5 = (1,2),\ S_6 = (2,0),\ S_7 = (2,1),\ S_8 = (2,2).$$

States $S_2$, $S_3$, and $S_7$ are inessential states of the chain. The other 6 states form a set of essential states. This set is divided into two subsets:

$$G_1 = \{S_0, S_4, S_8\},\ G_2 = \{S_1, S_5, S_6\}.$$

The chain goes from any state of the set $G_1$ to a state of the set $G_2$, and from any state of the set $G_2$ to a state of the set $G_1$. If the system is in a state of the set $G_1$, then there is no competition. If the system is in a state of the set $G_2$, then there is a competition. Thus, $h = \frac{1}{2}$, $w_1 = w_2 = \frac{3}{4}$.

**Theorem 1.** *Suppose* $M = 2$. *Then the greatest possible value of the competitions intensity is equal to* $\frac{1}{2}$, *and the least possible value of the tape velocity is equal to* $\frac{3}{4}$.

*Proof.* It follows from Lemma 2 that, if $M = 2$, then plans of particles can be given such that $h = \frac{1}{2}$, $v = \frac{3}{4}$. If $M = 2$, then plans cannot be given such that the competitions intensity is more than $\frac{1}{2}$. Indeed, if there is a competition at moment of time $T$, then there will be no competition at moment of time $T + 1$. Theorem 1 has been proved.

*3.2. Phase Pendulums*

Suppose $N = M = 2$, and we define the plan $a^{(2)}$, shifting plan $a^{(1)}$ for a fixed number of positions.

**Lemma 3.** *Suppose*

$$a^{(i)} = 0.\left(a_1^{(i)}a_2^{(i)}\dots a_m^{(i)}\right),\ i = 1, 2,$$

*and we get the plan $a^{(2)}$, shifting plan $a^{(1)}$ onto c positions $(0 \le c \le m-1)$ to the left, i.e.,*

$$a^{(2)}_{k+c} = a^{(1)}_k, \ k = 1, 2, \ldots, m,$$

*where addition is meant modulo m. Then the system comes to the state of synergy after a time interval with finite expectation.*

*Proof.* The system is a Markov chain (Feller, 1970), (Kemeny, & Snell, 1976), (Borovkov, 1986) with $m^2$ states. The system comes to the state of synergy with a positive probability, and the system cannot leave the state of synergy. Therefore a state is essential if and only if the system is in the state of synergy. The Markov chain comes from any inessential state to the set of essential states after a time interval with finite expectation. Lemma 3 has been proved.

**Theorem 2.** *Suppose*

$$a^{(i)} = 0.a^{(i)}_1 \ldots a^{(i)}_{l_i}(a^{(i)}_{l_i+1}a^{(i)}_{l_i+2} \ldots a^{(i)}_{l_i+m}), \ i = 1, 2,$$

*and we get the repeating part of $a^{(2)}$, shifting the repeating part of $a^{(1)}$ onto a fixed number of positions. Then the system comes to the state of synergy after a time interval with finite expectation.*

*Proof.* Both particles will read digits of repeating parts of their plans after a time interval with finite expectation. Theorem 2 follows from Lemma 3.

### 4. Functional Interpretation of Bipendulum

In Section 4, we consider functional representation of plans. We define variation of proper fraction .

Suppose the binary representation of the number $j/K$ is

$$\frac{j}{K} = 0.a_1 \ldots a_l(a_{l+1} \ldots a_{l+m}).$$

Suppose

$$\theta(i) = \begin{cases} 0, & a_i = a_{i+1}, \\ 1, & a_i \ne a_{i+1} \end{cases}$$

The *variation* of the representation of the number $\frac{j}{K}$ is defined as

$$V(j/K) = \frac{1}{m} \sum_{i=l+1}^{l+m} \theta(i),$$

where the addition, in indexes, is meant modulo $m$. The limit

$$Var(a) = \lim_{k \to \infty} \frac{1}{k} \sum_{i=1}^{k} \theta(i)$$

is called the *variation of the binary representation* of a real number $a \in [0, 1)$, $a = 0, a_1a_2 \ldots a_k \ldots$ if this limit exists. It is evident that $V(a) = Var(a)$ for rational number $a$. Suppose

$$Var_{\min} = \min\left(Var(a^{(1)}), Var(a^{(2)})\right).$$

**Theorem 3.** *Suppose $N = M = 2$. Then inequalities are true*

$$h \le 2Var_{\min}(2 + Var_{\min})^{-1},$$

$$v_1 = v_2 \ge 2(2 + Var_{\min})^{-1}.$$

*Proof.* If $\theta(i) = 0$ and a particle reads digit $a_i$ at time $T$, then the particle will be read digit $a_{i+1}$, $i = 1, 2, \ldots,$ at time $T+1$. If $\theta(i) = 1$, the particle reads digit $a_i$ at time $T$, and no competition takes place, then the particle will read digit $a_{i+1}$ at time $T+1$, $i = 1, 2, \ldots$. Suppose $\theta(i) = 1$, the particle reads digit $a_i$ at time $T$, and there is a competition at this time. Then, at time $T+1$, with probability $\frac{1}{2}$, the particle reads digit $a_i$, and with probability $\frac{1}{2}$, the particle reads digit $a_{i+1}$. Therefore, the average time of particle going from digit $a_i$ to digit $a_{i+1}$ is not greater than

$$(1 - Var_{\min}) \cdot 1 + Var_{\min} \cdot \frac{3}{2} = 1 + \frac{Var_{\min}}{2}.$$

Thus, $v_1 = v_2 \geq ((2 + Var_{min})/2)^{-1} = 2(2 + Var_{min})^{-1}$, $h = 2(1 - v_1) \leq 2Var_{min}(2 + Var_{min})^{-1}$. Theorem 3 has been proved.

**Corollary 2.** *Suppose $N = M = 2$. Then the following inequalities are true*

$$h_1 = h_2 \leq \begin{cases} 2Var_{\min}/(1 + Var_{\min}), & Var_{\min} \leq 2/3, \\ 1/2, & Var_{\min} \geq 2/3, \end{cases}$$

$$v_1 = v_2 \geq \begin{cases} 2/(2 + Var_{\min}), & Var_{\min} \leq 2/3, \\ 3/4, & Var_{\min} \geq 2/3. \end{cases}$$

## 5. Generalized Bernoulli Algebras on Proper Fractions

### *5.1. The Concept of Generalized Bernoulli Algebra*

Algebras of proper common fractions with respect to the operation, called Bernoulli shift, (Schuster, 1984), were considered in (Buslaev, Tatashev, 2016). If this operation is applied to a fraction, the fraction is multiplied by 2, and the integral part of the product is excluded. We introduce algebras with operation of Bernoulli shift. If this operation is applied, then the fraction is multiplied by the natural number $N$, and the integral part is excluded. Consider the set of proper fractions

$$G_K = \left\{ \frac{0}{K}, \frac{1}{K}, \frac{2}{K}, \ldots, \frac{K-1}{K} \right\}$$

with a unary operation

$$FB^N(x) = Nx - [Nx],$$

where $[Nx]$ is the integral part of $Nx$.

Suppose $W_K^N$ is the algebra with the operation $FB^N$ on the set $G_K$. The number $i/K$ $(0 \leq i \leq K-1)$ generates an algebra. Denote by $W_K^N(i)$ the subalgebra generated by the element $i/K$, $i = 0, 1, \ldots, K-1$.

Elements $i/K$ and $j/K$ are called *communicating* with each other if $j/K \in W_K^N(i)$ and $i/K \in W_K^N(j)$.

The element $i/K$ is called *inessential* if there exists an element $j/K$ such that $j/K \in W_K^N(i)$ and $i/K \notin W_K^N(j)$. The other elements are *essential.*

The set of essential elements can be divided into disjoint sets such that any two elements of the same set communicate with each other. These sets are called *classes of communicating essential elements* or *orbits.* If an element $i/N$ satisfies equation

$$FB^N\left(\frac{i}{K}\right) = i/K,$$

then this element is called *absorbing.*

If $i/K$ and $j/K$ communicate with each other, then $W_K^N(j) = W_K^N(i)$, $i, j = 1, 2, \ldots, K-1$. If $j/K \neq W_K^N(i)$ and $i/K \neq W_K^N(j)$, then $W_K^N(i)$ and $W_K^N(j)$ are disjoint subalgebras.

Identifying subalgebras $W_K^N(i)$ and $W_K^N(j)$ with each other for $W_K^N(i) = W_K^N(j)$, we divide the algebra $W_K^N$ into disjoint subalgebras. Each of these subalgebras contains a class of communicating essential elements and a set of inessential elements. This set of inessential elements can be empty. These subalgebras are called *connected components of the algebra.*

Consider a proper fraction $i/K$, $K \geq 1$, $0 \leq i < K$. The following theorem allows to find the $N$-ary representation of this fraction:

$$i/K = \sum_{j=1}^{\infty} a_j \cdot N^{-j},$$

where $a_j$ equals one of numbers $0, 1, \ldots, N-1$, $j = 1, 2, \ldots$

**Theorem, (Broido, Ilyina, 2006).** *Suppose*

$$b_0 = i/K,\ b_j = FB^N(b_{j-1}),\ j = 1, 2, \ldots$$

*Then $a_j = [Nb_{j-1}]$.*

This theorem gives an approach which allows to find the $N$-ary representation of proper fraction $b_0 = \frac{i}{K}$. The value of the first digit $a_1$ of the representation equals the integral part of $Nb_0$. The fractional part of the product is the value of $b_1$. The value of the digit $a_j$ equals the integral part of the product $b_{j-1}$ and $N$. The fractional part of this product equals the value of $b_j$, $j = 1, 2, \ldots$ Suppose numbers $l$ $(l \geq 1)$ and $m$ $(m \geq 1)$ are such that $b_l = b_{l+m}$, and there are no numbers $i$ and $j$ such that $i, j < l + m$ and $b_i = b_j$. Then we have $a_i = a_{i+m}$ for any $i \geq l$. Hence the $N$-ary representation of the proper fraction $s/K$ contains an aperiodic part with length $l - 1$ and a repeating part with length $m$. We write

$$\frac{s}{K} = 0.a_1 \ldots a_{l-1}(a_l \ldots a_{l+m-1}).$$

We have formulated an approach to find the binary representation of a proper fraction. This approach is similar to the approach of converting the decimal representation of a proper fraction to the $N$-ary representation, (Broido & Ilyina, 2006). It is easy to reconstruct the trajectory of the element $s/K$ of the algebra $W_K^N$ if we know the $N$-ary representation of the number $s/K$.

Some problems of number theory, related to $N$-ary representation of numbers, are considered in (Uteshev, Cherkasov, Shaposhnikov, 2001).

*5.2. Absorbing Elements of Algebra $W_K^N$*

We shall prove theorems about absorbing elements of algebras $W_K^N$.

**Lemma 4.** *If $N = 2$, then the element $\frac{0}{K}$ is a unique absorbing element of the algebra $W_K^2$.*

*Proof.* It is obvious that $\frac{0}{K}$ is an absorbing element of the algebra $W_K^N$ for any $N$ and $K$. Since, for $i = 1, \ldots, K - 1$,

$$\frac{i}{K} < \frac{2i}{K} < \frac{i}{K} + 1,$$

then there exist no other absorbing elements of the algebra $W_K^2$.

We shall give some examples. These examples show following. If $N \geq 3$, there exist algebras which do not contain elements not equal to $\frac{0}{N}$, and algebras which contain absorbing elements not equal to $0/N$.

*Example 1.* Algebra $W_5^3$ contains an absorbing element $\frac{0}{N}$ and an orbit with period 4.

*Example 3.* Algebra $W_6^3$ contains two absorbing elements $\frac{0}{6}$ and $\frac{3}{6}$. The other elements of the algebra are inessential.

**Lemma 5.** *Suppose $N = K + 1$. Then all elements of the subalgebra $W_K^N$ are absorbing.*

*Proof.* If $N = K + 1$, then we have

$$FB^N\left(\frac{j}{K}\right) = (K + 1)jK^{-1} - j = j/K, \ j = 0, 1, \ldots, K - 1.$$

*5.3. Structure of the Algebra $W_K^N$, Where $K$ is Coprime to $N$*

In Subsection 5.3, we prove theorems about generalized Bernoulli algebras in the case of coprime $K$ and $N$.

5.3.1 Preliminary Definitions and Results

Suppose the canonical representation of the number $K$, $K \geq 2$, has the form

$$K = p_1^{s_1} \ldots p_l^{s_l}, \tag{6}$$

where $2 \leq p_1 < \cdots < p_l$ are prime numbers, $s_1, \ldots, s_l$ are natural numbers. We write $K = \overline{p}^{\overline{s}}$, where $\overline{p} = (p_1, \ldots, p_l)$, $\overline{s} = (s_1, \ldots, s_l)$. Denote by $E(K)$ the Euler's function (Buchstab, 1972),

$$E(K) = \left(p_1^{s_1} - p_1^{s_1-1}\right) \ldots \left(p_l^{s_l} - p_l^{s_l-1}\right).$$

**Euler's theorem on numbers, (Buchstab, 1972).** *The value of $E(K)$ is equal to the number of positive integers that are coprime to $N$ and less than $K$*

This function can be also represented as

$$E(K) = K\left(1 - \frac{1}{p_1}\right)\left(1 - \frac{1}{p_2}\right) \ldots \left(1 - \frac{1}{p_l}\right).$$

**Euler's theorem on divisibility, (Buchstab, 1972).** *Suppose $N$ and $K$ are coprime natural numbers. Then $K$ is a divisor of $N^{E(K)} - 1$.*

Fermat's little theorem is a special case of Euler's theorem on divisibility.

**Fermat's little theorem, (Buchstab, 1972).** *If p is prime, and p is not a divisor of N, then p is a divisor of* $N^{p-1} - 1$.

Denote by $LCM(a_1, \ldots, a_L)$ the least common multiple of numbers $a_1, \ldots, a_l$. Suppose $K$ has the form (6).

*Generalized Euler's function* is the function $e(K)$, $e(1) = 1$,

$$e(K) = LCM\left(p_1^{s_1-1}(p_1 - 1), \ldots, p_l^{s_l-1}(p_l - 1)\right).$$

It is obvious that any divisor of $e(K)$ is a divisor of $E(K)$.

**The generalized Euler's divisibility theorem, (Buchstab, 1972).** *Let N and K be coprime positive integers. Then K is a divisor of* $N^{E(K)} - 1$.

According to this theorem, if $N$ and $K$ are coprime, then there exists a positive number $l$ such that $K$ is a divisor of $N^l - 1$. *The smallest positive integer l such that K is a divisor of* $N^l - 1$ *is called the order of K modulo N, (Buchstab, 1972). Denote by* $m(K, N)$ *the order of K modulo N.*

**The generalized Euler's order theorem, (Buchstab, 1972).** *Let N and K be coprime positive integers. Then* $m(K, N)$ *is a divisor of* $e(K)$.

The elements of the algebra $W_K^N$ can be divided into two classes. They are *the class of noncancelable fractions and the class of cancelable fractions*

$$W_K^N = W_K^{N,copr[ime]} + W_K^{N,canc[elable]}.$$

Consider the set of all elements $j/K$ of the algebra $W_K^N$ such that $j/K$ is a noncancelable fraction. The number of these elements is equal to the number of positive integers less than $K$ and coprime to $N$, i.e., this number is equal to Euler's function. Suppose

$$\delta(d, E(K)) = \begin{cases} \frac{E(K)}{m(K,N)}, & d = m(K, N), \\ 0, & d \neq m(K, N). \end{cases} \tag{7}$$

In accordance with section 5.1, an orbit of period $d$ is a set of elements of the algebra $W_K^N$

$$\frac{j_1}{K}, \frac{j_2}{K}, \ldots, \frac{j_d}{K},$$

$0 \leq j_i \leq K - 1$, $i = 1, \ldots, d$ such that

$$\frac{j_{i+1}}{K} = FB^N\left(\frac{j_i}{K}\right) \neq \frac{j_1}{K}, \quad i = 1, \ldots, d - 1,$$

$$\frac{j_1}{K} = FB^N\left(\frac{j_d}{K}\right).$$

Denote by $O(K, d)$ *the number of orbits with period d* ($d \geq 1$).

5.3.2 Simple Case

Suppose $K = p^s$, where $p \geq 2$ is a prime number, the number $p$ is not a divisor of $K$, $s$ is a natural number,

$$E(K) = p^s - p^{s-1}.$$

**Theorem 4.** *The following formula is true*

$$O(K, d) = \sum_{r=1}^{s} \delta(d, E(p^r)) = \sum_{r=1}^{s} \delta(m(p^r), E(p^r)).$$

The proof is based on Lemma 6.

**Lemma 6.** *Suppose* $r \leq s$, $j/p^r$ *is an element of the algebra* $W_{p^r}^N$ *such that it is a noncancelable fraction. Then this element belongs to an orbit of period* $m(p^r)$. *The set of cancelable fractions of the algebra* $W_{p^r}^N$ *is a subalgebra isomorphic to the algebra* $W_{p^{r-1}}^N$.

*Proof.* Suppose the number $a$ satisfies the equation

$$N^m \cdot \frac{j}{K} = \frac{j}{K} + a.$$

The number $a$ is natural if and only if the number $K$ is a divisor of $N^m - 1$. We take into account that $j$ and $K$ are coprime, and $j < K$. Applying to a noncancelable element the generalized Bernoulli shift operation $m(K, N)$ times, we obtain the same element. If the generalized Bernoulli shift is applied to this element less than $m(K, N)$ times, then the same element cannot be obtained. It is obviously that the subalgebra of cancelable fractions of the algebra $W^N_{p^r}$ and the algebra $W^N_{p^{r-1}}$ are isomorphic to each other. Lemma 6 has been proved.

5.3.3 General Case

Suppose $K$ is coprime to $N$.

**Lemma 7.** *Any element $j/K$ of the algebra $W^N_K$ belongs to an orbit, and, if $j/K \in W^{N,copr}_K$, i.e., the element $j/K$ is a noncancelable fraction, then the period of this orbit equals $m(K, N)$.*

*Proof.* Apply the Bernoulli shift operation to the element $j/K$ $i$ times. We obtain the same element if and only if the equation

$$N^i \cdot \frac{j}{K} = \frac{j}{K} + a, \tag{8}$$

contains natural number $a$. If $i = e(K)$ in (8), then the number $a$ is natural. We take into account that $K$ is coprime to $N$. Hence there exists an orbit with period not greater than $e(K)$ such that this orbit contains the element $j/K$. If the fraction $j/K$ is a noncancelable fraction, then $i = m(K, N)$ is the smallest value of $i$ such that the number $a$ is a natural number. From this Lemma 7 follows.

**Lemma 8.** *Let $K$ be a natural number, and $K$ is coprime to $N$. Then two elements $a \in W^{N,copr}_K$, $b \in W^{N,canc}_K$ of the algebra $W^N_K$ cannot belong to the same orbit.*

*Proof.* If an element is cancelable fraction, and we apply the generalized Bernoulli shift operation to this element any times, then we can obtain no noncancelable fraction. From this, Lemma 8 follows.

Denote by $\delta_{d,K}$ the number of orbits with period $d$ in the subalgebra $W^{N,copr}_K$. In accordance with Lemmas 7 and 8,

$$\delta_{m(p^r),p^r} = \frac{E(p^r)}{m(p^r, N)}.$$

If $d \neq m(p^r, N)$, then $\delta_{d,p^r} = 0$.

Suppose $R = R(N)$ is the set of vectors $r = (r_1, \dots, r_l)$ with integer nonnegative numbers, $0 \le r_1 \le s_1, \dots, 0 \le r_l \le s_l$, and at least one of numbers $r_1, \dots, r_l$ is positive.

**Theorem 5.** *Suppose $O(K, d)$ is the number of orbits with period $d$ ($d \ge 1$) in algebra $W^N_K$. Then the following formula is true*

$$O(K, d) = \sum_{\overline{r} \in R} \delta(d, \overline{p}^{\overline{r}}), \tag{9}$$

*where $\delta(d, p^r)$ is calculated in accordance with (7).*

*Proof.* If $r_1, r_2 \in R(K)$, then, for $K_1 = \overline{p}^{\overline{r}_1}$, $K_2 = \overline{p}^{\overline{r}_2}$, $K_1 \neq K_2$ we have

$$W^{N,copr}_{K_1} \cap W^{N,copr}_{K_2} = \emptyset.$$

Therefore no orbit can belong to both $W^{N,copr}_{K_1}$ and $W^{N,copr}_{K_2}$. From this fact and Lemmas 7 and 8, Theorem 5 follows.

*5.4. Algebra $W^N_K$ in the Case of $K = N^h K_1$, Where $K_1$ and $N$ are Coprime*

In the subsection 5.4, we assume that $K$ equals $K_1$ multiplied by $N^h$, where $K_1$ and $N$ are coprime. We shall prove a theorem about the form of generalized of Bernoulli algebras.

**Theorem 6.** *Let $h$ be a natural number,*

$$K = N^h K_1,$$

*where $K_1 = p_1^{s_1} \dots p_l^{s_l}$, $p_1 < \cdots < p_L$ are prime numbers, and none of these numbers is not a divisor of $N$. Then the following is true.*

*1) Algebras $W^N_K$ and $W^N_{K/N^h}$ contain the same number of orbits with each period.*

*2) The algebra $W^N_K$ contains $K_1$ essential elements. The other $K - K_1$ elements are nonessential elements.*

*3) The graph of algebra $W^N_K$ contains, as its subgraphs, $K_1$ trees, and the depth of each of these tree equals $h$. Each of $K_1$ essential elements of the algebra $W^N_K$ is the root of precisely one tree, and the degree of the root equals 1. The degree of* *other vertices, for any tree, is equal to N except the vertices of the level h. Any tree contains $N^h$ elements. One of these elements is the root of the tree, and there are $N^{i-1}$ elements that are vertices of the level i for any $i = 1, \ldots, h$.*

The proof of Theorem 6 is similar to proofs of Theorems 5 – 7 in (Buslaev, & Tatashev, 2016), where the case of $N = 2$ was considered.

### *5.5. Algebras $W_K^N$ and $W_d^N$, Where d is a Divisor of N*

Consider algebras $W_K^N$ and $W_d^N$, where $d$ is a divisor of the number $N$. In Subsection 5.5, we prove theorems that alow to compare the algebras $W_K^N$ and $W_d^N$.

**Theorem 7.** *Let d be a divisor of the number K. If the element $i/d$ of the algebra $W_d^N$ belongs to the orbit with period m, then the element $\frac{K}{d}\frac{i}{K}$ of the algebra $W_K^N$ belongs to an orbit with period m. If the elements $i/d$ and $j/d$ of the algebra $W_d^N$ belong to the same orbit, then elements $\frac{K}{d}\frac{i}{K}$, $\frac{K}{d}\frac{j}{K}$ of the algebra $W_K^N$ also belong to the same orbit. If the elements $i/d$ and jd of the algebra $W_d^N$ belong to different orbits, then the elements $\frac{K}{d}\frac{i}{K}$ and $\frac{K}{d}\frac{j}{K}$ of the algebra $W_K^N$ also belong to the different orbits. If the element $i/d$ of the algebra $W_d^N$ is inessential, then the element $\frac{K}{d}\frac{i}{K}$ of the algebra $W_K^N$ is also inessential. Elements $\frac{K}{d}\frac{i}{K}$ and $\frac{K}{d}\frac{j}{K}$ of the algebra $W_K^N$ belong to the same connected component if and only if the elements $i/d$ and $j/d$ of the algebra $W_d^N$ also belong to the same connected component.*

*Proof.* The value of the element $\frac{K}{d}\frac{i}{K}$ of the algebra $W_K^N$ and the value of element $i/d$ of the algebra $W_d^N$ are the same for any $i = 1, 2, \ldots, d-1$. From this, Theorem 7 follows.

**Corollary 3.** *Let d be a divisor of the number K. Then the number of orbits with period m, the number of connected components and the number of elements of the algebra $W_K^N$ are not less than in the algebra $W_d^N$.*

## 6. Rational Pendulums and Generalized Bernoulli Algebras

In section 6 we consider connection between generalized Bernoulli algebras and real valued pendulums. We show that the algebraic construction of rational logistic plans has an effect on properties of real valued pendulums.

### *6.1. Generalized Bernoulli Algebra and Real Valued Pendulum*

Consider a real valued pendulum with rational particles plans given by the rational numbers $a^{(1)} = s_1/k_1$, $a^{(2)} = s_2/k_2$, $0 \le s_i < k_i$, $i = 1, 2$. Suppose $k$ is the least common multiple of numbers $k_1$, $k_2$, and

$$a^{(i)} = \frac{j_i}{k},\ 0 \le j_i < k,\ i = 1, 2. \tag{10}$$

The generalized Bernoulli algebra $W_k^N$ contains all subalgebras (orbits), satisfying initial conditions (10).

### *6.2. Subalgebras, Orbits, and Velocity of Turing Tapes*

To each initial condition $a^{(i)}$, $i = 1, \ldots, M$, assign vector of elements $b^{(i)}$, $i = 1, \ldots, M$, where $a^{(i)} = b^{(i)}$ if $a^{(i)}$ is an element of the orbit of a subalgebra, or the first element of the orbit of the subalgebra on the path if $a^{(i)}$ is an inessential element.

**Lemma 9.** *Suppose $M \ge 2$, and $a^{(i)}$, $i = 1, \ldots, M$, are plans. Then, with probability 1, numerical characteristics of the real valued pendulum are the same as in the case of plans $b^{(i)}$, $i = 1, \ldots, M$, and the elements of the pendulum, belonging to the same orbit, can be not taken into account.*

The proof is evident.

**Theorem 8.** *Suppose $M \ge 2$, and plans of particles $a^{(i)}$, $i = 1, \ldots, M$, belong to the same subalgebra of the algebra $W_k^N$. Then the system comes to the state of synergy after a time interval with finite expectation.*

*Proof.* Suppose $M = 2$. We can get the periodic part of plan $a^{(2)}$, shifting the periodic part of the plan $a^{(1)}$ onto a fixed number of positions. There will be a finite number on competitions before the the system comes to the state of synergy, or the periodic parts of the plan $a^{(1)}$ and $a^{(2)}$ coincide after a time interval with finite expectation, and the system also come to the state of synergy. From this the theorem follows.

In the general case, each particle will read digits of periodic parts of plans after an instant with finite expectation. Suppose that, from time $T_1$, the set of competing particles, containing the particle with the least index, wins each competition. The particle $P_1$ does not lose competitions. The particle $P_2$ can lose competitions with the particle $P_1$. However there can be no more than $d-1$ competitions of these particles. The particle $P_2$ cannot compete with the particle $P_1$ after a finite instant $T_2$. Similarly, there are $T_3, \ldots, T_M$ ($T_1 \le \cdots \le T_M$) such that, after time $T_i$ none of particles $P_1, \ldots, P_i$ can compete with any other particle of this set. There will be no competitions after time $T_M$. Thus, with positive probability, the system comes to the state of synergy after time bounded above. From this, the theorem follows.

*6.3. The Velocity of Turing Tapes with Plans from Different Subalgebras*

Suppose there 2 particles and plans of particles belong to different subalgebras.

Particles will read repeating parts of their plans after a time interval with finite expectation. Therefore it is sufficient to know to which subalgebras plans belong.

Let $(i_1, i_2)$ be the system state such that the particle $P_j$ reads the $i_j$-th digit of the period, $j = 1, 2$.

We introduce Markov chain $X(t')$ obtained from the original chain as follows. The process is considered only at instants $t'$ such that competitions take place at these instants. New time scale is obtained from the original chain, excluding all instants such that there are not any competitions at these instants.

We can find transition probabilities between all states. There are no more than two positive elements at each row of the transition probability matrix. Break of the chain is the state of synergy. The time of going to the next state is calculated. This time depends on what state will be the next state. The problem of calculation of transition probabilities is connected with analysis of linear Diophantine equations with two variables. When we have found to what class of communicating essential states the initial state belongs, we find the system of equations for steady state probabilities and calculate the steady probabilities. When we have found steady state probabilities, we calculate the average duration between competitions, and calculate the velocity of particles tapes.

Following theorems give conditions of synergy in the case such that the plans of particles belong to different algebras.

We introduce the operation +,* of bitwise addition and multiplication of binary representations of rational plans $a^{(i)}$. Then the equality $a^{(1)} * a^{(2)} = a^{(1)}$ is true if the set of positions with 'ones' of number $a^{(1)}$ belongs to the set of positions with 'ones' of the number $a^{(2)}$. Denote this relation by $a^{(1)} \leq a^{(2)}$. From $a^{(1)} + a^{(2)} = 1$ it follows that $a^{(1)} * a^{(2)} = 0$.

**Theorem 9.** *Suppose $M = 2$, $N = 2$*

$$a^{(1)} + a^{(2)} = 1.$$

*Then the system comes to the state of synergy after a time interval with finite expectation.*

*Proof.* Suppose

$$a^{(j)} = 0.(a_1^{(j)} \dots a_d^{(j)}),\ j = 1, 2;$$

$b^{(1)} = 0.(b_1^{(1)} \dots b_d^{(1)})$, where $b_i^{(1)} = a_{i+1}^{(1)}$, $i = 1, \dots, d$ (addition is meant modulo $d$), i.e., we get the plan $b^{(1)}$, shifting the plan $a^{(1)}$ onto a position to left. Consider bipendulum with plans $b^{(1)}$ and $a^{(2)}$. If $a_i^{(2)} = a_{i+1}^{(2)}$ (addition is meant modulo $d$), then the first competition cannot take place at time $T = i$, $i = 1, \dots, d$. If $a_i^{(2)} \neq a_{i+1}^{(2)}$, then $b_i^{(1)} = a_{i+1}^{(1)} = a_i^{(2)}$, and, therefore, the first competition also does not take place at time $T = i$. Thus the bipendulum is in the state of synergy from the initial instant. From this and Lemma 3, it follows that the original bipendulum comes to the state of synergy after a time interval with finite expectation. Theorem 9 has been proved.

**Theorem 10.** *Suppose $M = 2$, $N = 2$*

$$a^{(1)} \leq a^{(2)}.$$

*Then the system comes to the state of synergy after a time interval with finite expectation.*

The proof is evident.

*Remark 1.* The following example shows that the synergy can take place for plans given by fractions with coprime denominators.

Suppose $N = M = 2$,

$$a^{(1)} = \frac{1}{7} = 0.(001),\ a^{(2)} = \frac{4}{9} = 0.(011100).$$

The number $a^{(1)}$ can be represented in the form $a^1 = 0.(001001)$. Therefore no competition can take place at the first 6 steps. Hence no competition can take place in the future.

*Remark 2.* The following example shows the following. It is possible that plans of all particles are given by fractions with the same denominator such that this denominator is a prime number, and the system cannot come to the state of synergy.

Suppose $N = M = 2$,

$$a^{(1)} = \frac{3}{73} = 0.(000010101),\ a^{(2)} = \frac{13}{73} = 0.(001011011).$$

The system cannot come to the state of synergy.

Indeed, if we shift the repeating part of a plan onto any number of positions, then a competition takes place after a finite time.

*6.4. Interpretation of Real Valued Pendulum in Terms of Bernoulli Algebra*

We can assume, in the general case, that there are $N$ vertices and $M$ particles, then the plans of the particles $P_j$, are given by fractions such that these fractions generate orbits. This means that the process of cyclic shift of the periodic sequence

$$a^{(j)} = \frac{s_j}{k_j},\ 0 \le s_j < k_j,\ j = 1, \ldots, M,$$

takes place.

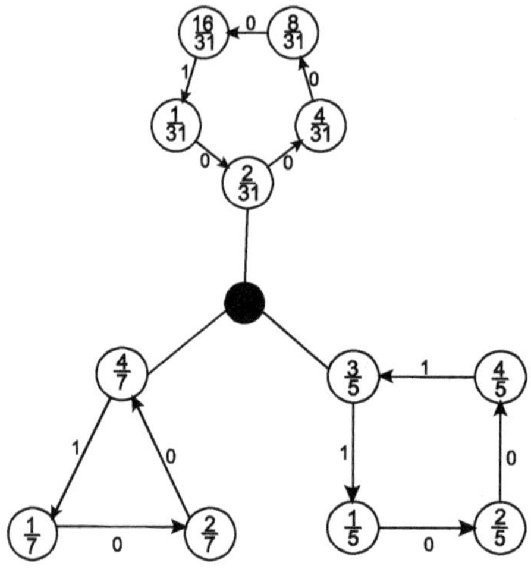


Figure 4. Bipendilum with orbits, $M = 3, N = 2$

We introduce the system such that this system is equivalent to real valued pendulum, Figures 4, 5. There are $M$ particles $P_1, \ldots, P_M$ and $k$ vertices, where $k$ is the least common multiple of numbers $k_1, \ldots, k_M$. Each vertex corresponds to an element of the algebra $W_k^N$. The initial state is given such that the particle $P_j$ is in the vertex, corresponding to an element of the subalgebra generated by the element $a^{(j)} = \frac{s_j}{k_j}, 0 \le s_j < k_j, j = 1, \ldots, M$. The vertex such that this vertex corresponds to the element with representation

$$\alpha = 0.(a_1, \ldots, a_m)$$

is labeled with the value of digit $a_m$. At each discrete instant, the particle located in the vertex, corresponding to the element $x$, goes to the vertex corresponding to the element $y = FB^N(x)$, if no competition takes place. A competition takes place if, for some $i, j$ $(0 \le i, j \le N-1)$, there are particles, trying to come from a vertex with the label $i$ to a vertex with the label $j$, and particles, trying to come from a vertex with the label $j$ to a vertex with the label $i$. If a competition takes place, then

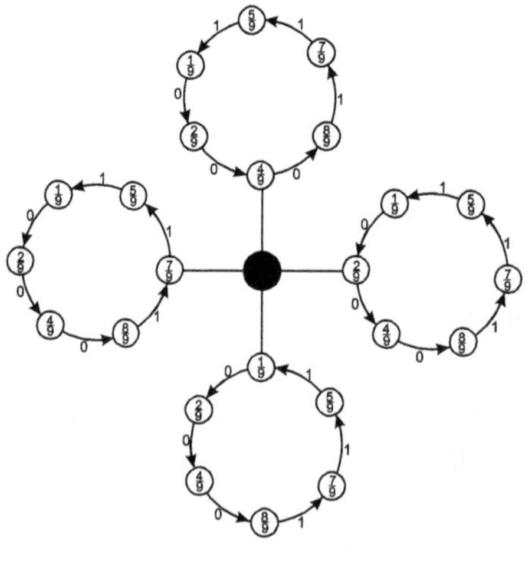

Figure 5. Phase bipendilum,$M = 4, N = 2$

only particles, winning the competition, move. Particles, losing the competition, do not move. Each particle moves along an orbit of the subalgebra $W_K^N$. If, at present time $T$, the particle is located in the vertex, corresponding to the element

$\alpha = 0.(a_1, \ldots, a_m)$, then, at this instant, in the real valued pendulum, the representation of this element is written on the particle tape. At this instant, the particle of the real valued pendulum reads the digit $a_1$, and therefore tries to come to the vertex $V_{a_1}$. The preceding digit that the particle read is $a_m$. Therefore the particle is located in the vertex $V_{a_m}$ at present time.

The velocity of particle, in the introduced system, and the velocity of particle tape, in the real valued pendulum, are the same.

**7. Random Walks Real Valued Pendulums**

Suppose that, at each time $T$, we choose the index of the vertex such that the particle tries to come to this vertex. The particle must be located in the same vertex at next instant with probability $q$, $0 < q < 1$. Any other vertex is chosen with probability $0 < p < 1$, $(N-1)p + q = 1$. The attempt of particle to come to another vertex is realized if no competition takes place, or the particle wins the competition. Competitions take place in the cases such that in these cases competitions take place in the real valued pendulum with planned movement. Rules of competitions resolution are the same. Particles, losing a competition, do not move, and do not repeat the attempt. The index of the next vertex is chosen again for each particle.

**Theorem 11.** *Suppose M=2. Then, with probability 1, the intensity of competitions in random pendulum equals*

$$\frac{p^2(1 - q^2 - (N-1)p^2)}{1 + 2pq - q^2}.$$

*Proof.* Consider a Markov chain with two states. The chain is in the state $G_1$ if both particles are in the same vertex. The chain is in the state $G_2$ if particles are in different vertices.

Denote by $p_{ij}$ the probability of the transition from the state $G_i$ to the state $G_j$, $i, j = 1, 2$.

If the chain is in the state $G_1$, then there is no competition. With probability $q^2 + (N-1)p^2$, the particles will be located in the same vertex at time $T + 1$. With probability $q^2 + (N-1)p^2$, the particles will be located in different vertices at time $T + 1$. Therefore, we have

$$p_{11} = q^2 + (N-1)p^2, \; p_{12} = 1 - q^2 - (N-1)p^2.$$

If the chain is in the state $G_2$ at the time $T$, then with probability $p^2$ there is a competition and the chain will be in the state $G_1$ at time $T+1$. With probability $(N-2)p^2$, particles come to the same vertex, and this vertex is not any vertex containing one of these particles. The system will be in state $G_1$ in this case. With probability $2pq$, one of particles does not change vertex, and the other particle comes to the same vertex. In this case, the chain also comes to the state $G_1$. Hence,

$$p_{21} = (N-1)p^2 + 2pq, \; p_{22} = 1 - (N-1)p^2 - 2pq.$$

The chain consists of two communicating states. There exist steady state probabilities. Denote by $p_i$ the steady probability of the state $G_i$, $i = 1, 2$. Steady probabilities satisfy the equation

$$(1 - q^2 - (N-1)p^2)p_1 = ((N-1)p^2 + 2pq)p^2,$$

$$p_1 + p_2 = 1.$$

Solution of this system is

$$p_1 = \frac{(N-1)p^2 + 2pq}{1 + 2pq - q^2}, \; p_2 = \frac{1 - q^2 - (N-1)p^2}{1 + 2pq - q^2}.$$

If the chain is in the state $G_1$ at present time, then there is no competition. If the chain is in the state $G_2$, then a competition takes place with probability $p^2$. The steady state probability of competition at current time equals

$$p^2 p_2 = \frac{p^2(1 - q^2 - (N-1)p^2)}{1 + 2pq - q^2}.$$

Suppose $H(T)$ is number of competitions in the time interval $(0, T)$. With probability 1, $\frac{H(T)}{T}$ tends as $T \to \infty$ to the steady probability of a competition at present time, (Borovkov, 1984). From this, the theorem follows.

Suppose $p = q = \frac{1}{N}$; then, from Theorem 11, we have the following statement.

**Corollary 1.** *Suppose each next vertex is chosen equiprobably. Then, with probability 1, the intensity of competitions in random pendulum equals* $\frac{N-1}{N(N^2+1)}$.

If $N = 2$, and $p = q = \frac{1}{2}$, in accordance with Theorem 11, we have $h = \frac{1}{2}$. This value of competitions intensity, in the case of bipendulum, has been found in (Kozlov, Buslaev, & Tatashev, 2015c), (Buslaev, & Tatashev, 2015).

## 8. Computer Simulation of Irrational Pendulums

### *8.1. Random Sequences of Digits and Random Pendulum*

A number is called *normal* to base $N$ if every sequence of $k$ consecutive digits of $N$-ary representation appears with limiting probability $N^{-k}$, (Becher & Figueira, 2002). As it is noted in (Becher & Figueira, 2002), there exists widely conjectured opinion that the fundamental constants, like $\sqrt{2}$, $\pi$, $e$, are normal to every $q \geq 2$. However this statement has not been proved. The same hypothesis is also supposed to be true (Becher & Figueira, 2002) on the irrational algebraic numbers.

In (Kolmogorov, & Uspenskii, 1987), it was considered in what case a sequence of digits in $N$-ary representation of a real number can be regarded as if the value of every next digit is chosen with probability $1/N$ independent of the other digits of this plan and other plans. It is assumed that the number is defined with aid of an algorithm. In (Kolmogorov, & Uspenskii, 1987), approaches have been developed that allow to characterize the sequences, behaving as random sequences. However this problem is very hard to solve, and no exhaustive solution of this problem has been found.

It is obvious that normality of a digits sequence to any base is a necessary condition for the sequence can be regarded as a random sequence.

Suppose that a sequence of digits in $N$-ary representation of real valued number plans can be regarded as if every next digit is chosen with probability $1/N$ independent of the other digits of this plan and other plans. Behavior of this real valued pendulum is similar to the behavior of the random valued pendulum, considered in Section 7, except the following. Particles, losing competitions, try to come to the same vertex again. However the intensity of competitions will be the same as in the case of random pendulum. A stochastic process can be introduced for this real valued pendulum. This stochastic process is similar to the Markov chain considered in the proof of Theorem 11. Transition probabilities of this stochastic process do not depend on that whether a competition takes place at preceding instant or not, and, hence, the stochastic process is a Markov chain. Thus, if $M = 2$, then, in accordance with Corollary 3, the intensity of competitions equals

$$h = \frac{N-1}{N(N^2+1)},$$

and the particles tapes velocity equals

$$w_1 = w_2 = 1 - \frac{h}{2} = \frac{2N^3 + N + 1}{2N(N^2+1)}.$$

If $N = 2$, i.e., in the case of bipendulum, we have

$$h = \frac{1}{10}, \; w_1 = w_2 = \frac{19}{20}.$$

We shall describe results of simulation experiments. In these experiments plans was given with aid of well-known irrational constants.

### *8.2. Irrational Plans*

Rational plans are related to periodic conditions and irrational plans are related to chaotic conditions of movement.

If plans of particles are irrational numbers, then the system cannot be described with finite Markov chain.

Simulation experiments have been implemented. The plans were irrational numbers such as $\sqrt{2}(mod\ 1)$, $\sqrt{3}(mod\ 1)$, $\pi - 3$, $\sqrt{5}(mod\ 1)$, Figure 6.

### *8.3. Chaotic Behavior of the System in the Case of Irrational Plans*

Let us describe results of experiments in the case $N = M = 2$ (bipendulum). The results of experiments show that the velocity of particles is equal to $\frac{19}{20}$ as in the case of the chaotic bipendulum. The simulation experiments are stable in the case of rational plans. The simulation experiments can be unstable in the case of irrational plans.

### *8.4. Phase Pendulums*

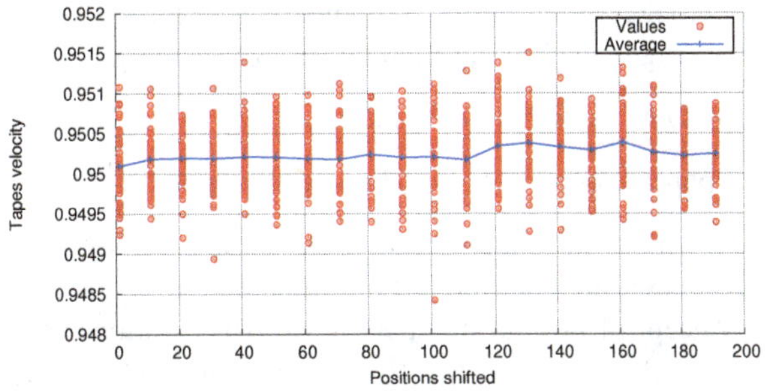


Figure 6. Logistic pendulum $\sqrt{2}(\text{mod}\,1) - \sqrt{3}(\text{mod}\,1)$

Let us get the plan $a_2$, shifting plan $a_1$ onto a fixed number of positions. If plans are rational numbers, then the system comes to the state of synergy after a time interval with a finite expectation. If plans are irrational numbers, then the system comes to the state of synergy, with probability 1, after a time interval interval $T_{syn}$ with infinite expectation.

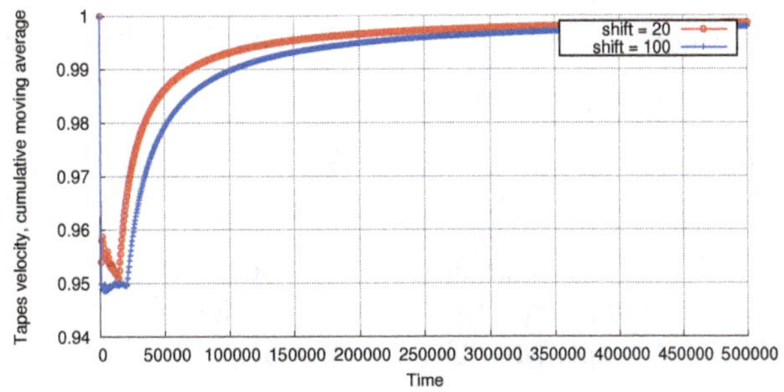


Figure 7. Phases shift $\sqrt{2}$ mod 1

The behavior of the bipendulum has been investigated with plans $a_1 = \sqrt{2}$ and $a_2$ such that we get $a_2$, shifting $a_1$ onto a fixed number of positions. The dependence of the average velocities on the time interval $(0, T)$ is shown in Figure 7. We suppose that the system comes to the state of synergy with probability 1 for a time interval. However the duration of this interval can be large.

## 9. Comments, and Further Research

*(9.1)* General transport-logistical problem is described in article (Buslaev, & Tatashev, 2015).

*(9.2)* Rational pendulum as $N = M = 2$ (bipendulum) is introduced considered in (Kozlov, Buslaev, & Tatashev, 2015c).

*(9.3)* Algebras with rational logistical plans are described in (Buslaev, & Tatashev, 2016). The Bernoulli algebra divides the triangle of rational numbers into subalgebras. If the plans belong to the same algebra, then the tape velocity of the bipendulum equals 1 *(the synergy)*.

*(9.4)* There is remarkable distinction between behavior of rational pendulums and behavior of irrational pendulums. In particular, the estimation of irrational bipendulum velocity over half a million characters is equal to the random walk bipendulum velocity.

*(9.5)* Irrational phase bipendulum converges also to synergy but with remarkably lower velocity.

*(9.6)* It is interesting to find exact lower bound of a real valued pendulum numeric characteristic when quantities of vertices and particles are equal to each other. In our paper, this problem has been solved in the case of 2 particles and any number of vertices.

*(9.7)* It is interesting to find the intensity competitions in the case of random pendulum with any number of particles. In our paper, this problem has been solved in the case of 2 particles.

*(9.8)* It is proved from random walk theory that on two dimension lattice any point is reached with 100% probability during infinite expectation time. In case of a random walk on a lattice with dimension more than two, the particle returns to given point with probability less than one. We may assume that, with positive probability, phase irrational pendulum with more than two particles is not synchronized for finite time.

*(9.9)* We use constructive approaches to determine irrational numbers. Problems of algorithmic construction of irrational numbers was considered in (Becher & Figueira, 2002), (Kolmogorov, & Uspenskii, 1987).

*(9.10)* Limits (1) –(3) do not exist for some irrational bipendulum .

*(9.11)* The model, considered in our paper, can be interpreted as a traffic model.

*(9.12)* We introduce the following classification of plans. The plan can be given by a fraction without non-repeating part in the simplest case. In more general case, there are both a non-repeating part and repeating part in the fraction which determines the plan. We use the concept of normal numbers to classify irrational plans. Normality of plans is not a sufficient condition for the particles velocity to equal $\frac{2N^3+N+1}{2N(N^2+1)}$. The sequences of plan symbols must seem independent.

*(9.13)* The dynamical system, studied in our paper, is similar to a physical system of interacting pendulums. This physical system is considered in (Mandelstam, 1972).

**References**

Becher V., & Figueira, S. (2002). An example of a computable absolute normal numbers. *Theoretical Computer Science* (Vol. 270, pp. 947 – 958).

Borovkov, A.A. (1986). *Probability theory* (2th ed.) Moscow: Nauka. (In Russian.)

Broido, W.P., & Ilyina, O.P. (2006). *Computer and systems architecture.* Saint Petersburg, Piter. (In Russian.)

Buchstab, A.A. (1972). *Number theory.* Moscow, Nauka. (In Russian.)

Buslaev, A.P., & Tatashev, A.G. Bernoulli algebra on common fractions and generalized oscillations (2016). *Journal of Mathematics Research.* (Vol. 8, no. 3, pp. 82 – 93). http://dx.doi.org/10.5539/jmr.v8n3p82

Buslaev, A.P., & Tatashev, A.G. (2015). Generalized real numbers pendulums and transport logistic applications. In *New Developments in Pure and Applied Mathematics* (pp. 388 – 392). Vienna. www.inase.org/library/2015/vienna/bypaper/MAPUR/MAPUR-63.pdf

Feller, W. (1970). *An introduction to probability theory and its applications. Vol. 1.* New York, John Willey.

Kemeny, J.G., & Snell, J.L. (1976). *Finite Markov chains.* New York, Heidelberg, Tokyo: Springer Verlag.

Kolmogorov, A.N., & Uspenskii, V.A. (1987). Algorithms and randomness *Theory of Probability and its Applications* (Vol. 32, no. 3, pp. 389 – 412).

Kozlov V.V., Buslaev A.P., &Tatashev A.G. (2015a). A dynamical communication system on a network. Journal of Computational and Applied Mathematics (Vol. 275, pp. 247 – 261). http://dx.doi.org/10.1016/j.cam.2014.07.026

Kozlov V. V., Buslaev A. P.,& Tatashev A. G. (2015b) Monotonic walks on a necklace and coloured dynamic vector. International Journal of Computer Mathematics (Vol. 92, no. 9, pp. 1910 – 1920). http://dx.doi.org/1080/00207160.2014/915964

Kozlov, V.V., Buslaev, A.P., & Tatashev, A.G. (2015c). On real-valued oscillations of a bipendulum. *Applied Mathematics Letters* (Vol. 46, pp. 44 – 49). http://dx.doi.org/10.1016/j.aml.2015.02.003

Kozlov V.V., Buslaev A.P.,& Tatashev A.G., Yashina M.V. (2014) Monotonic walks of particles on a chainmail and coloured matrices. *Proceedings of the 14th International Conference on Computational and Mathematical Methods in Science and Engineering,* CMSSE 2014, Cadiz Spain, June 3 – 7 2014, vol. 3, pp. 801 – 805.

Mandelstam, L.I. (1972). Lectures on oscillation theory. Moscow: Nauka, 1972. (In Russian.)

Schuster, H.G. (1984). *Deterministic chaos: An introduction.* Weinheim, Physik–Verlag.

Uteshev, A.Ju., & Cherkasov, T.M., & Shaposhnikov, A.A. (2001). *Digits and codes.* Saint Petersburg, Saint Petersburg University. (In Russian.)

# 8

# Augmented Stabilized and Galerkin Least Squares Formulations

Rakesh Ranjan[1,2], Yusheng Feng[1,2] & Anthony Theodore Chronopoulos[1,3]

[1] Center for Simulation, Visualization and Real-Time Prediction (SiViRT), USA

[2] Department of Mechanical Engineering, University of Texas, San Antonio, USA

[3] Department of Computer Science, University of Texas, San Antonio, USA

Correspondence: Rakesh Ranjan, E-mail: ranrakesh@gmail.com

**Abstract**

We study incompressible fluid flow problems with stabilized formulations. We introduce an iterative penalty approach to satisfying the divergence free constraint in the Streamline Upwind Petrov Galerkin (SUPG) and Galerkin Least Squares (GLS) formulations, and prove the stability of the formulation. Equal order interpolations for both velocities and pressure variables are utilized for solving problems as opposed to div-stable pairs used earlier. Higher order spectral/*hp* approximations are utilized for solving two dimensional computational fluid dynamics (CFD) problems with the new formulations named as the Augmented SUPS (ASUPS) and Augmented Galerkin Least Squares (AGLS) formulations. Excellent conservation of mass properties are observed for the problem with open boundaries in confined enclosures. Inexact Newton Krylov methods are used as the non-linear solvers of choice for the problems studied. Faithful representations of all fields of interest are obtained for the problems tested.

**Keywords:** ASUPS/AGLS formulations, spectral methods, stabilized methods, incompressible Flow, inexact Newton Krylov method.

## 1. Introduction

Computational fluid dynamics problems are usually solved with the finite volume and finite difference approaches. Finite element procedures have not been able to enjoy the same popularity as finite volume based methods because of ill-conditioning of the mixed weak form Galerkin formulation, satisfaction of the restrictive *inf-sup* condition or the LBB condition (Ladyzhenskaya-Babuska-Brezzi) condition to ensure coercivity of the functional, along with inherent central difference approximation of lower order finite element methods that become unstable at moderate to high Reynolds numbers. Various methods have been proposed to alleviate these difficulties for solving CFD problems with finite element procedures. The methods that have gained popularity are based on Penalty finite element methods (Reddy, 1982), Least Squares type approaches (Pontaza and Reddy, 2003), and Stabilized Finite element methodologies (Brooks and Hughes, 1980, 1982; Hughes and Brooks, 1982). Least Squares finite element methods have issues with extensive loss of mass in contraction regions in addition to increase of the degrees of freedom to be solved with introduction of vorticity or stresses in two dimensional problems (Ranjan and Reddy, 2012). Vorticity and stresses are auxillary variables and often not of interest in flow computations. Penalty finite element formulations interpret the incompressibility as a constraint in the Navier-Stokes equations and involve solution of velocity fields as the only degrees of freedom. Penalty finite element methods have been used for solving incompressible flow however they often require a very high value of the Penalty parameter to ensure a divergence free velocity field and for open boundary problems the Penalty parameters can often lead to a very ill-conditioned global matrix which is not amenable for solution with iterative solvers. These problems prohibit the usage of such procedures for solutions of large scale problems. Stabilized finite element methods on the other hand provide a variational setting for the Navier-Stokes equations and provide a viable alternative for solving incompressible flow problems in primitive variables. Stabilization methods for solving advection-dominated problems were introduced by Hughes and Brook (1979, 1982) and Brooks and Hughes (1980, 1982). The optimal upwinding schemes provide nodally exact solutions to advection diffusion problems in one dimension (Brooks and Hughes, 1982) and generalizations to multi-dimensions were also outlined. The streamline upwind finite element method added an extra term to the under-diffused Galerkin finite element formulation, which greatly improved the conditioning of the resulting stiffness matrix.

Streamline Upwind Petrov Galerkin (SUPG) finite element method was first proposed by Brooks and Hughes (1982). The SUPG formulation with additional pressure stabilization (PSPG) first appeared in Hughes et al. (1986) for the Stokes problem and was later generalized to the solution of Navier-Stokes equations. The SUPS (Streamline Upwind and Pressure Stabilization) (Bazilevs et al, 2012) formulation provided a variational setting for the solution of incompressible flow problems, at the same time that they admit equal order interpolations for the velocity and pressure variables. In addition to removing the requirement of satisfaction of the LBB criterion they also improve the conditioning of the discrete form. Pressure stabilization yields a better conditioned matrix, consequently facilitating the use of iterative solvers. An excellent presentation of stabilized finite element methods for solving incompressible flow equations has been provided in (Bazilevs et al, 2012) and the references therein. An earlier report highlighted results obtained from the augmented SUPS formulation for incompressible flow (Ranjan et al, 2016c) for solving both two dimensional and three dimensional problems. Further study on effective scalable pre-conditioners for solving incompressible flow problems, and introduction of inexact Newton-Krylov methods for solving spectral method based CFD problems with the augmented SUPS formulation have been provided in Ranjan et al. (2016b,c).

Lower order finite element methods require extensive meshing in areas of high gradients of the evolving fields. Spectral/hp finite element methods alleviate the extensive meshing requirements and further provide spectrally accurate results. With the advent of higher order spectral/hp finite element methods (Karniadakis and Sherwin, 1999), it was realized that very accurate numerical solutions could be obtained with coarser meshes for solving complicated problems in both structures and computational fluid dynamics areas. Similar to $hp$ finite element methods the spectral/$hp$ element methods use orthogonal polynomials (Legendre or Chebyshev based), which offer orthogonality and better conditioning as compared to Lagrange based expansions for higher than $4^{th}$ order polynomials. In due considerations of the advantages of spectral/$hp$ finite element procedures we employ spectral/$hp$ methods for solving incompressible flow problems.

In literature artificial compressibility and Penalty finite element methods have been proposed as closely related ideas which have known to provide superior convergence behavior for Navier-Stokes equations. The Penalty method which leads to the solution of a perturbed problem has been proved to provide solutions that are not too far from the original problem and the distance between the two solutions goes to zero as the penalty parameter $\hat{\epsilon}$ goes to zero (Ern and Guermond, 2004). Even in commercial codes artificial compressibility is often utilized to obtain solutions to the stiff linear system generated from the Navier-Stokes equations when there are convergence problems during solution. A related idea to the Penalty finite element method is the iterative penalty finite element method. Iterative penalty methods have been used for solving incompressible flow problems by Codina (Codina et al, 1994; Codina Rovira et al, 1992). Discontinuous pressure and bilinear velocity div-stable pairs were utilized for solving incompressible flow problems. Recovery of the post computed pressures from a discontinuous cell based field to a continuous field for presentation purposes was obtained via a $L_2$ least squares projection from the center point values to the quadrature points while post processing. They eliminated the pressure from the solution and solved for the velocities only. Usage of unequal interpolation which are LBB stable requires a complex coding structure as compared to equal order interpolations if pressure is carried as a primary variable in the formulation. Further a discontinuous pressure field projected to the finite element space with a least squares projection provides a smooth pressure field at the quadrature points at a price of loss of consistency of the momentum equations. In contrast, in this work we solve for the coupled monolithic problem carrying both pressures and velocities as primary variables in the formulations. We utilize an equal order interpolation for both velocities and pressures in the formulations which is consistent with the governing differential equations being solved. We attempt to bring the advantages of stabilized finite element formulations with iterative penalization of the divergence free constraint to improve the velocity-pressure coupling in solving incompressible flow problems. We name the formulations as Augmented Stabilized SUPS (ASUPS) and Augmented Galerkin Least Squares (AGLS) formulations for solving incompressible flow equations. We provide the stability analysis of both formulations and demonstrate for benchmark problems the formulations provide better non-linear convergence. With this improvement in the formulations we solve some benchmark CFD problems and report results. Some of the advantages of the present formulation are as follows;

- It circumvents the need for artificial mass matrix enhancement with the predictor corrector step while solving incompressible flow problems. The true mass matrix is utilized for solving all problems.
- It allows the usage of standard monolithic solvers in conjugation with inexact newton krylov methods for solving problems.
- Non linear convergence is achieved with the Navier- Stokes in their natural form. This is attributed to an adequate addressing of the velocity-pressure coupling as opposed to a total loss of stability of the discrete form with SUPG formulations.
- Allows the use of alpha family of approximation for stepping the equations in time in a space time decoupled framework.
- More scales in the problem are resolved than are possible with an artificially enhanced mass matrix in effect accessing a larger bandwidth of fine scale problem.

We introduce some notation for the following discussion and proofs.

- The penalty parameter, $\hat{\epsilon}$ is a small positive constant.
- Lower case letters $(i, j, k, l, m, n)$ to denote positive integers (not bold).
- Utilize the lower case letter $h$ to denote the grid-size (not bold).
- We denote by $R^N$ the Euclidean space of dimensions $N = (2, 3)$.
- $(C_k, C_k', C', C'', C_{k1}...., C_{k5})$ denote positive constants independent of viscosity and of any element diameter $h_K$.

This paper is organized as follows. In section 2, we outline the incompressible Navier-Stokes equations and introduce the iterative penalty method for relaxing the divergence free condition. In section 3 we perform the stability analysis of the ASUPS formulation. In section 4 we perform the stability analysis of the AGLS formulation. Spatial discretization with the spectral/*hp* finite element formulation is introduced in section 5. Section 6 provides details on the non-linear solution techniques and time stepping algorithms. Section 7 provides an array of sample CFD problems solved with the ASUPS and AGLS formulations respectively. Section 8 contains numerical experiments, and Section 9 concludes the paper.

## 2. Incompressible Flow Equations and Penalty Techniques

Let us consider the flow of a Newtonian fluid with density $\rho$, and viscosity $\mu$. Let $\Omega \subset R^N$ and $t \in [0, T]$ be the spatial and temporal domains respectively, where $N$ is the number of space dimensions. Let $\Gamma$ denote the boundary of $\Omega$. The equations of fluid motion that describe the unsteady Navier-Stokes governing incompressible flows are provided by

$$\frac{\partial \mathbf{u}}{\partial t} + \mathbf{u} \cdot \nabla \mathbf{u} + \nabla p - \frac{1}{Re} \nabla^2 \mathbf{u} = \mathbf{f} \text{ on } \Omega \; \forall \; t \in [0, T] \tag{1}$$

$$\nabla \cdot \mathbf{u} = 0, \text{ on } \Omega \; \forall \; t \in [0, T] \tag{2}$$

where, $\mathbf{u}$, $\mathbf{f}$, and $p$ are the velocity, body force, and pressure respectively. $Re$ denotes the Reynolds number, $\nu = \frac{\mu}{\rho}$ is the kinematic viscosity of the fluid, $p$ is the pressure and $\mathbf{u}$ is the fluid velocity. The part of the boundary at which the velocity is assumed to be specified is denoted by $\Gamma_g$:

$$\mathbf{u} = \mathbf{g} \text{ on } \Gamma_g \; \forall t \in \; [0, T] \tag{3}$$

Penalty method relaxes the incompressibility condition as

$$\frac{\partial \mathbf{u}}{\partial t} + \mathbf{u} \cdot \nabla \mathbf{u} + \nabla p - \frac{1}{Re} \nabla^2 \mathbf{u} = \mathbf{f} \text{ on } \Omega \; \forall \; t \in [0, T] \tag{4}$$

$$\nabla \cdot \mathbf{u} = -\hat{\epsilon} p, \text{ on } \Omega \; \forall \; t \in [0, T] \tag{5}$$

In the above equation 5 $\hat{\epsilon}$ is a small positive constant. Penalty procedures are known to improve the convergence of the mixed finite element formulation. However, they can still provide an unstable computation for very low values of the penalty parameter. Iterative penalty methods relax this condition on the penalty parameter and allow for larger values while maintaining stability of the formulation (Dai and Cheng, 2008). Iterative penalty thus introduces a regularization step to enforce the incompressibility constraint. We employ the formulation of this method as outlined in (Dai and Cheng, 2008; Lin et al, 2004; Xiao-liang and Shaikh, 2006). The governing equations take the form

$$\frac{\partial \mathbf{u}^k}{\partial t} + \mathbf{u}^\mathbf{k} \cdot \nabla \mathbf{u}^\mathbf{k} + \nabla p^k - \frac{1}{Re} \nabla^2 \mathbf{u}^k = \mathbf{f} \text{ on } \Omega \; \forall \; t \in [0, T] \tag{6}$$

$$\nabla \cdot \mathbf{u}^\mathbf{k} = -\hat{\epsilon}(p^k - p^{k-1}), \text{ on } \Omega \; \forall \; t \in [0, T] \tag{7}$$

where, $k$ denotes the regularization step. The natural boundary conditions associated with eq (1) are the conditions on the stress components, and these are the conditions assumed to be imposed on the remaining part of the boundary,

$$\mathbf{n} \cdot \sigma = \mathbf{h} \text{ on } \Gamma_h \; \forall t \in [0, T] \tag{8}$$

where $\Gamma_g$ and $\Gamma_h$ are the complementary subsets of the boundary $\Gamma$, or $\Gamma = \Gamma_g \cup \Gamma_h$. In eq (8) we denote the outward normal by $\mathbf{n}$, the stress tensor by $\sigma$ and the traction components by $\mathbf{h}$. As the initial condition, a divergence free velocity field, $\mathbf{u}_0(\mathbf{x})$, is imposed. Error estimates for the initial value of the iterative penalty method obtained from eq.4-5 for Stokes equations (dropping the non-linearity) have been provided in (Carey and Krishnan, 1982; Oden et al, 1982; Stenberg,

1990). Similar error estimates of the iterative penalty method eq. 6-7 for the Stokes problem have been provided in (Xiao-liang and Shaikh, 2006).

## 3. Stability Analysis of ASUPS Formulation

Let us study the stability of the SUPS finite element formulation resulting from the iterative regularization of the divergence free constraint. We prove the stability of the formulation based on the initial value of the algorithm (detailed below). We consider the steady state form of the equations for analysis. The linearized Navier-Stokes equations are defined on a bounded domain $\Omega \subset R^N, N = 2, 3$, with a polygonal or polyhedral boundary $\Gamma$. Let $C_h$ denote a partition of $\Omega$ into elements consisting of triangles or convex quadrilaterals. Let us denote $L^2(\Omega)$ as the space of square integrable functions in $\Omega$, and $L_0^2$ the space of $L_2$ functions with zero mean value in $\Omega$. The $L_2$ inner product in $\Omega$ is $(.,.)$. Further, we also employ $(.,.)_K$ the $L^2$ inner product and $||.||_{0,K}$ the $L_2$ norm in the element domain $K$. Let $C^0(\Omega)$ be the space of continuous functions in $\Omega$. $H_0^1$ denotes the Sobolev space of functions with square-integrable value and derivatives in $\Omega$ with zero value on the boundary $\Gamma$. $H^1$ norm of a function $u$ is

$$||\mathbf{u}||_{H^1} = \left\{ \int_\Omega \left( |\mathbf{u}(\mathbf{x})|^2 + |\mathbf{u}'(\mathbf{x})|^2 \right) dx \right\}^{\frac{1}{2}} \tag{9}$$

Further we denote the triangular or quadrilateral decomposition of the domain by the following (Ciarlet, 2002; Johnson, 2012; Oden and Reddy, 2012)

$$R_m(K) = \begin{cases} P_m(K), & \text{if K is a triangle or tetrahedron} \\ Q_m(K), & \text{if K is a quadrilateral or hexahedron} \end{cases} \tag{10}$$

where for each integer $m \geq 0$, $P_m$ and $Q_m$ have the usual meaning.

Let the finite element spaces which we wish to work with be given by

$$\mathbf{V}_h = \{\mathbf{w} \in H_0^1(\Omega)^N | V_K \in R_k(K)^N, K \in C_h\} \tag{11}$$

$$P_h = \{p \in C^0(\Omega) \cap L_0^2(\Omega) | p_K \in R_l(K), K \in C_h\} \tag{12}$$

where the integers $k$ and $l$ are greater than or equal to 1.

### *3.1 Consistency*

Momentarily reverting back the notation introduced earlier let us denote $(u, p)$ the solution of equation 1-2 and $(\mathbf{u}_{\hat{\epsilon}}, p_{\hat{\epsilon}})$ the solution to 6-7. It has been shown the following error estimates hold for the weak finite element formulation of the iterative penalty method (Xiao-liang and Shaikh, 2006)

$$||\mathbf{u} - \mathbf{u}_{\hat{\epsilon}}^k||_1 + ||p - p_{\hat{\epsilon}}^k||_0 \leq C_0 \left( \inf_{\mathbf{u} \in \mathbf{V}_h} ||\mathbf{u} - \mathbf{u_h}||_1 + \inf_{p \in P_h} ||p - p_h||_0 + \hat{\epsilon} ||p - p_{\hat{\epsilon}}^{k-1}||_0 \right) \tag{13}$$

where the constant $C_0$ is independent of $h$ and $\hat{\epsilon}$. For the above estimates we define $H^s(\Omega)$ to be the standard *Sobolev* spaces with corresponding inner products $(.,.)_{s,\Omega}$ and norm $||.||_{s,\Omega}$. The formulation is consistent since the pressure converges iteratively. Further since the SUPS formulation requires addition of scaled residuals of the momentum and the continuity equations it retains consistency upon addition of these terms.

### *3.2 Stability Analysis*

Steady-state regularized equations are written as;

$$(\nabla \mathbf{u})\mathbf{a} - 2\nu \nabla \cdot \varepsilon(\mathbf{u}) + \nabla p = \mathbf{f} \ \ in \ \ \Omega \tag{14}$$

$$\nabla \cdot \mathbf{u} = -\hat{\epsilon}(p^k - p^{k-1}) \ \ in \ \ \Omega \tag{15}$$

$$\mathbf{u} = \mathbf{0} \ \ on \ \ \Gamma \tag{16}$$

where $\mathbf{a}$ is a given velocity field, $\mathbf{u}$ is the unknown velocity, $p$ is the pressure, $\nu$ the viscosity, and $\varepsilon(\mathbf{u})$ is the symmetric part of the velocity gradient and $\mathbf{f}$ is the body force. The initial value of the algorithm follows by replacing the regularization of the incompressiblity with a one-step process as

$$\nabla \cdot \mathbf{u} = -\hat{\epsilon}(p - p_0) \ \ in \ \ \Omega \tag{17}$$

Assuming the given data $\mathbf{a}(\mathbf{x})$ the linearised velocity field and $\nu(x)$ viscosity satisfy;

$$\nabla \cdot \mathbf{a}(\mathbf{x}) = 0$$
$$\nu(x) = \nu, \ \nu \geq 0 \tag{18}$$

The finite element formulation we wish to study can be written as: find $\mathbf{u}_h \in \mathbf{V}_h$ and $p_h \in P_h$ such that

$$B(\mathbf{u_h}, p_h; \mathbf{w}, q) = F(\mathbf{w}, q), (\mathbf{w}, q) \in \mathbf{V_h} \times P_h \tag{19}$$

with

$$B(\mathbf{u}, p; \mathbf{w}, q) = ((\nabla\mathbf{u})\mathbf{a}, \mathbf{w}) + 2\nu(\varepsilon(\mathbf{u}), \varepsilon(\mathbf{w})) - (\nabla \cdot \mathbf{w}, p) + (\nabla \cdot \mathbf{u}, q) + \hat{\epsilon}(p, q) + \sum_{K \in C_h} ((\nabla\mathbf{u})\mathbf{a} + \nabla p - 2\nu(\nabla \cdot \varepsilon(\mathbf{u})), \tau(x, Re_K(\mathbf{x}))((\nabla\mathbf{w})\mathbf{a} + \nabla q))_K \tag{20}$$

We obtain:

$$F(\mathbf{w}, q) = (\mathbf{f}, \mathbf{w}) + \sum_{K \in C_h} (\mathbf{f}, \tau(x, Re_K(\mathbf{x}))((\nabla\mathbf{w})\mathbf{a} + \nabla q))_K + \hat{\epsilon}(p_0, q) \tag{21}$$

We note that the stabilization parameter is bounded in each domain $K$ by a constant. Be it defined;

$$\tau(\mathbf{x}, Re_K(\mathbf{x})) = \frac{h_K}{2|\mathbf{a}(\mathbf{x})|_{\hat{p}}}, \quad Re_K(\mathbf{x}) \geq 1$$
$$\tau(\mathbf{x}, Re_K(\mathbf{x})) = \frac{m_k h_K^2}{8\nu}, \quad 0 \leq Re_K(\mathbf{x}) < 1 \tag{22}$$

where we define $Re_K(\mathbf{x})$ as

$$Re_K(\mathbf{x}) = \frac{m_k h_K |\mathbf{a}(\mathbf{x})|_{\hat{p}}}{4\nu} \tag{23}$$

and

$$|\mathbf{a}(\mathbf{x})|_{\hat{p}} = \left(\Sigma_{i=1}^{N} |a_i(\mathbf{x})|^{\hat{p}}\right)^{\frac{1}{\hat{p}}}, \quad 1 \leq \hat{p} < \infty$$
$$|\mathbf{a}(\mathbf{x})|_{\hat{p}} = max_{i=1,N}|a_i(\mathbf{x})|, \quad \hat{p} = \infty \tag{24}$$

Obtaining a common bound on $\tau(Re_K(\mathbf{x}))$, with no restrictions on $Re$ we arrive at

$$\tau(\mathbf{x}, Re_K(\mathbf{x})) \leq \frac{m_k h_K^2}{8\nu} \tag{25}$$

where

$$m_k = min\{\frac{1}{3}, 2C_k\} \tag{26}$$

Here, $m_k$ and $C_k$ are two positive constants with $m_k$ defined above.

Lemma I: For stability of the formulation we require

$$B(\mathbf{u}, p; \mathbf{u}, p) \geq \left(\frac{1}{2} C_K^{''} \nu ||\varepsilon(\mathbf{u})||_0^2 + \frac{1}{2} ||\tau^{\frac{1}{2}}(\mathbf{x}, Re_K(\mathbf{x}))((\nabla\mathbf{u})\mathbf{a} + \nabla p))||_0^2\right) + \hat{\epsilon}||p||_0^2 \tag{27}$$

where $C_K^{''}$ is a positive constant independent of $h_K$.

Proof: From integration by parts

$$((\nabla\mathbf{u})\mathbf{a}, \mathbf{u}) = 0, \quad \mathbf{u} \in \mathbf{V}_h \tag{28}$$

since $\mathbf{u} = \mathbf{0}$ on $\Gamma$ for $\mathbf{u} \in V_h$. Therefore we have:

$$B(\mathbf{u}, p; \mathbf{u}, p) = 2\nu(\varepsilon(\mathbf{u})), \varepsilon(\mathbf{u})) + \hat{\epsilon}(p, p) + \sum_{K \in C_h} ((\nabla\mathbf{u})\mathbf{a} + \nabla p - 2\nu(\nabla \cdot \varepsilon(\mathbf{u})), \tau(\mathbf{x}, Re_K(\mathbf{x}))((\nabla\mathbf{u})\mathbf{a} + \nabla p))_K \tag{29}$$

Thus we only need a measure of:

$$\sum_{K \in C_h} (2\nu\nabla \cdot \varepsilon(\mathbf{u}), \tau((\nabla\mathbf{u})\mathbf{a} + \nabla p))_K \tag{30}$$

Applying Schwarz inequality:

$$\sum_{K\in C_h}(2\nu\nabla\cdot\varepsilon(\mathbf{u}),\tau((\nabla\mathbf{u})\mathbf{a}+\nabla p))_K \le \sum_{K\in C_h}\left[\|(2\nu\nabla\cdot\varepsilon(\mathbf{u})\|_0^2\,\|\tau((\nabla\mathbf{u})\mathbf{a}+\nabla p)\|_0^2\right]^{\frac{1}{2}} \tag{31}$$

Utilizing the bound on the stabilization parameter and value of $m_k$ eq. 26:

$$\sum_{K\in C_h}(2\nu\nabla\cdot\varepsilon(\mathbf{u}),\tau((\nabla\mathbf{u})\mathbf{a}+\nabla p))_K \le \sum_{K\in C_h}\left[C_k h_K^2\nu\|(\nabla\cdot\varepsilon(\mathbf{u})\|_0^2\,\|\tau^{\frac{1}{2}}((\nabla\mathbf{u})\mathbf{a}+\nabla p)\|_0^2\right]^{\frac{1}{2}} \tag{32}$$

Inverse estimates provides us element level bounds on the derivatives as follows (Ern and Guermond, 2004)

$$\sum_{K\in C_h} C_k' h_K^2\|\nabla\cdot\varepsilon(\mathbf{u})\|_0^2 \le \|\varepsilon(\mathbf{u})\|_0^2 \tag{33}$$

In addition:

$$(x,y)\le\frac{1}{2}((x,x)+(y,y)) \tag{34}$$

Let us denote a positive constant $\beta$ such that

$$\beta=\left(\frac{C_k}{C_k'}\right)^{\frac{1}{2}} \tag{35}$$

Combining all the above:

$$\sum_{K\in C_h}(2\nu\nabla\cdot\varepsilon(\mathbf{u}),\tau((\nabla\mathbf{u})\mathbf{a}+\nabla p))_K \le \sum_{K\in C_h}\beta^2\frac{\nu}{2}\|(\varepsilon(\mathbf{u})\|_0^2+\frac{1}{2}\|\tau^{\frac{1}{2}}((\nabla\mathbf{u})\mathbf{a}+\nabla p)\|_0^2 \tag{36}$$

Finally we obtain:

$$\begin{aligned}B(\mathbf{u},p;\mathbf{u},p) \ge{}& 2\nu(\varepsilon(\mathbf{u})),\varepsilon(\mathbf{u}))+\hat{\epsilon}(p,p)+\|\tau^{\frac{1}{2}}(\mathbf{x},Re_K(\mathbf{x}))((\nabla\mathbf{u})\mathbf{a}+\nabla p))\|_0^2\\ &-\nu\frac{\beta^2}{2}\|(\varepsilon(\mathbf{u})\|_0^2-\frac{1}{2}\|\tau^{\frac{1}{2}}((\nabla\mathbf{u})\mathbf{a}+\nabla p)\|_0^2\end{aligned} \tag{37}$$

Thus we obtain the following

$$B(\mathbf{u},p;\mathbf{u},p)\ge\left(2-\frac{\beta^2}{2}\right)\nu\|\varepsilon(\mathbf{u})\|_0^2+\frac{1}{2}\|\tau^{\frac{1}{2}}(x,Re_K(\mathbf{x}))((\nabla\mathbf{u})\mathbf{a}+\nabla p))\|_0^2+\hat{\epsilon}\|p\|_0^2 \tag{38}$$

Restrictions due to the stability analysis of the formulation require

$$2-\frac{\beta^2}{2}\ge 0 \tag{39}$$

which provides

$$\beta\le 2 \tag{40}$$

Since the regularization steps present the same problem with previous values of the pressures stability is further guaranteed across the regularization steps.

## 4. Stability Analysis of AGLS Formulation

We perform the stability analysis of the Galerkin Least Squares (GLS) formulation next. We consider the steady form of the equations. Let the weighting spaces which we wish to work with be given by (as defined above)

$$\mathbf{V}_h=\{\mathbf{w}\in H_0^1(\Omega)^N|\mathbf{V}_K\in R_k(K)^N, K\in C_h\} \tag{41}$$

$$P_h=\{p\in C^0(\Omega)\cap L_0^2(\Omega)|p_K\in R_l(K), K\in C_h\} \tag{42}$$

where the integers $k$ and $l$ are greater than or equal to 1.

*4.1 Consistency*

Momentarily reverting back the notation introduced earlier let us denote $(u, p)$ the solution of equation 1-2 and $(\mathbf{u}_{\hat{\epsilon}}, p_{\hat{\epsilon}})$ the solution to 6-7. It has been shown the following error estimates hold for the weak finite element formulation of the iterative penalty method (Xiao-liang and Shaikh, 2006)

$$||\mathbf{u} - \mathbf{u}^k_{\hat{\epsilon}}||_1 + ||p - p^k_{\hat{\epsilon}}||_0 \leq C_0 \left( \inf_{\mathbf{u} \in \mathbf{V}_h} ||\mathbf{u} - \mathbf{u_h}||_1 + \inf_{p \in P_h} ||p - p_h||_0 + \hat{\epsilon} ||p - p^{k-1}_{\hat{\epsilon}}||_0 \right) \tag{43}$$

where the constant $C_0$ is independent of $h$ and $\hat{\epsilon}$. For the above estimates we define $H^s(\Omega)$ to be the standard *Sobolev* spaces with corresponding inner products $(.,.)_{s,\Omega}$ and norm $||.||_{s,\Omega}$. The formulation is consistent since the pressure converges iteratively. Further since the GLS formulation requires addition of scaled residuals of the momentum and the continuity equations it retains consistency upon addition of these terms. The following convergence estimates were proved in (Lin, 1997).

$$||\mathbf{u} - \mathbf{u}_{\hat{\epsilon}}||_{\mathbf{H}^1} \leq M\hat{\epsilon}^k \tag{44}$$

and the iteratively regularized pressure as

$$\left( \int_0^T ||p - p_{\hat{\epsilon}}||_2 dt \right)^{\frac{1}{2}} \leq M\hat{\epsilon}^k \tag{45}$$

From the above equation we can see we only need to assume the penalty parameter $\hat{\epsilon} < 1$ but not necessarily choose it very small since by iteration the error $O(\hat{\epsilon}^k)$ can reach any accuracy that we desire. Further since we choose $\hat{\epsilon}$ to be not very small the formulation is less stiff and more stable.

*4.2 Stability Analysis*

Steady-state linearised equations are written as;

$$(\nabla \mathbf{u})\mathbf{a} - 2\nu\nabla \cdot \varepsilon(\mathbf{u}) + \nabla p = \mathbf{f} \;\; in \;\; \Omega \tag{46}$$

$$\nabla \cdot \mathbf{u} = -\hat{\epsilon}(p - p_0) \;\; in \;\; \Omega \tag{47}$$

$$\mathbf{u} = \mathbf{0} \;\; on \;\; \Gamma \tag{48}$$

where $\mathbf{a}$ is a given velocity field, $\mathbf{u}$ is the unknown velocity, $p$ is the pressure, $\nu$ the viscosity, and $\varepsilon(\mathbf{u})$ is the symmetric part of the velocity gradient and $\mathbf{f}$ is the body force vector. Assuming the given data $\mathbf{a(x)}$ the linearised velocity field and $\nu(x)$ viscosity satisfy;

$$\nabla \cdot \mathbf{a(x)} = 0$$
$$\nu(x) = \nu, \;\; \nu \geq 0 \tag{49}$$

The finite element formulation we wish to study can be written as: find $\mathbf{u}_h \in \mathbf{V}_h$ and $p_h \in P_h$ such that

$$B(\mathbf{u}, p; \mathbf{w}, q) = F(\mathbf{w}, q), (\mathbf{w}, q) \in \mathbf{V_h} \times P_h \tag{50}$$

with

$$\begin{aligned} B(\mathbf{u}, p; \mathbf{w}, q) = {} & ((\nabla \mathbf{u})\mathbf{a}, \mathbf{w}) + 2\nu(\varepsilon(\mathbf{u})), (\varepsilon(\mathbf{w})) - (\nabla \cdot \mathbf{w}, p) \\ & + (\nabla \cdot \mathbf{u}, q) + \hat{\epsilon}(p, q) + \delta(\nabla \cdot \mathbf{u}, \nabla \cdot \mathbf{w}) + \\ \sum_{k \in C_h} ((\nabla \mathbf{u})\mathbf{a} + \nabla p - 2\nu(\nabla \cdot \varepsilon(\mathbf{u})), & \tau(x, Re_K(x))((\nabla \mathbf{w})\mathbf{a} + \nabla q - 2\nu(\nabla \cdot \varepsilon(\mathbf{w}))_K \end{aligned} \tag{51}$$

In the above equation 51 the parameter $\delta$ is the least squares stabilization parameter on the incompressibility constraint. We obtain:

$$F(\mathbf{w}, q) = (\mathbf{f}, \mathbf{w}) + \sum_{k \in C_h} (f, \tau(x, Re_K(x))((\nabla \mathbf{w})\mathbf{a} + \nabla q - 2\nu(\nabla \cdot \varepsilon(\mathbf{w}))_K + \hat{\epsilon}(p_0, q) \tag{52}$$

We note that the stabilization parameter is bounded in each domain $K$ by a constant. Be it defined;

$$\tau(\mathbf{x}, Re_K(\mathbf{x})) = \frac{h_K}{2||\mathbf{a(x)}||_{pn}}, \;\; Re_K(\mathbf{x}) \geq 1$$
$$\tau(\mathbf{x}, Re_K(\mathbf{x})) = \frac{m_K h_K^2}{8\nu}, \;\; 0 \leq Re_K(\mathbf{x}) < 1 \tag{53}$$

Obtaining a common bound on $\tau(Re_K(\mathbf{x}))$, with no restrictions on $Re_K$ we obtain

$$\tau(\mathbf{x}, Re_K(\mathbf{x})) \leq \frac{m_k h_K^2}{8\nu} \tag{54}$$

where

$$m_k = min\{\frac{1}{3}, 2C_k\} \tag{55}$$

Lemma II: For stability of the formulation we require

$$B(\mathbf{u}, p; \mathbf{u}, p) \geq \frac{1}{2}(C_{'}\nu\varepsilon||\mathbf{u}||)_0^2 + \frac{1}{2}C_{''}||\tau^{\frac{1}{2}}(x, Re_K(x))((\nabla\mathbf{u})\mathbf{a} + \nabla p))||_0^2) + ||\delta^{\frac{1}{2}}\nabla\cdot\mathbf{u}||_0^2] + \hat{\epsilon}||p||_0^2 \tag{56}$$

where $C_{'}$ and $C_{''}$ are positive constants independent of $h_K$.

Proof: From integration by parts

$$((\nabla\mathbf{u})\mathbf{a}, \mathbf{u}) = 0, \quad \mathbf{u} \in \mathbf{V}_h \tag{57}$$

since $\mathbf{u} = \mathbf{0}$ on $\Gamma$ for $\mathbf{u} \in \mathbf{V}_h$. Therefore we have:

$$B(\mathbf{u}, p; \mathbf{u}, p) = 2\nu(\varepsilon(\mathbf{u})), \varepsilon(\mathbf{u})) + \hat{\epsilon}(p, p) + \delta(\nabla\cdot\mathbf{u}, \nabla\cdot\mathbf{u}) + \sum_{K\in C_h}((\nabla\mathbf{u})\mathbf{a} + \nabla p - 2\nu(\nabla\cdot\varepsilon(\mathbf{u})), \tau(x, Re_K(x))((\nabla\mathbf{u})\mathbf{a} + \nabla p - 2\nu(\nabla\cdot\varepsilon(\mathbf{u}))_K \tag{58}$$

The above statement proves the stability of the AGLS formulation. To derive estimates from this analysis we further examine the statement.

Classical definition of the stabilization parameter assumes a relationship of the form Eq. 22. Additional terms in the parameter dependent on the constant viscosity of the medium are traditionally defined as in (Ranjan et al, 2016a,b) $\tau = \tau(h, \nu, \mathbf{a})$. In the ensuing discussion we define the stabilization parameter to be the maximum value of this parameter obtained across all elements $K$ and assume a constant viscosity in the domain. Let us denote this value of the stabilization parameter as $\tau_{sup}$. For this constant value of the parameter we obtain:

$$\begin{aligned} B(\mathbf{u}, p; \mathbf{u}, p) = {} & 2\nu(\varepsilon(\mathbf{u})), \varepsilon(\mathbf{u})) + \hat{\epsilon}(p, p) + \delta(\nabla\cdot\mathbf{u}, \nabla\cdot\mathbf{u}) + \\ & \sum_{K\in C_h}\tau_{sup}((\nabla\mathbf{u})\mathbf{a} + \nabla p), (\nabla\mathbf{u})\mathbf{a} + \nabla p))_K + \\ & \sum_{K\in C_h} 4\nu^2\tau_{sup}(\nabla\cdot\varepsilon(\mathbf{u}), \nabla\cdot\varepsilon(\mathbf{u}))_K - \\ & 4\nu\sum_{K\in C_h}\tau_{sup}((\nabla\cdot\varepsilon(\mathbf{u}), (\nabla\mathbf{u})\mathbf{a} + \nabla p))_K \end{aligned} \tag{59}$$

Thus we only need a measure of:

$$\sum_{K\in C_h}(4\nu\nabla\cdot\varepsilon(\mathbf{u}), \tau_{sup}((\nabla\mathbf{u})\mathbf{a} + \nabla p))_K \tag{60}$$

Applying Cauchy-Schwarz (utilizing positivity of $\nu$):

$$\sum_{K\in C_h}(4\nu\nabla\cdot\varepsilon(\mathbf{u}), \tau_{sup}((\nabla\mathbf{u})\mathbf{a} + \nabla p))_K \leq \sum_{K\in C_h}\left[||(4\nu\nabla\cdot\varepsilon(\mathbf{u})||_{0,K}^2 \, ||\tau_{sup}((\nabla\mathbf{u})\mathbf{a} + \nabla p)||_{0,K}^2\right]^{\frac{1}{2}} \tag{61}$$

Utilizing the bound on the stabilization parameter and value of $m_k$ eq. 55:

$$\sum_{K\in C_h}(4\nu\nabla\cdot\varepsilon(\mathbf{u}), \tau_{sup}((\nabla\mathbf{u})\mathbf{a} + \nabla p))_K \leq \sum_{K\in C_h}\left[4C_{k1}h_k^2\nu||(\nabla\cdot\varepsilon(\mathbf{u})||_{0,K}^2 \, ||\tau_{sup}^{\frac{1}{2}}((\nabla\mathbf{u})\mathbf{a} + \nabla p)||_{0,K}^2\right]^{\frac{1}{2}} \tag{62}$$

Inverse estimates (on element level integrals) provides us element level bounds on the derivatives as follows (Ern and Guermond, 2004)

$$\sum_{K\in C_h} C_{k2}h_k^2||\nabla\cdot\varepsilon(\mathbf{u})||_{0,K}^2 \leq ||\varepsilon(\mathbf{u})||_{0,K}^2 \tag{63}$$

Let us denote by the positive constant $C_{k3} = \frac{C_{k1}}{C_{k2}}$ Therefore it follows:

$$\sum_{K \in C_h} (4\nu\nabla \cdot \varepsilon(\mathbf{u}), \tau_{sup}((\nabla\mathbf{u})\mathbf{a} + \nabla p))_K \leq \sum_{K \in C_h} \left[ 4C_{k3}\nu||(\varepsilon(\mathbf{u})||^2_{0,K} \, ||\tau^{\frac{1}{2}}_{sup}((\nabla\mathbf{u})\mathbf{a} + \nabla p)||^2_{0,K} \right]^{\frac{1}{2}} \tag{64}$$

In addition we have:

$$(x, y) \leq \frac{1}{2}((x, x) + (y, y)) \tag{65}$$

Combining all the above we get:

$$\sum_{K \in C_h} (4\nu\nabla \cdot \varepsilon(\mathbf{u}), \tau_{sup}((\nabla\mathbf{u})\mathbf{a} + \nabla p))_K \leq \sum_{K \in C_h} C_{k3}||(\varepsilon(\mathbf{u})||^2_{0,K} + \nu C_{k3}||\tau^{\frac{1}{2}}_{sup}((\nabla\mathbf{u})\mathbf{a} + \nabla p)||^2_{0,K} \tag{66}$$

This leads to:

$$\begin{aligned} B(\mathbf{u}, p; \mathbf{u}, p) \geq\ & 2\nu(\varepsilon(\mathbf{u})), \varepsilon(\mathbf{u})) + \hat{\epsilon}(p, p) + \delta(\nabla \cdot \mathbf{u}, \nabla \cdot \mathbf{u}) + \\ & \sum_{K \in C_h} \tau_{sup}(\mathbf{x}, Re_K(\mathbf{x}))((\nabla\mathbf{u})\mathbf{a} + \nabla p), (\nabla\mathbf{u})\mathbf{a} + \nabla p))_K + \\ & \sum_{K \in C_h} 4\nu^2\tau_{sup}(\nabla \cdot \varepsilon(\mathbf{u}), \nabla \cdot \varepsilon(\mathbf{u}))_K - \\ & \left( \sum_{K \in C_h} C_{k3}||(\varepsilon(\mathbf{u})||^2_{0,K} + \nu C_{k3}||\tau^{\frac{1}{2}}_{sup}((\nabla\mathbf{u})\mathbf{a} + \nabla p)||^2_{0,K} \right) \end{aligned} \tag{67}$$

Thus we obtain the following

$$\begin{aligned} B(\mathbf{u}, p; \mathbf{u}, p) \geq\ & (2\nu - C_{k3})||\varepsilon(\mathbf{u})||^2_0 + ||\hat{\epsilon}^{1/2}p||^2 + ||\delta^{1/2}\nabla \cdot \mathbf{u}||^2 + \\ & (1 - \nu C_{k3})||\tau^{1/2}_{sup}((\nabla\mathbf{u})\mathbf{a} + \nabla p)||^2_0 + ||2\nu\tau^{1/2}_{sup}\nabla \cdot \varepsilon(\mathbf{u})||^2_0 \end{aligned} \tag{68}$$

Further from Poincare inequality we obtain

$$C_{k4}h_k^2||\nabla \cdot \varepsilon(\mathbf{u})||^2 \geq ||\varepsilon(\mathbf{u})||^2 \tag{69}$$

Therefore we obtain

$$\begin{aligned} B(\mathbf{u}, p; \mathbf{u}, p) \geq\ & ||\varepsilon(\mathbf{u})||^2_0(2\nu - C_{k3}) + ||\hat{\epsilon}^{1/2}p||^2 + ||\delta^{1/2}\nabla \cdot \mathbf{u}||^2 + \\ & (1 - \nu C_{k3})||\tau^{1/2}_{sup}((\nabla\mathbf{u})\mathbf{a} + \nabla p)||^2_0 + \frac{4\nu^2}{h_k^2}C_{k5}\tau_{sup}||\varepsilon(\mathbf{u})||^2_0 \end{aligned} \tag{70}$$

Assimilating common terms we obtain

$$\begin{aligned} B(\mathbf{u}, p; \mathbf{u}, p) \geq\ & ||\varepsilon(\mathbf{u})||^2_0 \left( (2\nu - C_{k3}) + \frac{4\nu^2}{h_k^2}\tau_{sup}C_{k5} \right) + ||\hat{\epsilon}^{1/2}p||^2 + ||\delta^{1/2}\nabla \cdot \mathbf{u}||^2 + \\ & (1 - \nu C_{k3})||\tau^{1/2}_{sup}((\nabla\mathbf{u})\mathbf{a} + \nabla p)||^2_0 \end{aligned} \tag{71}$$

where we define the constant $C_{k5} = \frac{1}{C_{k4}}$. For the above inequality to satisfy eq. 56 we require

$$\begin{aligned} (1 - \nu C_{k3}) \geq 0 \\ \left( (2\nu - C_{k3}) + \frac{4\nu^2}{h_k^2}\tau_{sup}C_{k5} \right) \geq 0 \end{aligned} \tag{72}$$

The restrictions on viscosity from the above equations are obtained as (maximum value of viscosity)

$$C_{k3} \leq \frac{1}{\nu} \tag{73}$$

The second term from the stability estimate obtains

$$2\nu\left(1+\frac{2\nu}{h_k^2}\tau_{sup}C_{k5}\right)\geq C_{k3} \tag{74}$$

Note, in the derivation of the above restriction on the stabilizaton parameter we did not use the definition of the parameter given by Eq. 54. This is an important note, since the estimate provided by Eq. 74 follows directly from the stability analysis for the maximum value of the stabilization parameter inside the domain.

**5. Spectral/*hp* Finite Element Formulation**

A class of *p*-type elements known as the *spectral elements* use the Lagrange polynomial through the zeroes of the Gauss-Lobatto polynomials (Karniadakis and Sherwin, 1999; Ranjan, 2011; Ranjan and Reddy, 2009). The spectral finite element approximation is described as follows. The primary variables are each approximated as

$$\Delta^e = \sum_{j=1}^{n}\Delta_j\psi_j \tag{75}$$

where $\psi_j$ are the nodal expansions, which are provided by the following one-dimensional $\hat{C}^0$ spectral nodal basis as mentioned in Karniadakis et al. (1999).

$$\psi_i(\xi) = h_i^e(\xi) = \frac{(\xi-1)(\xi+1)L_n'(\xi)}{n(n+1)L_n(\xi_i)(\xi-\xi_i)} \tag{76}$$

where $\Delta_j$ are the nodal values due to the Kronecker delta property of the spectral basis. $L_n = P_n{}^{(0,0)}$ is the Legendre polynomial of order $n$ and $\xi_i$ denote the location of the roots of $(\xi-1)(\xi+1)L_n'(\xi) = 0$ in the interval $[-1,+1]$. The element stiffness matrices (derived in the following section) were obtained based on the Gauss-Lobatto- Legendre (GLL) rules. Gauss-Lobatto-Legendre rule includes both end points of the interval, that is, $\xi_i = \pm 1$. The points and weights are respectively listed as

$$\xi_i = \begin{cases} -1 & \text{if } i=0, \\ \xi^{1,1}_{i-1,Q-2} & \text{if } i=1,\ldots.,Q-2 \\ 1 & \text{if } i=Q-1>0. \end{cases} \tag{77}$$

The weights for the GLL quadrature rule are obtained as follows

$$w_i^{0,0} = \frac{2}{Q(Q-1)[L_{Q-1}(\xi_i)]^2} \quad i=0,1\ldots,Q-1 \tag{78}$$

In the formulae above, $L_Q(\xi)$ is the Legendre polynomial $L_Q(\xi) = P_Q^{0,0}(\xi)$. The zeroes of the Legendre polynomial do not have an analytical form and are commonly tabulated or evaluated separately for input in the computational framework. First and second derivatives of the Legendre polynomials are required in the formulation. The first derivative differentiation matrix is provided as

$$d_{ij} = \begin{cases} -\frac{Q(Q-1)}{4} & \text{if } i=j=0, \\ \frac{L_{Q-1}(\xi_i)}{L_{Q-1}(\xi_j)}\frac{1}{\xi_i-\xi_j} & i\neq j, 0\leq i,j\leq Q-1 \\ 0 & \text{if } 1\leq i=j\leq Q-2, \\ \frac{Q(Q-1)}{4} & \text{if } i=j=Q-1 \end{cases} \tag{79}$$

The differentiation operation utilizing the differentiation matrix is obtained

$$\frac{d}{dx}\Delta^e = \sum_{j=1}^{n} d_{ij}\Delta_j \tag{80}$$

The computation of higher order derivatives follows the approach for the computation of the first derivative. One can compute the entries of the $q^{th}$ order differentiation matrix $d_{ij}^{(q)}$ by evaluating the $q^{th}$ derivative of the Legendre polynomial at the quadrature points. Another convenient formulae for the higher order derivative is

$$d_{ij}^{(q)} = (d_{ij})^q \tag{81}$$

For the purposes of the current discussion we will be only concerned with the first and second derivatives.

The construction of the two dimensional spectral basis follows a tensor product with nodal expansions in either direction. Thus the construction of the tensor product follows

$$\phi_{ij}(\xi_1, \xi_2) = \psi_i(\xi_1)\psi_j(\xi_2) \tag{82}$$

These functions are defined on the domain (master element)

$$\bar{\Omega} = \{(\xi_1, \xi_2)| - 1 \leqslant \xi_1, \xi_2 \leqslant 1\} \tag{83}$$

which will be transformed by affine mapping to each element $\Omega_e$.

**6. Finite Element Formulation**

Let the spectral/*hp* spectral element spaces which we wish to work with be given by

$$\mathbf{W}^{hp} = \{\mathbf{w} \in H_0^1(\Omega)^N | V_K \in R_k(K)^N, K \in C_h\} \tag{84}$$

$$P^{hp} = \{p \in C^0(\Omega) \cap L_0^2(\Omega) | p_K \in R_l(K), K \in C_h\} \tag{85}$$

where the integers $k$ and $l$ are greater than or equal to 1. These spaces are defined using nodal spectral/*hp* basis provided by Legendre polynomials.

We consider the dimensional form of the Navier-Stokes equations for the development of the ASUPS formulation. Consider a spectral/*hp* element discretization of $\Omega$ into subdomains, $\Omega_e$, $k = 1, 2, ...., n_{el}$ where $n_{el}$ is the number of spectral/*hp* elements into which the domain is divided. Based on this discretization, for velocity and pressure, we define the trial discrete function spaces $\mathfrak{I}_u^{hp}$ and $\mathfrak{I}_p^{hp}$, and the weighting function spaces; $\mathbf{W}_u^{hp}$ and $P_p^{hp}$ defined above. These spaces are selected, by taking the Dirichlet boundary conditions into account, as subspaces of $\left[H^{1h}(\Omega)\right]^{n_{el}}$ and $\left[H^{1h}(\Omega)\right]$, where $\left[H^{1h}(\Omega)\right]$ is the finite-dimensional function space over $\Omega$. We write the stabilized Galerkin formulation of eq. 6-7 as follows: Find $u^{hp} \in \mathfrak{I}_u^{hp}$ and $p^{hp} \in \mathfrak{I}_p^{hp}$ such that $\forall$ $\mathbf{w}^{hp} \in \mathbf{W}_u^{hp}$, $q^{hp} \in P_p^{hp}$

$$\begin{aligned}
&\int_\Omega \mathbf{w}^{hp}\rho \cdot \left(\frac{\partial \mathbf{u}^{hp}}{\partial t} + \mathbf{u}^{hp} \cdot \nabla \mathbf{u}^{hp} - \mathbf{f}\right) d\Omega - \int_\Omega \mathbf{w}^{hp} \cdot \nabla \cdot \sigma(p^{hp}, \mathbf{u}^{hp}) d\Omega + \\
&\int_\Omega q^{hp} \nabla \cdot \mathbf{u}^{hp} d\Omega + \int_\Omega \epsilon q^{hp} p^{k,hp} d\Omega + \sum_{e=1}^{n_{el}} \int_{\Omega^e} \frac{1}{\rho} \tau \left(\rho \mathbf{u}^{hp} \cdot \nabla \mathbf{w}^{hp} + \nabla q^{hp}\right) \cdot \\
&\qquad \left[\rho \frac{\partial \mathbf{u}^{hp}}{\partial t} + \rho\left(\mathbf{u}^{hp} \cdot \nabla \mathbf{u}^{hp} - f\right) - \nabla \cdot \sigma(p^{hp}, \mathbf{u}^{hp})\right] d\Omega^e + \\
&\qquad \sum_{e=1}^{n_{el}} \int_{\Omega^e} \delta \nabla \cdot \mathbf{w}^{hp} \rho \nabla \cdot \mathbf{u}^{\mathbf{hp}} d\Omega^e = \int_{\Gamma_h} \mathbf{w}^{hp} \cdot h^{hp} d\Gamma
\end{aligned} \tag{86}$$

As can be seen from Eqn. 86, four stabilizing terms are added to the standard Galerkin formulation. In Eqn. 86 the first three terms and the right hand side constitute the classical Galerkin formulation of the problem. The surface integrals on the right are obtained from the weak form development as the natural boundary conditions. The first series of element-level integrals comprise the SUPG and PSPG stabilization terms, which are added to the variational formulation (Brooks and Hughes, 1982; Hughes et al, 1986). Two extra terms are added to the variational formulation. The new integral adds a term $(\psi_i \psi_j)$ to the stabilized finite element formulation. A similar term is added to the forcing function which depends on last known regularization step value of the pressure. The regularization step and the non-linear iteration can be handled as a single iteration in the new formulation. Prescriptions on determinations of the stabilization parameters based on inverse estimates have been utilized by Gervasio (Gervasio and Saleri, 1998). In their paper the higher order stabilization parameter (with applications to spectral methods) was defined.

$$\tau(\mathbf{x}) = \frac{h_k}{2|u_h^n|_{pn} {N_d}^2} \xi(Re_K(\mathbf{x})) \tag{87}$$

where

$$Re_K(\mathbf{x}) = \frac{m|u_h^n|_{pn} h_k}{2\nu {N_d}^2} \tag{88}$$

$$\xi(Re_K(\mathbf{x})) = \begin{cases} Re_K(\mathbf{x})) & Re_{x(\mathbf{x}))} \leq 1, \\ 1 & 1 \leq Re_x(\mathbf{x})) \end{cases} \tag{89}$$

The second stabilization term on the continuity equation is known as the least squares incompressibility constraint (LSIC) term (Tezduyar and Sathe, 2003). This term is provided as

$$\delta_k = \frac{\lambda h_k |u_h^n|_{pn}}{N_d{}^2} \xi(Re_K(\mathbf{x})) \tag{90}$$

In this paper we resort to the stabilization parameters defined by Eq. 87-90. In the above equations $N_d$ denotes the number of degrees of freedom in each element. The constant $m_k$ is defined above. The $p$ norm of the velocity is defined

$$|\mathbf{u_h^n}(\mathbf{x})|_{pn} = \begin{cases} (|(u_h^n)_1(\mathbf{x})|^{pn} + |(u_h^n)_2(\mathbf{x})|^{pn})^{\frac{1}{pn}} & 1 \leq pn \leq \infty, \\ max_{i=1,2}|(u_h^n)_i(\mathbf{x})| & pn = \infty \end{cases} \tag{91}$$

We used the notation $p$ to denote pressure and $N$ to denote the space dimensions earlier. To avoid any confusion with notation we denote the $p$ norm with $pn$ and $N_d$ as the number of degrees of freedom in each element.

The spatial discretization of Eq. 86 leads to a system of non-linear ordinary differential equations,

$$[\mathbf{M}(\mathbf{v}) + \mathbf{M}_{\kappa_1}(\mathbf{v})]\,\dot{\mathbf{v}} + [\mathbf{N}(\mathbf{v}) + \mathbf{N}_{\kappa_1}(\mathbf{v})]\,\mathbf{v} + [\mathbf{K} + \mathbf{K}_{\kappa_1}]\,\mathbf{v} - [\mathbf{G} + \mathbf{G}_{\kappa_1}]\,\mathbf{p} = [\mathbf{F} + \mathbf{F}_{\kappa_1}] \tag{92}$$

$$[\mathbf{M}_{\kappa_2}(\mathbf{v})]\,\dot{\mathbf{v}} + \mathbf{G^T}\mathbf{v} + \mathbf{N}_{\kappa_2}(\mathbf{v})\mathbf{v} + \mathbf{K}_{\kappa_2}\mathbf{v} + \mathbf{G}_{\kappa_2}\mathbf{p} = E \tag{93}$$

Here, $\mathbf{v}$ is the vector of unknown nodal values of $\mathbf{v^{hp}}$, and $p$ is the vector of nodal values of $p^{hp}$. The matrices $\mathbf{N}(\mathbf{v})$, $\mathbf{K}$, and $\mathbf{G}$ are derived, respectively, from the advective, viscous, and pressure terms. The vectors $\mathbf{F}$ and $\mathbf{E}$ are due to boundary conditions. The subscripts $\kappa_1$ and $\kappa_2$ identify the SUPG and PSPG contributions, respectively. The various matrices forming the discrete finite element equations are provided. The definition of the terms in the stiffness matrices (after linearization) are outlined below;

$$(\mathbf{M}(\mathbf{v}) + \mathbf{M}_{\kappa_1}(\mathbf{v})) = \int_\Omega \left[\rho \mathbf{w}^{hp} \frac{\partial \mathbf{u}^{hp}}{\partial t} + \sum_{e=1}^{n_{el}} \kappa_1 \rho \frac{\partial \mathbf{u}^{hp}}{\partial t}\right] d\Omega \tag{94}$$

$$(\mathbf{N}(\mathbf{v}) + \mathbf{N}_{\kappa_1}(\mathbf{v})) = \int_\Omega \left[\rho \mathbf{w}^{hp} \cdot (\mathbf{u}^{hp} \cdot \nabla \mathbf{u}^{hp}) + \sum_{e=1}^{n_{el}} \kappa_1 \cdot (\mathbf{u}^{hp} \cdot \nabla \mathbf{u}^{hp})\right] d\Omega \tag{95}$$

$$(\mathbf{K} + \mathbf{K}_{\kappa_1} + \mathbf{K}_{\kappa_\delta}) = \int_\Omega \left[\nabla \cdot \mathbf{w}^{hp} \cdot 2\mu\varepsilon(\mathbf{u}^{hp}) + \sum_{e=1}^{n_{el}} \kappa_1 \cdot 2\mu\varepsilon(\mathbf{u}^{hp}) + \kappa_\delta \rho \nabla \cdot \mathbf{u}^{hp}\right] d\Omega \tag{96}$$

$$(\mathbf{G} + \mathbf{G}_{\kappa_1}) = \int_{\Omega^e} \nabla \cdot \mathbf{w}^{hp} p^{hp} d\Omega + \sum_{e=1}^{n_{el}} \int_{\Omega^e} \kappa_1 \nabla \cdot p^{hp} d\Omega \tag{97}$$

$$\mathbf{M}_{\kappa_2}(\mathbf{v}) = \sum_{e=1}^{n_{el}} \kappa_2 \rho \frac{\partial \mathbf{u}^{hp}}{\partial t} d\Omega \tag{98}$$

$$\mathbf{G}^T = \int_{\Omega^e} q^{hp} \nabla \cdot \mathbf{u}^{hp} d\Omega \qquad \mathbf{G}_{\kappa_2} = \sum_{e=1}^{n_{el}} \kappa_2 \cdot \nabla \cdot p d\Omega + \int_{\Omega^e} \epsilon q^{hp} p^{k,hp} d\Omega \tag{99}$$

$$\mathbf{N}_{\kappa_2}(\mathbf{v}) = \sum_{e=1}^{n_{el}} \kappa_2 \cdot (\mathbf{u}^{hp} \cdot \nabla \mathbf{u}^{hp}) \rho d\Omega \quad \mathbf{K}_{\kappa_2} = \int_{\Omega^e} \sum_{e=1}^{n_{el}} \kappa_2 \cdot \nabla \cdot 2\mu\varepsilon(\mathbf{u})\Omega \tag{100}$$

And the forcing functions are defined as

$$\mathbf{F} + \mathbf{F}_{\kappa_1} = \int_{\Omega^e} \left\{\mathbf{w}^{hp} \rho \cdot \mathbf{f} + \sum_{e=1}^{n_{el}} \kappa_1 \cdot \rho \mathbf{f}\right\} d\Omega \tag{101}$$

$$\mathbf{E} = \int_{\Omega_e} \sum_{e=1}^{n_{el}} \kappa_2 \cdot \rho \mathbf{f} d\Omega \tag{102}$$

The stabilization of the Galerkin formulation presented in this paper is a generalization of the SUPG/PSPG/LSIC formulation, employed thus far for solving incompressible flow problems. In the equations above the parameters $\kappa$ are defined as

$$\kappa_1 = \tau(\mathbf{u}^{hp} \cdot \nabla \mathbf{w}^{hp}) \quad \kappa_2 = \frac{1}{\rho}\tau(\nabla q^{hp}) \quad \kappa_3 = \delta \nabla \cdot \mathbf{w}^{hp} \tag{103}$$

With such stabilization procedures described above, it is possible to use elements that have equal order interpolation functions for the velocity and pressure, and are otherwise unstable (Hughes et al, 1986). There are various ways to linearise the non-linear terms in stabilized finite element methodology. We store the tangent stiffness matrices. The least squares term was set to zero in the present case.

The Galerkin Least Squares (GLS) formulation adds a least squares term involving the momentum equation to the Galerkin finite element formulation. The GLS stabilization is a more general stabilization approach that includes the essence of the SUPG and PSPG type stabilizations as subcomponents of its formulation. This approach has been successfully applied to Stokes flows, compressible flows (Shakib et al, 1991), and incompressible flows at finite Reynolds numbers (Tezduyar, 1992). In the GLS stabilization of incompressible flows, the stabilizing terms added are obtained by minimizing the sum of the squared residual of the momentum equation integrated over each element domain. Consequently similar to the SUPG and PSPG stabilizations, because the stabilizing terms involve the residual of the momentum equation as a factor, the GLS stabilized formulation is consistent. For time-dependent problems, a strict implementation of the GLS stabilization technique necessitates finite element discretization in both space and time, and leads to a space-time coupled finite element formulation of the problem. A slightly liberal implementation of the GLS formulation requires discretization in space only. This formulation still retains the consistency of the governing differential equations since we seek the solution of the partial differential equation minus the time dependent term in the stabilization process. Space-time formulations have been known to generate extensive linear systems with the accompanying lethargy in solution procedures for a fully coupled simulation. In the interest of saving computational efforts in obtaining the solutions for the Navier-Stokes equations we will employ a space-time decoupled GLS formulation.

To formalize the GLS formulation for incompressible flow we introduce some notation. Let the spatial domain $\Omega$ be divided into non- overlapping subdomains $\Omega_e$, where $e$=1, 2, 3, ..*nel* where *nel* is total number of elements in the domain. Based on this discretization of the velocity and pressures we define the trial function spaces $\mathbf{\Gamma}_u^{hp}$ and $\Gamma_p^{hp}$ and the weighting function spaces $\mathbf{W}_u^{hp}$ and $W_p^{hp}$. These function spaces are selected by taking the Dirichlet boundary conditions into account as subsets of the finite dimensional space $H^{1h}(\Omega)$. The GLS stabilized finite element formulation is written as follows: Find $\mathbf{u} \in \mathbf{\Gamma}_u^{hp}$ and $p \in \Gamma_p^{hp}$ such that for $\mathbf{w} \in \mathbf{W}_u^{hp}$ and $q \in W_p^{hp}$

$$\begin{aligned}&\int_\Omega \mathbf{w^{hp}}\rho \cdot \left(\frac{\partial \mathbf{u^{hp}}}{\partial t} + \mathbf{u^{hp}} \cdot \nabla \mathbf{u^{hp}} - \mathbf{f}\right) d\Omega - \int_\Omega \mathbf{w^{hp}} \cdot \nabla \cdot \sigma(p^{hp}, \mathbf{u^{hp}}) d\Omega + \\ &\int_\Omega q^{hp} \nabla \cdot \mathbf{u^{hp}} d\Omega + \sum_{e=1}^{n_{el}} \int_{\Omega^e} \tau \left[\mathbf{u^{hp}} \cdot \nabla \mathbf{w^{hp}} - \frac{1}{\rho}\nabla \cdot \sigma(q^{hp}, \mathbf{w^{hp}})\right] \cdot \\ &\left[\rho\left(\frac{\partial \mathbf{u^{hp}}}{\partial t} + \mathbf{u^{hp}} \cdot \nabla \mathbf{u^{hp}} - \mathbf{f}\right) - \nabla \cdot \sigma(p^{hp}, \mathbf{u^{hp}})\right] d\Omega^e = \int_{\Gamma_h} \mathbf{w^{hp}} \cdot \mathbf{h^{hp}} d\Gamma\end{aligned} \tag{104}$$

The above formulation requires the specification of the stabilization parameter. The stabilization parameter is obtained from the stability analysis of the formulation performed earlier. Classically it has been derived from a multi-dimensional generalization of the parameter derived in Shakib (1989) in one dimension. In the following developments it is also specified based on the advection and diffusion limits. The classical definition of the stabilization parameter is provided as

$$\tau_{GLS} = \left[\left(\frac{2}{\Delta t}\right)^2 + \left(\frac{4\nu}{h^2}\right)^2 + \left(\frac{2\|u^h\|}{h}\right)^2\right]^{-1/2} \tag{105}$$

This classical stabilization parameter is designed to provide stability to the formulation.

We employ the decoupled space-time formulation in the interest of lower compute times. We seek to obtain solutions to the incompressible Navier-Stokes with a modification of the original GLS formulation augmented with an iterative penalization of the incompressibility constraint. This augmentation is found to provide better non linear convergence histories as compared to the classical GLS formulation used to date. The idea is similar to artificial compressibility methods which have been known to improve the convergence rates for stiff problems obtained from the discretization of Navier-Stokes equations. In the iteratively penalized formulation however we maintain consistency with a regularization of the incompressibility equation. The space-time decoupled GLS stabilized finite element formulation is written as follows: $\mathbf{u} \in \mathbf{\Gamma}_u^{hp}$ and $p \in \Gamma_p^{hp}$ such that for $\mathbf{w} \in \mathbf{V}_u^{hp}$ and $q \in V_p^{hp}$

$$\int_\Omega \mathbf{w^{hp}}\rho \cdot \left(\frac{\partial \mathbf{u^{hp}}}{\partial t} + \mathbf{u^{hp}} \cdot \nabla \mathbf{u^{hp}} - \mathbf{f}\right) d\Omega - \int_\Omega \mathbf{w^{hp}} \cdot \nabla \cdot \sigma(p^{hp}, \mathbf{u^{hp}}) d\Omega +$$
$$\int_\Omega q^{hp} \nabla \cdot \mathbf{u^{hp}} d\Omega + \sum_{e=1}^{n_{el}} \int_{\Omega^e} \tau \left[\mathbf{u^{hp}} \cdot \nabla \mathbf{w^{hp}} - \frac{1}{\rho} \nabla \cdot \sigma(q^{hp}, \mathbf{w^{hp}})\right] \cdot$$
$$\left[\rho\left(\frac{\partial \mathbf{u^{hp}}}{\partial t} + \mathbf{u^{hp}} \cdot \nabla \mathbf{u^{hp}} - \mathbf{f}\right) - \nabla \cdot \sigma(p^{hp}, \mathbf{u^{hp}})\right] d\Omega^e + \int_\Omega \epsilon q^{hp} p^{hp,k} d\Omega =$$
$$\int_{\Gamma_h} \mathbf{w^{hp}} \cdot \mathbf{h^{hp}} d\Gamma \tag{106}$$

Two extra terms are added to the variational formulation when compared to the GLS formulation. The new integral adds a term $(\psi_i \psi_j)$ to the stabilized finite element formulation. The regularization step and the non-linear iteration can be handled as a single iteration in the augmented formulation. Augmentation is achieved with the relaxation of the divergence free condition via iterative penalization. We use higher order spectral elements where GLS and SUPS formulations are distinct.

The spatial discretization of Eq. 106 leads to a set of non-linear ordinary differential equations,

$$[\mathbf{M}(\mathbf{u}) + \mathbf{M}_{\kappa_1}(\mathbf{u}) + \mathbf{M}_{\kappa_3}(\mathbf{u})]\,\dot{\mathbf{u}} + [\mathbf{N}(\mathbf{u}) + \mathbf{N}_{\kappa_1}(\mathbf{u}) + \mathbf{N}_{\kappa_3}(\mathbf{u})]\mathbf{u} +$$
$$[K + K_{\kappa_1} + \mathbf{K}_{\kappa_3}]\mathbf{u} - [G + G_{\kappa_1} + G_{\kappa_3}]\,p = [F + F_{\kappa_1}] \tag{107}$$

$$[\mathbf{M}_{\kappa_2}(\mathbf{u})]\,\dot{\mathbf{u}} + \mathbf{G^T u} + \mathbf{N}_{\kappa_2}(\mathbf{u})\mathbf{u} + \mathbf{K}_{\kappa_2}\mathbf{u} + \mathbf{G}_{\kappa_2}\mathbf{p} = \mathbf{E} \tag{108}$$

Here, $\mathbf{u}$ is the vector of unknown nodal values of $\mathbf{u^{hp}}$, and $p$ is the vector of nodal values of $p^{hp}$. The matrices $\mathbf{N}(\mathbf{u})$, $\mathbf{K}$, and $\mathbf{G}$ are derived, respectively, from the advective, viscous, and pressure terms. The vectors $\mathbf{F}$ and $\mathbf{E}$ are due to boundary conditions and the iterative penalization procedure. The subscripts $\kappa_1$, $\kappa_2$, $\kappa_3$ identify the SUPG, PSPG and GLS contributions, respectively. The various matrices forming the discrete finite element equations are provided. The definition of the terms in the stiffness matrices are outlined below;

$$(\mathbf{M}(\mathbf{u}) + \mathbf{M}_{\kappa_1} + \mathbf{M}_{\kappa_3}(\mathbf{u})) = \int_\Omega \left[\rho \mathbf{w^{hp}} \frac{\partial \mathbf{u^{hp}}}{\partial \mathbf{t}} + \sum_{e=1}^{n_{el}} \kappa_1 \rho \frac{\partial \mathbf{u^{hp}}}{\partial t} + \kappa_3 \rho \frac{\partial \mathbf{u}^{hp}}{\partial t}\right] d\Omega \tag{109}$$

$$(\mathbf{N}(\mathbf{u}) + \mathbf{N}_{\kappa_1}(\mathbf{u}) + \mathbf{N}_{\kappa_3}(\mathbf{u})) =$$
$$\int_\Omega \left[\rho \mathbf{w^{hp}} \cdot (\mathbf{u^{hp}} \cdot \nabla \mathbf{u^{hp}}) \sum_{e=1}^{n_{el}} \kappa_1 \cdot (\mathbf{u^{hp}} \cdot \nabla \mathbf{u^{hp}}) + \kappa_3 (\mathbf{u}^{hp} \cdot \nabla \mathbf{u}^{hp})\right] d\Omega \tag{110}$$

$$(\mathbf{K} + \mathbf{K}_{\kappa_1} + \mathbf{K}_{\kappa_3}) = \int_\Omega \left[\nabla \cdot \mathbf{w^{hp}} \cdot 2\mu\varepsilon(\mathbf{u^{hp}}) + \sum_{e=1}^{n_{el}} \kappa_1 \cdot 2\mu\varepsilon(\mathbf{u^{hp}}) - \kappa_3 \nu\mu\Delta \mathbf{w}^{hp} \Delta \mathbf{u}^{hp}\right] d\Omega \tag{111}$$

$$(\mathbf{G} + \mathbf{G}_{\kappa_1} + \mathbf{G}_{\kappa_3}) = \int_{\Omega^e} \nabla \cdot \mathbf{w^{hp}} p^{hp} d\Omega + \sum_{e=1}^{n_{el}} \int_{\Omega^e} \left(\kappa_1 \nabla p^{hp} + \kappa_3 \nabla p^{hp}\right) d\Omega \tag{112}$$

$$\mathbf{M}_{\kappa_2}(\mathbf{u}) = \sum_{e=1}^{n_{el}} \kappa_2 \rho \frac{\partial \mathbf{u^{hp}}}{\partial t} d\Omega \tag{113}$$

$$\mathbf{G}^T = \int_{\Omega^e} q^{hp} \nabla \cdot \mathbf{u^{hp}} d\Omega \qquad \mathbf{G}_{\kappa_2} = \sum_{e=1}^{n_{el}} \kappa_2 \cdot \nabla \cdot p d\Omega + \int_{\Omega^e} \epsilon q^{hp} p^{k,hp} d\Omega \tag{114}$$

$$\mathbf{N}_{\kappa_2}(\mathbf{u}) = \sum_{e=1}^{n_{el}} \kappa_2 \cdot (\mathbf{u^{hp}} \cdot \nabla \mathbf{u^{hp}}) \rho d\Omega \quad \mathbf{K}_{\kappa_2} = \int_{\Omega^e} \sum_{e=1}^{n_{el}} \kappa_2 \cdot \nabla \cdot 2\mu\varepsilon(\mathbf{u})\Omega \tag{115}$$

And the forcing functions are defined as

$$\mathbf{F} + \mathbf{F}_{\kappa_1} = \int_{\Omega^e} \left\{ \mathbf{w^{hp}}\rho \cdot \mathbf{f} + \sum_{e=1}^{n_{el}} (\kappa_1 \cdot \rho\mathbf{f} + \kappa_3\rho\mathbf{f}) \right\} d\Omega \tag{116}$$

$$\mathbf{E} = \int_{\Omega_e} \sum_{e=1}^{n_{el}} \kappa_2 \cdot \rho\mathbf{f} d\Omega \tag{117}$$

The stabilization of the Galerkin formulation presented in this paper is the GLS formulation for solving incompressible flow problems. In the equations above the parameters $\kappa$ are defined as

$$\kappa_1 = \tau(\mathbf{u^{hp}} \cdot \nabla\mathbf{w}^{hp}) \quad \kappa_2 = \frac{1}{\rho}\tau(\nabla q^{hp}) \quad \kappa_3 = -\tau\nu\Delta\mathbf{w}^{hp} \tag{118}$$

With such stabilization procedures described above, it is possible to use elements that have equal order interpolation functions for the velocity and pressure, and are otherwise unstable (Hughes et al, 1986). We utilize the GLS formulation for presenting results to benchmark incompressible flow problems in a following section.

**7. Non Linear Solvers and Time Stepping Algorithms**

Let us consider the discrete non-linear problem obtained from the Navier-Stokes equations by;

$$\mathbf{F}(\mathbf{\Delta}) = \mathbf{0} \tag{119}$$

where $\mathbf{F}(\mathbf{\Delta}) : R^N \to R^N$ is a non-linear mapping with the following properties:

(1) There exists an $\mathbf{\Delta}^* \in R^N$ with $F(\mathbf{\Delta}^*) = 0$.

(2) F is continuously differentiable in a neighborhood of $\mathbf{\Delta}^*$.

(3) $F'(\mathbf{\Delta}^*)$ is non singular.

In the particular case of solution of incompressible flow problems, $\mathbf{\Delta} = \{\mathbf{u}, p\}$ denotes the vector of nodal variables for the velocity field and pressures over the domain of interest. For moderate to high Reynolds numbers the non-linear convective term dominates and the solution procedures have to be able to accommodate the strong non-linearities.

We assume that $\mathbf{F}(\mathbf{\Delta})$ is continuously differentiable in $\Re^N$ where $N$ is the number of spatial dimensions. Let us denote the Jacobian matrix by $\mathbf{J} \in \Re^N$. Classical Newton's method for solving the non-linear problem can be enunciated as (Elias et al, 2004, 2006)

$$\boxed{\begin{array}{c} \textit{for } k = 0 \textit{ until convergence} \\ \textit{solve } \mathbf{J}(\mathbf{\Delta_k})\mathbf{s_k} = -\mathbf{F}(\mathbf{\Delta_k}) \\ \textit{set } \mathbf{\Delta_{k+1}} = \mathbf{\Delta_k} + \mathbf{s_k} \end{array}}$$

Newton's method was designed for solving problems where the initial guess is very close to the non-linear solution $\mathbf{\Delta}^*$. Amongst the disadvantages of Newton's method are the requirements that the linear system be solved very accurately which can be computationally expensive. Computing an exact solution using a direct method can also be prohibitively expensive if the number of unknowns is large and may not be justified when $\mathbf{\Delta}_k$ is far from the solution. In spectral computations where there is high computational costs involved with quadratures it is imperative to resort to efficient procedures for solving the discrete systems of equations. Thus one might prefer to compute some approximate solution leading to the following Inexact Newton Algorithm

$$\boxed{\begin{array}{c} \textit{for } k = 0 \textit{ until convergence} \\ \textit{find some } \xi_k \in [0, 1\} \textit{ and } s_k \textit{ that satisfy} \\ ||\mathbf{F}(\mathbf{\Delta_k}) + \mathbf{F}'(\mathbf{\Delta_k})\mathbf{s_k}|| \leq \eta_k ||\mathbf{F}(\mathbf{\Delta_k})|| \\ \textit{set } \mathbf{\Delta_{k+1}} = \mathbf{\Delta_k} + \mathbf{s_k} \end{array}}$$

or a damped version of the final update to the new solution vector as; set $\mathbf{\Delta_{k+1}} = \mathbf{\Delta_k} + \alpha\mathbf{s_k}$ where, $\alpha$ is the damping parameter. For some $\eta_k$ where $|| \cdot ||$ is the norm of choice. This new formulation automatically allows the use of an iterative linear solver method, one first chooses $\eta_k$ and then applies the iterative solver to the algorithm until an $\mathbf{s_k}$ is determined for which the residual norm satisfies the convergence criterion. In this context, $\eta_k$ is often called the forcing term, since its role is to force the residual of the above equation to be sufficiently small. In general the non-linear forcing sequence needs to be specified to drive the solution towards the solution of the non-linear problem as;

$$\frac{||\mathbf{r_k}||}{||\mathbf{F}(\mathbf{\Delta_k})||} \leq \eta_k \tag{120}$$

The $||\mathbf{r_k}||$ is defined as the residual for the non-linear problem. In our implementations we use the element by element (EBE) BiCGSTAB method to compute the $\mathbf{s_k}$. In the EBE- BiCGSTAB implementation we store the element stiffness matrices. The amount of storage required for the problem varied as $O(e_p \times \left(ndf \times (p+1)^2\right))$. Here, $e_p$ is the number of elements stored in each processor, $ndf$ is the number of degrees of freedom in each element, and $p$ is the polynomial approximation of the spectral element. There exist several choices for the evaluation of the forcing function $||\eta_k||$ as mentioned in Eisenstat and Walker (1996). Amongst the choices is the usage of a constant value of $\eta_k = 10^{-4}$. We stored the element tangent stiffness matrix in our computational framework.

The derivations of the convergence behavior for a generic non-linear problem have been outlined in Dembo et al. (1982). The basic requirement for the convergence of the iterative method is that $\mathbf{F}$ is continuously differentiable in the neighbourhood of $\mathbf{\Delta}^* \in R^N$ for which $\mathbf{F}(\mathbf{\Delta}^*)$ is non singular and that $\mathbf{F}'$ is Lipschitz continuous at $\mathbf{x}^*$ with constant $\lambda$, i.e.

$$||\mathbf{F}'(\mathbf{\Delta}) - \mathbf{F}'(\mathbf{\Delta}^*)|| \leq \lambda||\mathbf{x} - \mathbf{x}^*|| \tag{121}$$

The construction of the tangent stiffness matrix required for the algorithm follows standard procedures in finite element analysis. In our implementations we store the tangent element stiffness matrices.

We employ the generalized $\alpha$ time integration method to solve the discretized equations. The generalized $\alpha$ method for Navier-Stokes equations for incompressible flows was first proposed by Jansen et al. (2000). In this method the stepping in time for the variables are approximated as

$$\dot{\mathbf{\Delta}}_{\mathbf{r}+\alpha_\mathbf{m}} = \dot{\mathbf{\Delta}}_\mathbf{r} + \alpha_\mathbf{m}(\dot{\mathbf{\Delta}}_{\mathbf{r+1}} - \dot{\mathbf{\Delta}}_\mathbf{r}) \tag{122}$$

$$\mathbf{\Delta}_{\mathbf{r}+\alpha_\mathbf{f}} = \dot{\mathbf{\Delta}}_\mathbf{r} + \alpha_f(\dot{\mathbf{\Delta}}_{\mathbf{r+1}} - \dot{\mathbf{\Delta}}_\mathbf{r}) \tag{123}$$

In the above equation $r$ denotes the time-step and $\mathbf{\Delta}$ is the vector of unknowns $\{\mathbf{u}, p\}$. The discrete momentum and continuity equations are collocated at intermediate values of the solution at every time step. The relationship between the nodal velocity degrees-of-freedom and their time derivatives is approximated by the discrete Newmark formula

$$\mathbf{\Delta_{r+1}} = \mathbf{\Delta_r} + \Delta t_n \left((1-\gamma)\dot{\mathbf{\Delta}}_\mathbf{r} + \gamma\dot{\mathbf{\Delta}}_\mathbf{r+1}\right) \tag{124}$$

Here $\alpha_m$, $\alpha_f$, and $\gamma$ are the real valued parameters chosen based on the stability and accuracy considerations. It was shown by Jansen et al. (2000) that second order accuracy in time is achieved provided that

$$\gamma = \frac{1}{2} + \alpha_m - \alpha_f \tag{125}$$

while unconditional stability is achieved provided that

$$\alpha_m \geq \alpha_f \geq \frac{1}{2} \tag{126}$$

To study the unsteady problem for the flow past a square obstruction we employ a space-time decoupled formulation. We begin by partitioning the time interval of interest $[0, T]$ into subintervals, or time steps. The time intervals $t_n$ and $t_{n+1}$ define the end points of the time $n^{th}$ time step and $\Delta t_n = t_{n+1} - t_n$ is the size of the $n^{th}$ time step. In general the time step may vary from step to step. A one-parameter family of second-order accurate and unconditionally stable algorithms may be obtained by setting $\gamma$ according to equation 125 and employing the following parametrization of the intermediate time steps

$$\alpha_m = \frac{1}{2}\left(\frac{3-\rho_\infty}{1+\rho_\infty}\right) \text{ and } \alpha_f = \left(\frac{1}{1+\rho_\infty}\right) \tag{127}$$

## 8. Numerical Examples

In this section we obtain solutions for some computational fluid dynamics problems with the formulations presented. First we consider the ASUPS formulation, and subsequent examples consider the AGLS formulation. In the first section we examine steady state conditions for the Kovasznay flow problem and skewed cavity problem. We provide agreement with analytical results for verification for the Kovaszany flow solutions. Steady state analysis results are presented for the flow past a square obstruction in a enclosure and unsteady state flow past a square cylinder. With the AGLS formulation we consider the performance for the driven cavity problem at aspect ratio of 2, and flow past two cylinders in tandem.

*8.1 ASUPS Formulation*

We consider first standard incompressible flow problems solved with the ASUPS formulation and follow with numerical examples solved with the AGLS formulation.

8.1.1 Kovasznay Flow

Benchmark problems in two dimensional computational fluid dynamics involve the solutions of Kovasznay flow among other problems. Analytical solutions have been provided for this flow to compare against numerical results by Kovasznay (1948). We study this standard two dimensional benchmark problem at Reynolds number of 40.

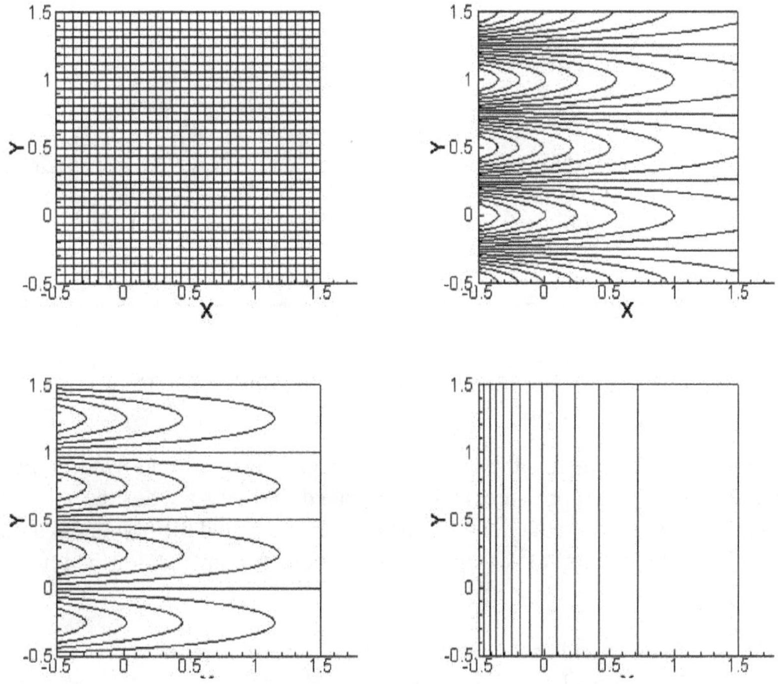


Figure 1. Koznavy Flow Re-40

The description of the problem requires to consider a two-dimensional domain $\Omega = [-0.5, 1.5] \times [-0.5, 1.5]$. Kovasznay provided the analytical solution to the problem. The solutions are provided by:

$$u(x,y) = 1 - e^{\lambda x} cos(2\pi y)$$
$$v(x,y) = \frac{\lambda}{2\pi} e^{\lambda x} sin(2\pi y)$$
$$p(x,y) = p_0 - \frac{1}{2} e^{2\lambda x} \tag{128}$$

where $\lambda = Re/2 - (Re^2/4 - 4\pi^2)^{1/2}$ and $p_0$ is the reference pressure (an arbitrary constant). Figure 1 presents the solution of the problem at $Re$ 40. The two dimensional mesh along with contour plots of $u$-velocity, pressure, and $v$-velocity, are presented in the figure (clockwise from left). We compare the *sup* norm between the analytical solutions and the numerical results. The *sup* norm was defined as

$$L_\infty = sup\, |x_{present} - x_{anal}| \tag{129}$$

The values of this $L_\infty$ norm was found to be 0.024170 and 0.002908 for the two components of the velocities and 0.032211 for the pressure. The non-linear tolerance for the present runs were kept at a reduction in $\mathbf{s_k}$ to a value less than $10^{-6}$ for generating the results in Figure 1. It should be noted that for an 'exact solve' for the problem solved with Picard method these numbers were 0.021667, 0.0028023, and 0.032027. For this comparison the $p$-level was set to 3. The Picard method solutions were obtained with iteration lagged linearisation. A Gauss elimination solver was used for obtaining the results for the Picard solution. Convergence study of the above formulation with increasing $p$-levels is provided in Figure 2.

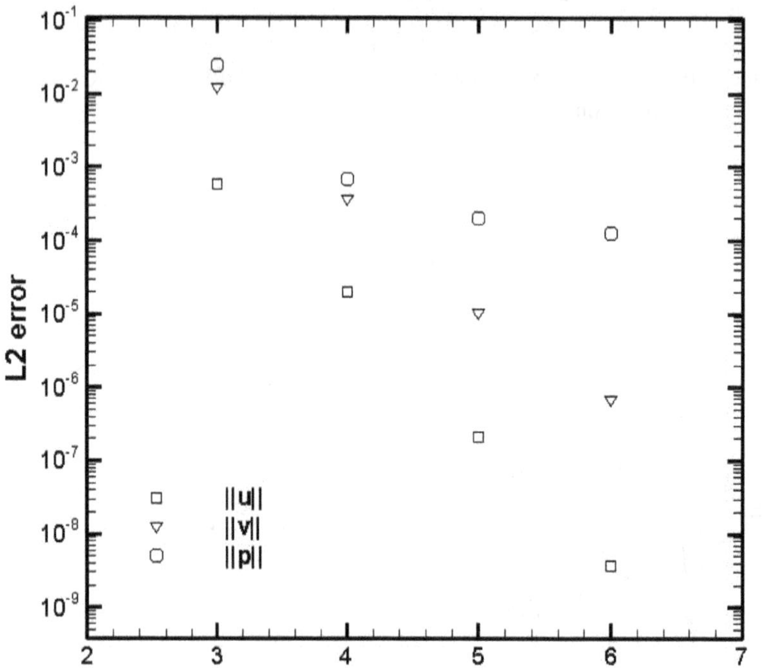


Figure 2. $p$ convergence of formulation Re-40, $\epsilon$=0.05

Obviously, the constant value of $\epsilon = 0.01$ affects the convergence with $p$-level beyond five (5). Further studies to obtain spectral convergence of the formulation (wrt pressures) with accurate design of $\epsilon$ are desired. For the problem considered the linear convergence tolerance was set at a reduction in the residual of $10^{-4}$ and the non linear tolerance was set at $10^{-9}$. For a p-level of three (3) the inexact algorithm took from 55-95 iterations to converge for the linearised discrete systems.

8.1.2 Skewed Cavity

Lid driven cavity problem for orthogonal grids has been studied extensively by researchers in the past (Ghia et al, 1982). The cavity problem for non-orthogonal grids has not been explored that extensively perhaps because of the oblique cavity orientation. In this section we test the spectral/*hp* Newton Krylov algorithm for solving the skewed cavity problem at angles of 30° and 45° respectively. We also provide the convergence history for both the SUPS and ASUPS formulations.

The length of the cavity was taken as unity, with the density of the fluid as 1, and lid velocity was taken as $U_L = 1$ respectively. The Reynolds number was defined based on the lid velocity $U_L$ and cavity length $L$. The viscosity was varied from 0.01 through 0.001 which obtained Reynolds numbers 100 and 1000 respectively. The angle of skew was defined as the inclination of the side walls of the cavity. Two different values of 30° and 45° were used. The reason for choosing the problem included an intricate study of the difference between different types of errors when performing numerical analysis. As mentioned in Demirdžić et al. (2005) who provide benchmark results for this problem, there are three sources of error in the study of any problem namely; (a) discretization errors: the difference between the exact solution of the correctly discretized equations on a given grid and the exact solution of the differential equation, (b) convergence errors: the difference between the exact and the approximate solution of the discretized equations left over after stopping the iterations. (c) algorithmic and programming errors: errors due to incorrect discretization, erroneous implementation of boundary conditions and programming errors etc. We assume only the first two sources of errors to dominate the numerical solutions. Demirdžić (Demirdžić et al, 2005) used the control volume approach for solving the equations on a 320 × 320 mesh.

In comparison we only utilized a 40 × 40 mesh with a $p_{level}$ of 6 in each element. Thus, we retain the high order spatial resolution offered by the spectral/*hp* methods at the same time we can resort to the non-linear solvers to reach benchmark solutions. The non-linear tolerance was set to $10^{-6}$ and the linear system solver tolerance was set to $10^{-4}$. We set the value of the regularization constant to $\epsilon = 0.0$ and 0.04 respectively. Figure 3 presents the streamline and the $v$ contour plots over the domain of the skewed cavity for Reynolds number of 100 with a skew angle of 30°. The corresponding

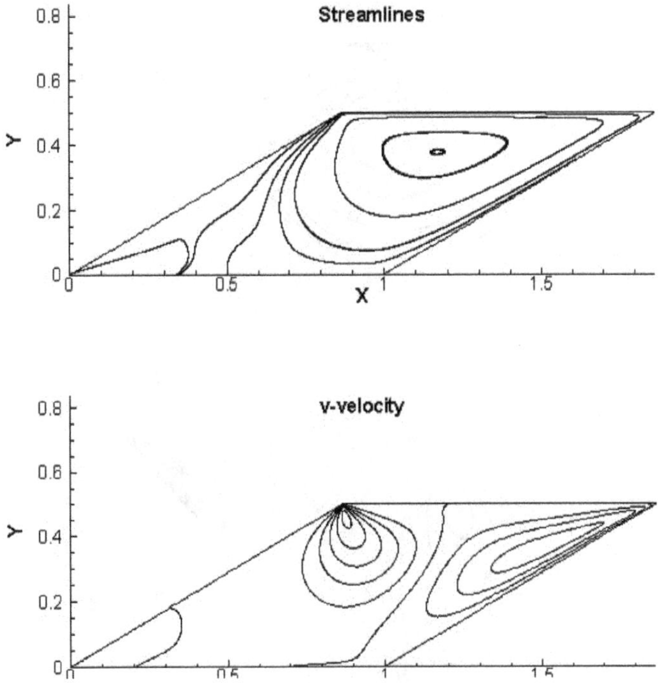


Figure 3. Contour plot of velocities Re-100, cavity skew=30

streamlines and $v$-contour plots have been presented in Figure 4 for a skew angle of 45° and Re-100. Figure 5 presents the streamlines and the $v$-contour plots for a skew angle of 45°. The agreement of the present results with benchmark results along the skewed vertical and horizontal centerline for an Re-100 is presented in Figure 6 for a skew angle of 30°. The agreement between the present results (lines) and benchmark results of Demirdžić (Demirdžić et al, 2005) (open squares) can be considered excellent.

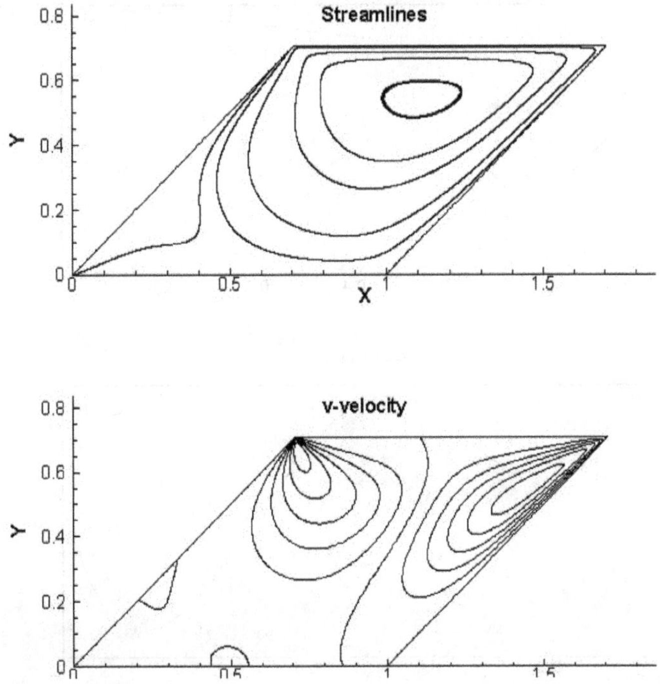


Figure 4. Contour plot of velocities Re-1000, cavity skew=45

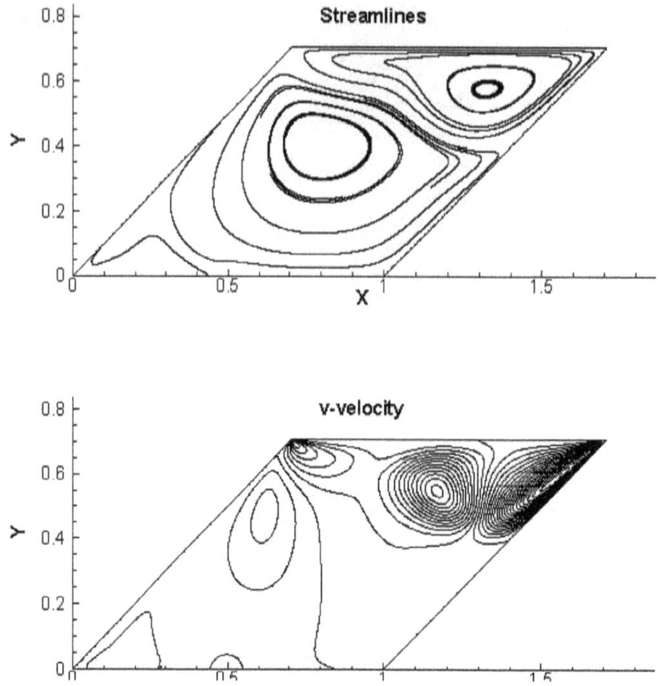


Figure 5. Contour plot of velocities Re-1000, cavity skew=45

Present results benchmarked with the skewed vertical and horizontal centerlines are presented in Figure 7 for Re-100 and skew angle of 45°. The non-linear convergence history for the SUPS formulation and the ASUPS formulation have been presented in Figure 8 and Figure 9 for the velocity and pressures respectively. We can discern from the figures that the ASUPS formulation performs better for this problem and converges more rapidly to the non-linear tolerance. In general excellent agreement was found for different skew angles and also differing Reynolds numbers for this problem.

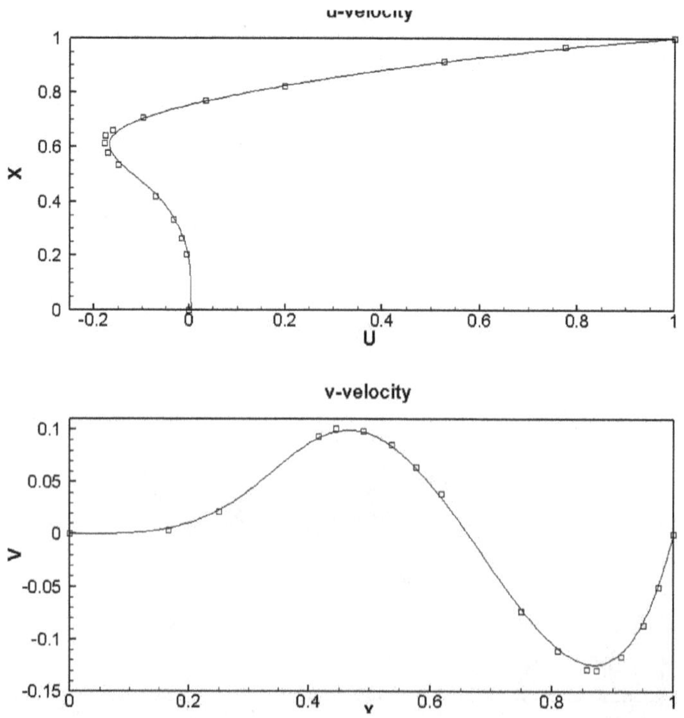


Figure 6. Centerline velocity at Re-100, cavity skew=30

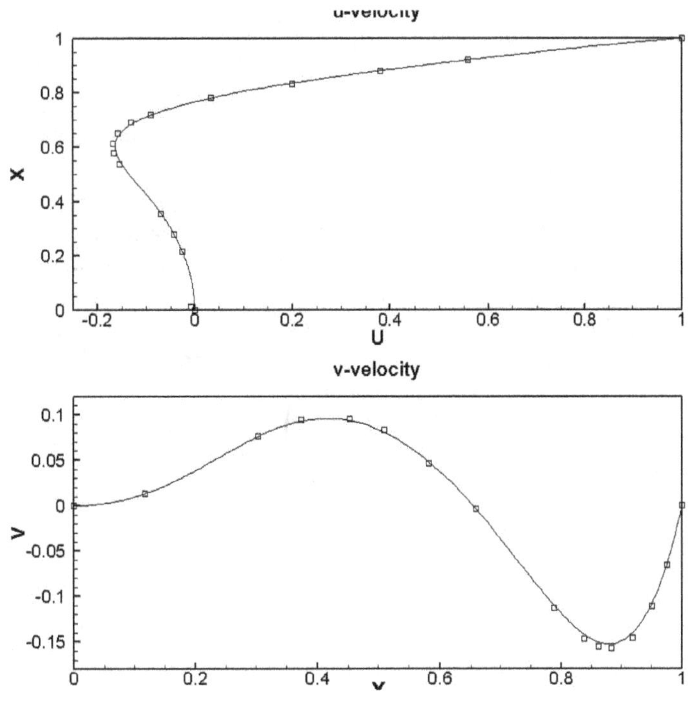


Figure 7. Centerline velocity at Re-100, cavity skew=45

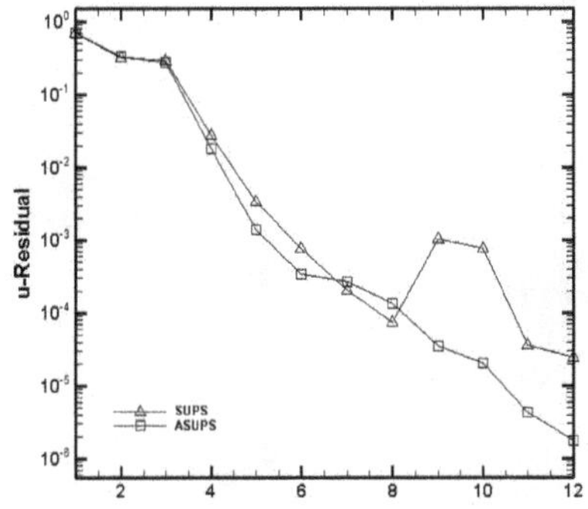


Figure 8. u-convergence

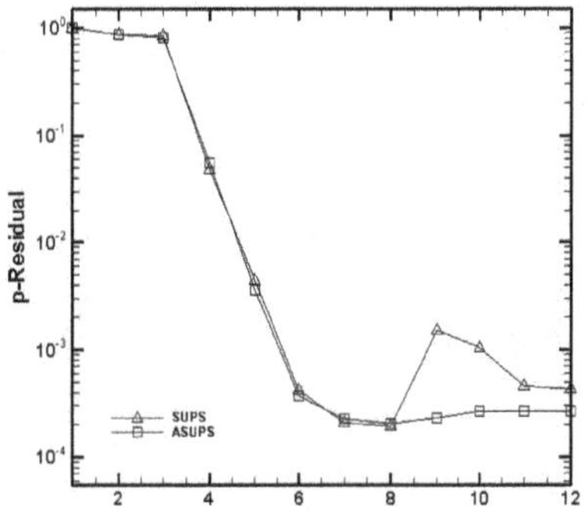


Figure 9. p-convergence

### 8.1.3 Flow Past a Square Obstacle

To further test robustness of the Inexact Newton Krylov algorithm for solving fluid flow problems in contraction regions we present the solutions for the problem of a square obstacle in a thin channel. We consider laminar incompressible flow for steady state analysis. The domain of interest was specified as a channel of length 6.0 units and height of 1 unit. We prescribe a square obstruction of length 0.40 and height 0.40 units, situated at x=1.4 inside the channel. The density of the fluid was taken as 1, with the fluid viscosity of 0.01. We specify uniform flow at the inlet of $u = 1$ and $v = 0$. Along the side and bottom walls and along the obstacle no-slip conditions of $u = v = 0$ was imposed. At the outflow boundary, the use of traction-free boundary conditions (zero normal and tangential stress) were imposed. The Reynolds number was considered based on the inlet height of the channel (taken as unity). We prescribed two different boundary conditions on the top wall for the problem. The first (BCI) entailed the specification of $u = 0$ and $v = 0$ for the case of a stationary top wall. The second boundary condition (BCII) allowed the top wall to move at a uniform velocity of $u = 1$ and $v = 0$. The second boundary condition (BCII) also tested the formulation for accurate resolution of the boundary layers for the top wall in motion. Both boundary value problems (BCI and BCII) were solved with the linear and non-linear solver tolerance set to $10^{-4}$.

Further, the length of the downstream reattachment point was found to be of larger length for the moving wall boundary condition. Figure 10 presents the development of the $u$ and $v$ contour plots for the stationary top wall. The development of the pressures and streamlines over the domain for this problem (BCI) are presented in Figure 11. As can be seen from Figure 11 the length of the downstream reattachment point was found to be of length 3.5 units. This is in agreement with the steady state solution of the problem solved in Laval (Laval and Quartapelle, 1990). It should be mentioned that in the above reference the upstream separation point and upstream recirculation region was not perceptible. However, due to the usage of spectral approximations we obtain more accurate resolution of the flow field in present simulation with perceptible development of a secondary circulation region upstream. The problem required from 1460 to 5452 iterations for reaching the linear convergence tolerance specified as $10^{-4}$ for the discrete linearised system generated solved with EBE-BJCGSTAB. The inexact Newton Krylov solved required from between 3-4 iterations. In the problem subject to different boundary conditions it was realized that most of the flow fields of interest occur at the sections of $x = 3$ and $x = 6$ respectively. Figure 12 presents the $U$ component of the velocities at the sections of $x = 3$ and $x = 6$ respectively for the stationary top wall with Reynolds number of 100. Figure 13 presents the contour plot of the $U$ and $V$ component of the velocity of the fluid when the top wall is set in motion with an $X$-component of the velocity 1. The development of the pressure and the streamline for this second boundary condition (BCII) is presented in Figure 14. From Figure 14 the length of the reattachment point downstream was found to be 4.0 units. Figure 15 presents the $U$ component of the velocity at the section $x = 3$ and $x = 6$ respectively for BCII with the top wall in motion.

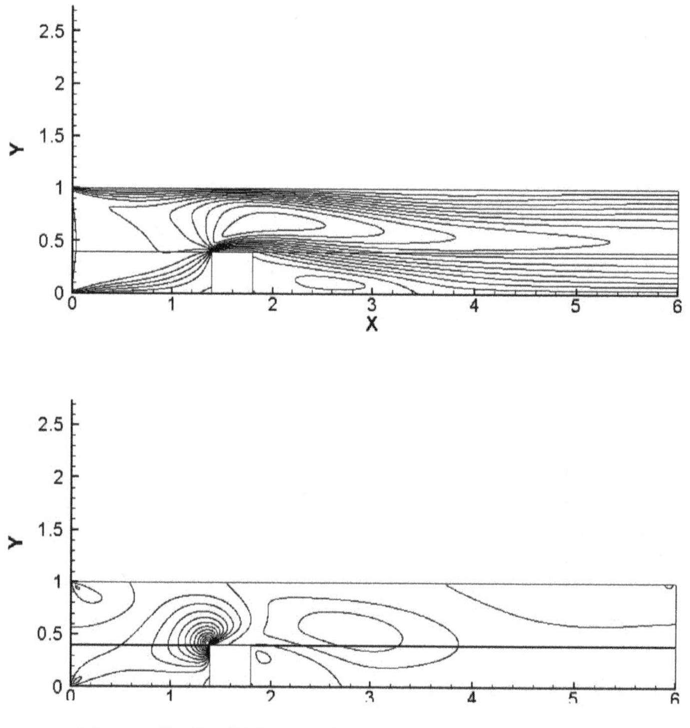


Figure 10. Re-100 $U$ and $V$ contour plots with No Slip.ps

8.1.4 Unsteady Flow Past a Square Cylinder

Let us consider two dimensional unsteady flow past a square obstruction. The Reynolds number is based on the free-stream velocity and side length of the square obstruction. The square cylinder of unit length ($L = 1.0$) is placed in a finite region $\bar{\Omega} = [0, 28] \times [0, 16]$. The center of the cylinder lies at $(x, y) = [8, 8]$. Thus the incoming flow is separated at a distance of $8L$ from the inlet of the channel. The square cylinder is situated symmetrically with respect to the $y$-coordinate.

The boundary conditions includes specification of the free stream velocity at the inlet of the channel of $u = 1$, and $v = 0$. At the top and bottom boundaries free stream velocity is imposed with a specification of $u = 1$, and $v = 0$. We consider the problem for a Reynolds number of Re-100 and set the regularization constant $\epsilon = 0.03$. In general we have found the iterative regularization constant to lie in the range of 0.01-0.05. This is also in conformity with recommendations for this constant found in literature (Dai and Cheng, 2008; Lin et al, 2004; Xiao-liang and Shaikh, 2006).

The outflow boundary condition include specification of traction free surfaces $t_x = 0$ and $t_y = 0$. The specification of pressure $p = 0$ completes the problem definition by anchoring the pressure for the simulations. The problem was discretized into a macro-mesh with 50 spectral elements along the x-and y-directions respectively. The connected model $\Omega^{hp}$ comprised of spectral elements with a uniform $p_{level}$ of 4 in each element. Previous numerical studies reported

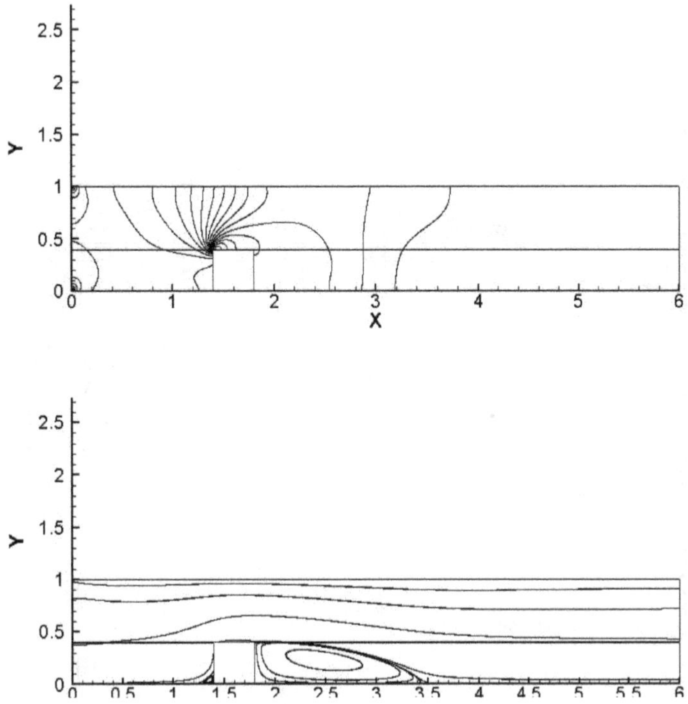


Figure 11. Re-100 Pressure and Streamlines with No Slip.ps

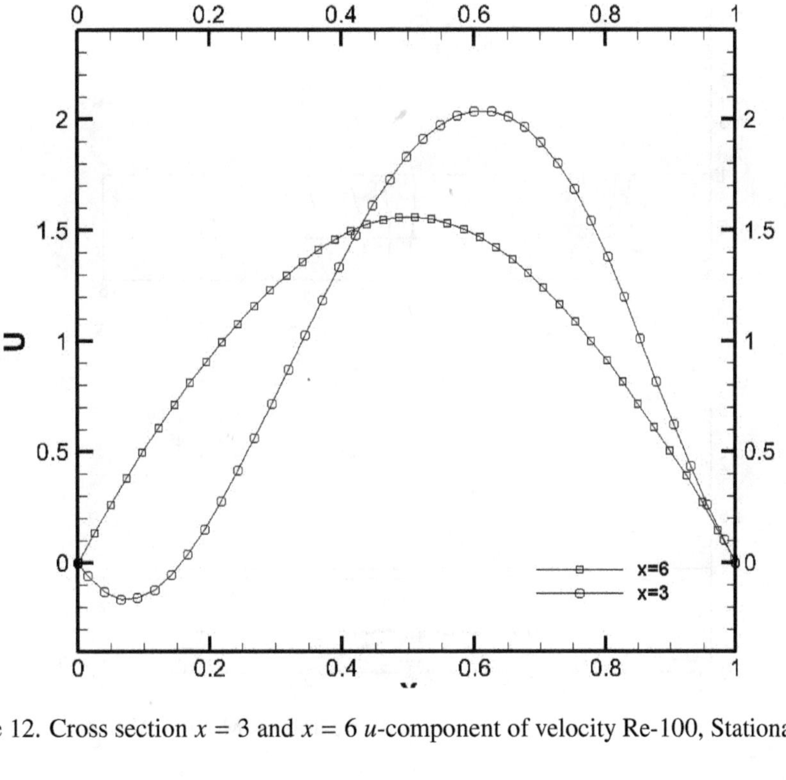


Figure 12. Cross section $x = 3$ and $x = 6$ $u$-component of velocity Re-100, Stationary Wall

significant dependence of the transient flow structures on the spatial resolution especially in the near-field of the square cylinder (Sohankar et al, 1998).

To address this issue we refine the mesh in the near vicinity of the square cylinder. To study the unsteady nature of the problem we employ a space-time decoupled finite element formulation. The temporal terms in the governing equations are represented by the generalized $\alpha$ family of approximation, which retain second order accuracy in time and allow for user controlled high-frequency damping by the single free integration parameter, $0.0 \leq \rho_{\infty}^{hp} \leq 1$. For $\rho_{\infty}^{hp} = 1$ the method

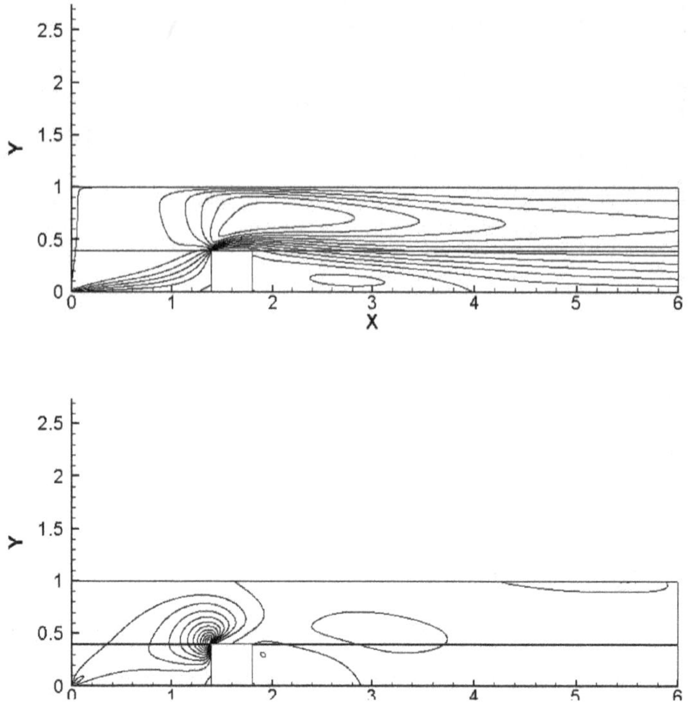


Figure 13. Re-100 $U$ and $V$ contour plots Moving Wall

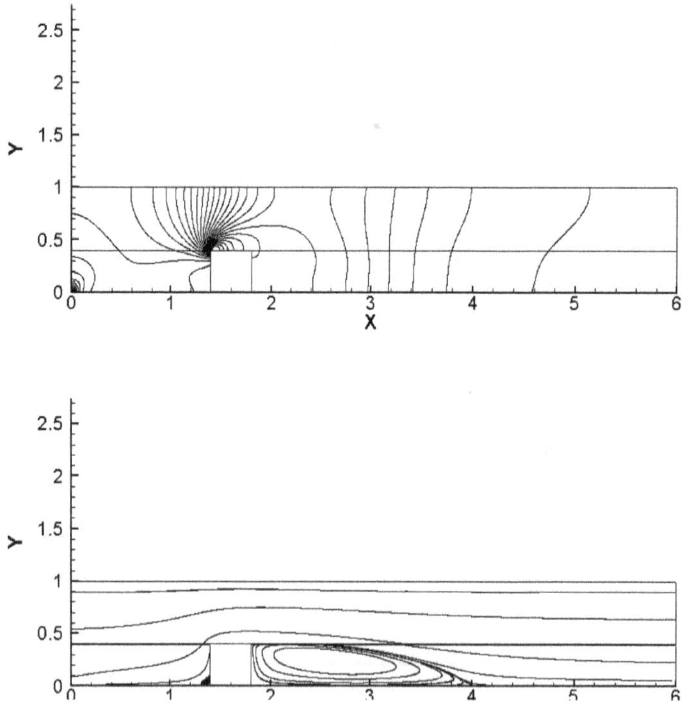


Figure 14. Re-100 Pressure and Streamlines Moving Wall

is identical to the trapezoidal rule. For the results presented, the choices $\rho_\infty^{hp} = 0.333$ and $\Delta t = 0.10$ were made. Even though the formulation is stable for larger time steps smaller time steps are required for accuracy reasons.

Figure 16 presents the $U$-component of the velocity for the flow past a square shaped cylinder once the unsteady nature of the flow is fully developed. The development of the Von-Karman vortex street is evident. The $V$-component of the velocity is shown in Figure 17. The time history of the lift coefficient has been presented in Figure 18. The lift coefficient exhibits a fluctuating component with the amplitude of 0.20. The transient evolution of the $U$, $V$-velocity component with

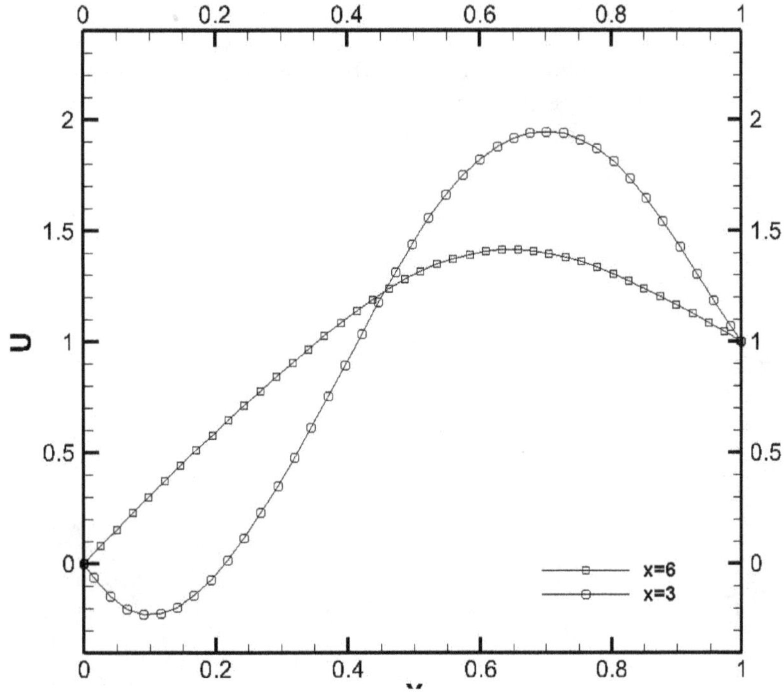


Figure 15. Cross section $x = 3$ and $x = 6$ $u$-component of velocity Re-100, Moving wall

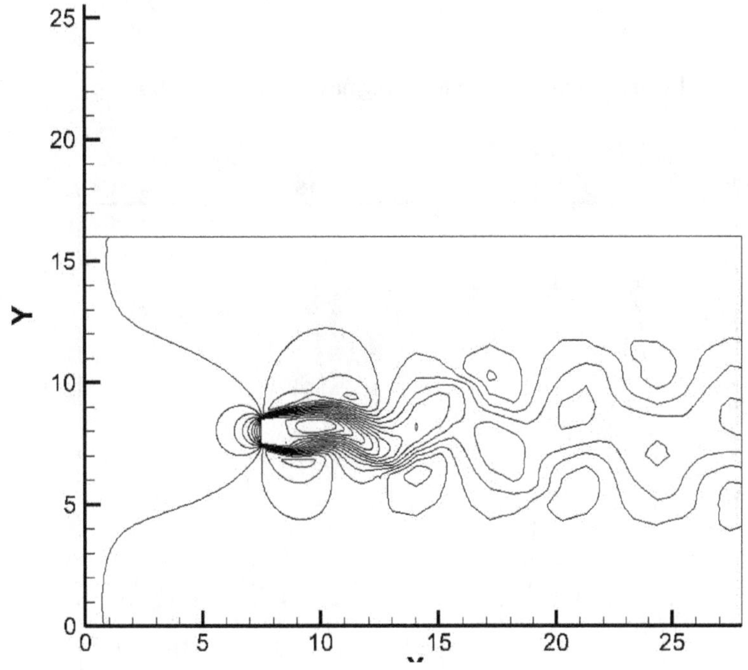


Figure 16. U-contour plots flow past a square Re-100

time at the location of a probe set at [10, 8] is presented in Figure 19. From the history of the lift coefficient we obtain a time period of vortex shedding of 7.4 time units. This time period allowed the transient evolution of the problem through 74 time steps for one shedding cycle. The time period provides a shedding frequency of 0.135. This shedding frequency is in agreement with the results presented in Sohankar et al. (1998) who reported a value of $St = 0.139$ and also Pavlov et al. (2000) who reported a value of 0.137. For checking the predictive capabilities of the formulation we verify the conservation of mass at the exit plane. The mass conservation obtained at the exit was divided by the inlet mass to obtain the ratio $m^{out}/m^{in}$. It is well known that most finite element formulations have issues with mass conservation. From figure 20 we notice the mass conservation is also periodic in nature with a very small amplitude. The mean value of the exit mass of the fluid vs. the inlet mass was found to be 0.998. This slight difference between the input and output mass can be also attributed to the inexact solve of the non-linear problem. It is evident there is negligible errors in conservation of mass which further verifies the predictive capabilities of the ASUPS formulation.

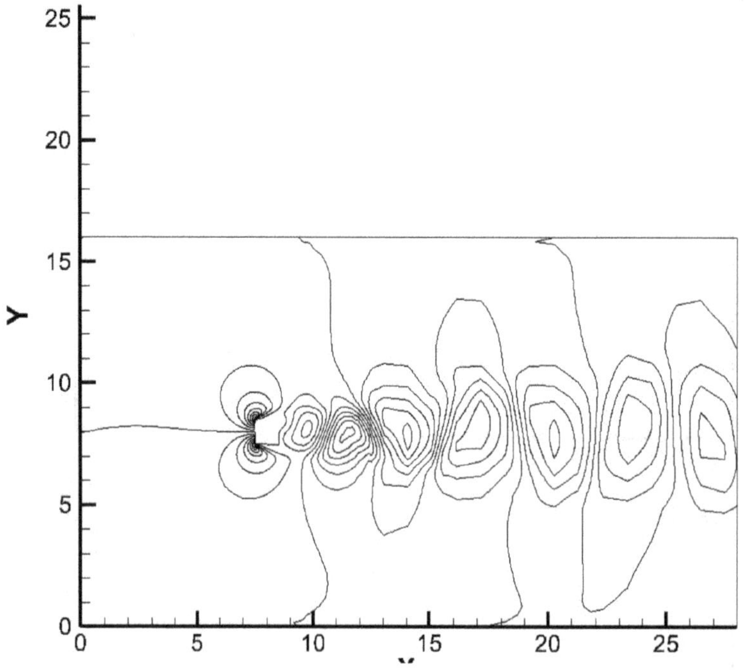


Figure 17. V-contour plots flow past a square Re-100

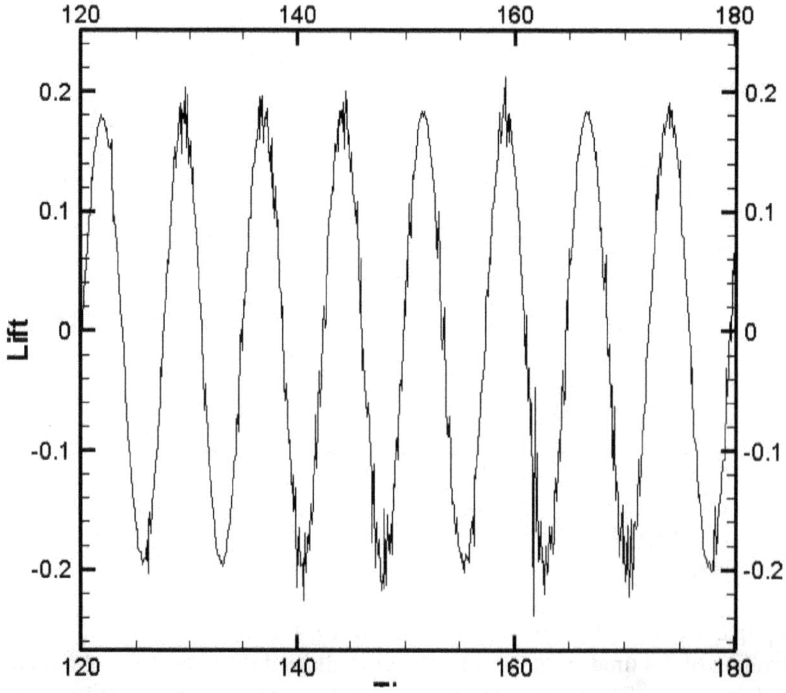


Figure 18. Fluctuating Lift Coefficient History

*8.2 AGLS Formulation*

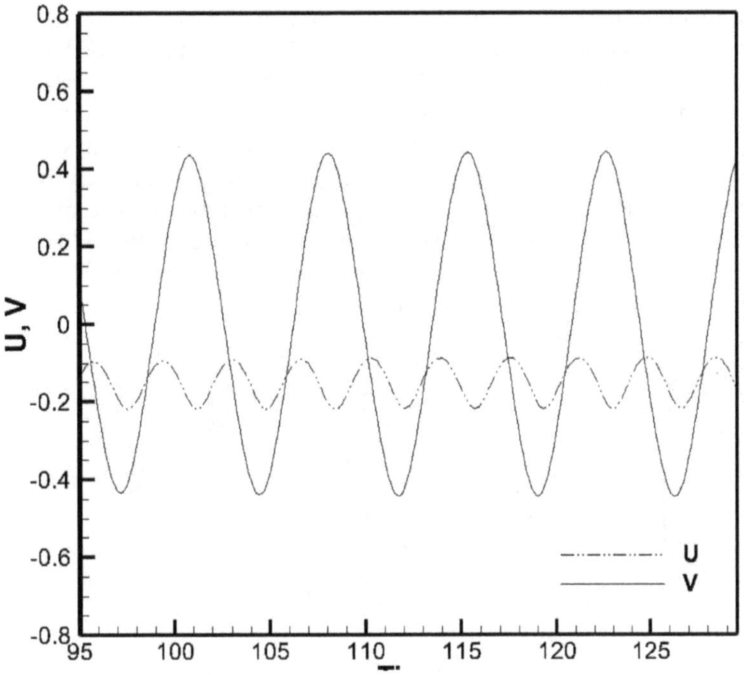


Figure 19. Time history of velocity at probe [10,8]

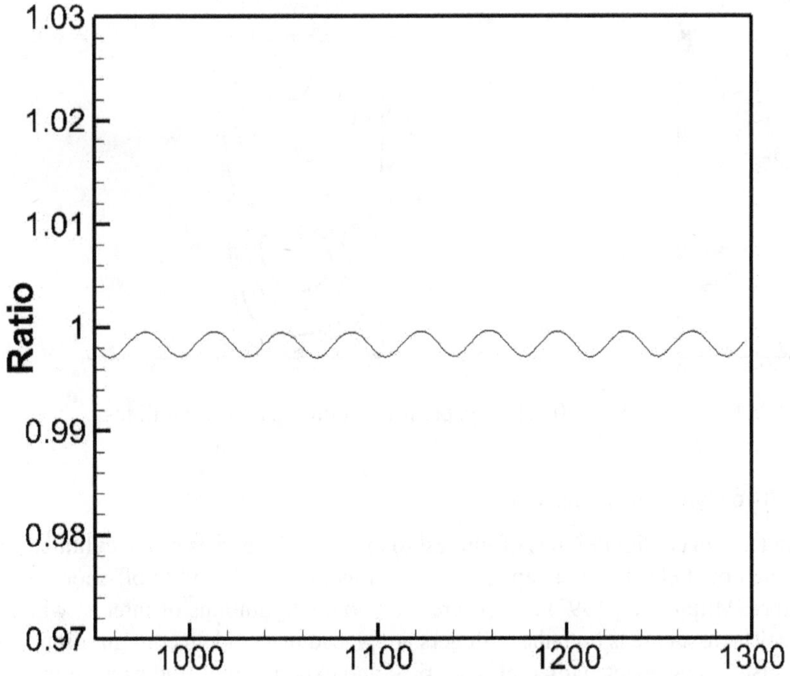


Figure 20. Mass conservation exit plane

In the following section we present numerical solutions obtained with the AGLS formulation. Both steady state and unsteady state examples are solved and validated.

8.2.1 Driven Cavity Aspect Ratio AR=2

Let us consider a benchmark problem in two dimensional incompressible flow computations, the driven cavity problem with an aspect ratio of 2. The specification of the problem is standard with the top surface of the cavity set in motion with an initial velocity of $u = 1$ and $v = 0$. No slip and no-penetration velocity boundary conditions are specified on the rest of the walls. A specification of the pressure to zero on the mid-bottom of the cavity completes the description of the problem. We consider the problem with an aspect ratio of AS=2 at a Reynolds number of 100. The dimensions of the cavity were taken as a unit length along the X direction and 2 units along Y. Reynolds number was based on the length of the cavity along the x-direction and the velocity of the driven surface. The problem was discretized into a macro-mesh of $50 \times 60$ elements with a $p_{level}$ of 3 in each element. A graded mesh was considered for analysis with a gradation towards the edges for capturing the boundary layers. The problem was subject to steady state analysis. Figure 21 presents the contour plot of the velocity and pressures over the domain for the problem. From the streamline plots it is evident two circulation zones develop inside the cavity. The problem was solved for two different Reynolds numbers of 100 and 400. The agreement of the centerline velocities at the mid-centerline are presented in Figure 22. The agreement of the present results with benchmark results of Gupta et al. (2005) is excellent.

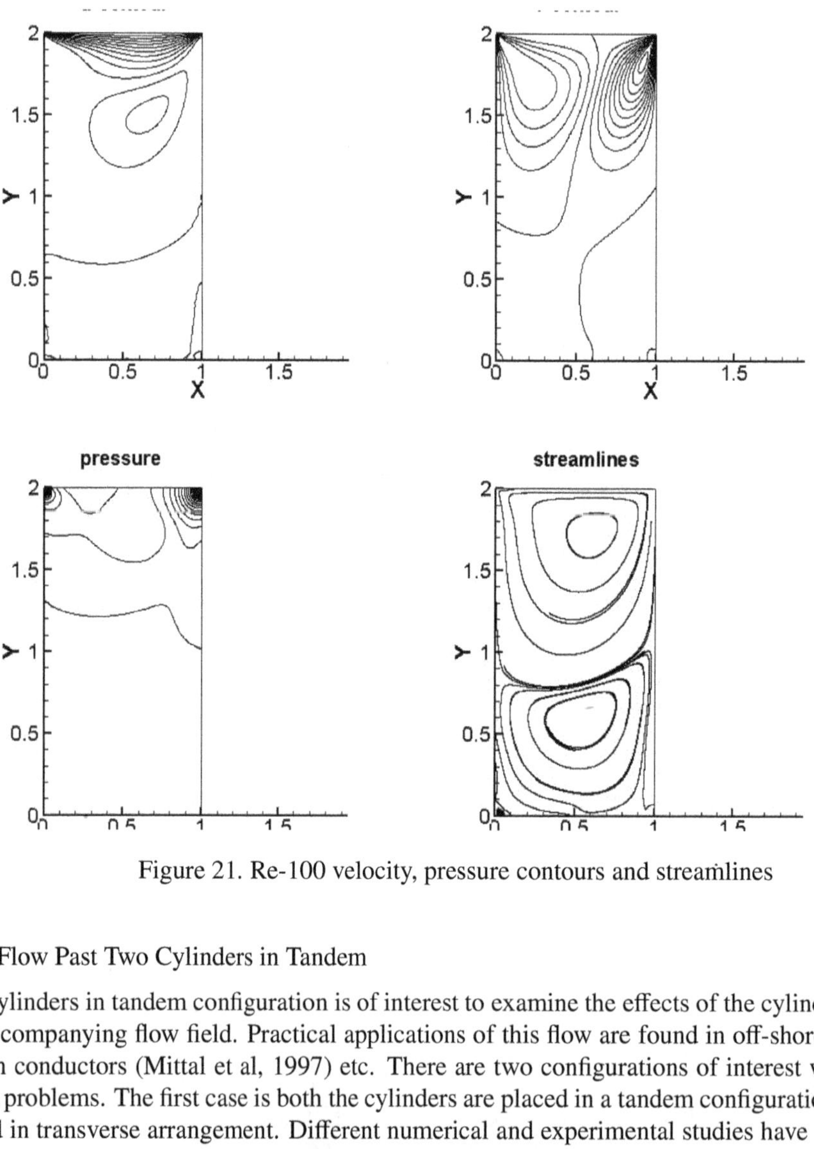


Figure 21. Re-100 velocity, pressure contours and streamlines

8.2.2 Unsteady Flow Past Two Cylinders in Tandem

Flow past two cylinders in tandem configuration is of interest to examine the effects of the cylinder interactions with each other and the accompanying flow field. Practical applications of this flow are found in off-shore structures, transmission cables with twin conductors (Mittal et al, 1997) etc. There are two configurations of interest while studying such solid fluid interaction problems. The first case is both the cylinders are placed in a tandem configuration, and the second is both cylinders placed in transverse arrangement. Different numerical and experimental studies have involved solving the flow past two cylinders in transverse configuration (Williamson, 1985). However flow past two cylinders inline with each other (in tandem) is relatively scarce. This problem has been examined in Mittal (Mittal et al, 1997) in detail. They considered the problem with a belt type boundary condition. Initial steady state flow field was perturbed with a clockwise and counter clockwise motion of the cylinders and the flow was allowed to develop.

In the following example we consider flow past two cylinders in tandem configuration without any initial perturbation of the flow field. This problem was carefully chosen as a sample problem to examine the robustness of the formulation for long time integration with the AGLS formulation with previously published results. We examine the flow interactions of the cylinders with each other and adjacent flow. The intent with an unperturbed initial condition was to be able to recover the unsteady solution obtained earlier for a similar problem. The diameters of the cylinders are denoted by $D$. The

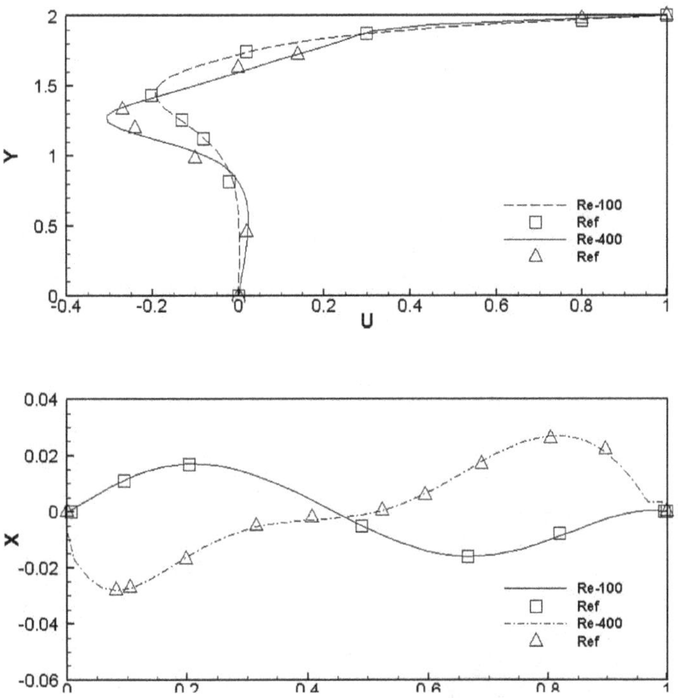


Figure 22. $x$ and $y$ component of velocities with benchmark results for $Re$ – 100 and 400

distance between the centers of both cylinders is denoted by $S$. It has been observed for cylinders in tandem configuration there exists a critical cylinder distance $S$ between the cylinders below which there is no distinct shedding observed behind the first cylinder. Even at this separation however there is formation of a vortex street behind the second cylinder. This distance has been reported as $S=3D$. We present the flow metrics observed for the cylinder separation distances of $5.0D$. For smaller distances lesser than $3D$ the flow perturbations at Reynolds number of 100 tend to reach steady state.

The problem domain was considered to be of length [32, 32]. The center of the first cylinder was placed at a separation of 5 units from the upstream boundary at the location [5, 16]. The second cylinder was placed at [10, 16]. The diameters of both cylinders were taken as 1. Free stream values of the velocity were specified on the top, bottom, and left faces of the computational domain. The exit face was specified traction free. The Reynolds number based on the diameter of the cylinders and the free stream value of the velocity was considered as Re-100. Figure 23 presents the contour plot of the velocity components at a specific time instant for the two cylinders in tandem. From the animation generated from the simulation the cylinders were found to be in anti-phase mode of vortex shedding. Contour plot of the pressure field has been presented in Figure 24. The obstruction in the flow downstream of the first cylinder prohibits an early development of the shedding phenomenon. For a single cylinder in free stream the Von-Karman vortex street is fully developed by the time instant t-50. However for the present case the development starts at around t-90. The values of the amplitude of the lift coefficient (Figure 25) and the Strouhal number for the first cylinder were found be close to the values for a single cylinder. The second cylinder which lies in the wake of the first cylinder was found to be significantly affected due to the presence of the first cylinder. The amplitudes of the lift coefficients for both cylinders were found be in agreement with the values presented for an initially perturbed flow condition by Mittal et al. (1997).

## 9. Conclusion

Augmented SUPS and Augmented Galerkin Least Squares formulations with iterative penalization of the incompressibility constraint were proposed for solving incompressible Navier- Stokes equations. Stability analysis was performed and better non-linear convergence history of the ASUPS formulation was demonstrated. The formulations were used for solving both steady state and unsteady state incompressible flow problems. Higher order/spectral methods provided the flexibility of obtaining exponentially accurate results with high $p_{levels}$ of resolutions on relatively coarse macro-meshes. We utilize inexact Newton Krylov methods for solving problems. No deterioration of results are observed for flow problems in confined enclosures with different boundary conditions. Excellent conservation of mass at the exit plane further verify the predictive capabilities of the formulations presented and the robustness of the solution algorithms.

**References**

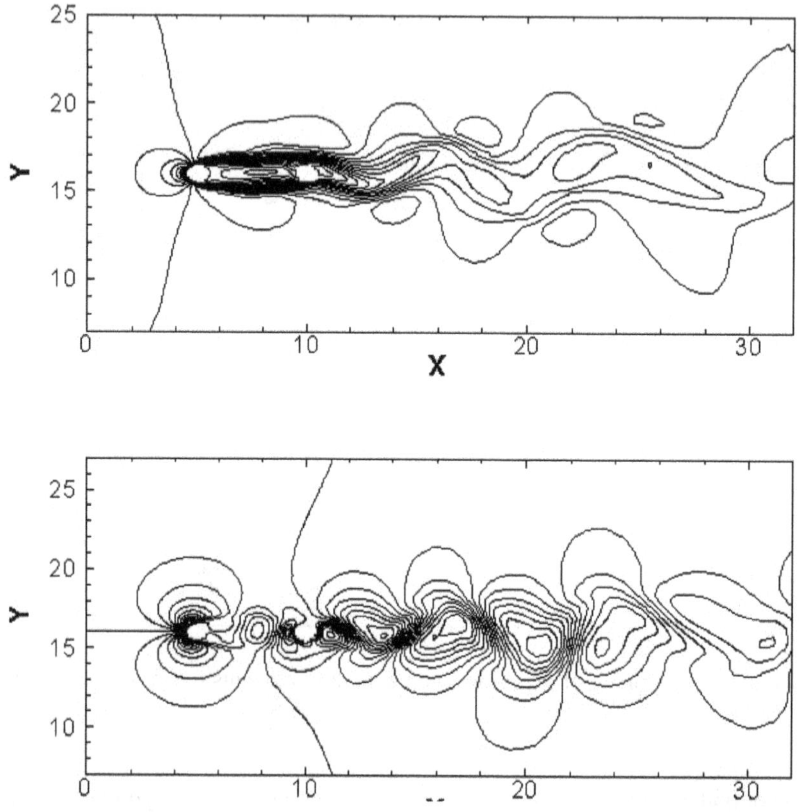


Figure 23. Contour plot of $U$ velocity for two cylinders in Tandem

Bazilevs, Y., Takizawa, K., & Tezduyar, T. E. (2012). Computational fluid-structure interaction: methods and applications. John Wiley & Sons.

Brooks, A., & Hughes, T. J. (1980). *Streamline upwind/petrov-galerkin methods for advection dominated flows.* In: Third Internat. Conf. of Finite Element Methods in Fluid Flow, Banff, Canada

Brooks, A. N., & Hughes, T. J. (1982). Streamline upwind/petrov-galerkin formulations for convection dominated flows with particular emphasis on the incompressible navier-stokes equations. *Computer methods in applied mechanics and engineering, 32*(1), 199-259.

Carey, G., & Krishnan, R. (1982). Penalty approximation of stokes flow. *Computer Methods in Applied Mechanics and Engineering, 35*(2), 169-206.

Ciarlet, P. G. (2002). The Finite Element Method for Elliptic Problems. *SIAM, 40.* https:/doi.org/10.1137/1.9780898719208

Codina R, Schäafer, U., Oñate, E. (1994). Mould filling simulation using finite elements. *International Journal of Numerical Methods for Heat & Fluid Flow, 4*(4), 291-310. https:/doi.org/10.1108/EUM0000000004108

Codina Rovira, R., et al. (1992). A finite element model for incompressible flous problems.

Dai, X., & Cheng, X. (2008). The iterative penalty method for stokes equations using q1-p0 element. *Applied Mathematics and Computation, 201*(1), 805-810.

Dembo, R, Eisenstat, S, Steihaug, T. (1982). Inexact newton methods. *SIAM Journal on Numerical analysis, 19*(2), 400-408. https:/doi.org/10.1137/0719025

Demirdžić, I., Lilek, Ž., Perić, M. (2005). Fluid flow and heat transfer test problems for non-orthogonal grids: Benchmark solutions. *International Journal for Numerical Methods in Fluids, 15*(3), 329-354.

Eisenstat, S., & Walker, H. (1996). Choosing the forcing terms in an inexact newton method. *SIAM Journal on Scientific Computing, 17*(1), 16-32. https:/doi.org/10.1137/0917003

Elias, R., Coutinho, A., & Martins, M. (2004). Inexact newton-type methods for non-linear problems arising from the supg/pspg solution of steady incompressible navier-stokes equations. *Journal of the Brazilian Society of Mechanical*

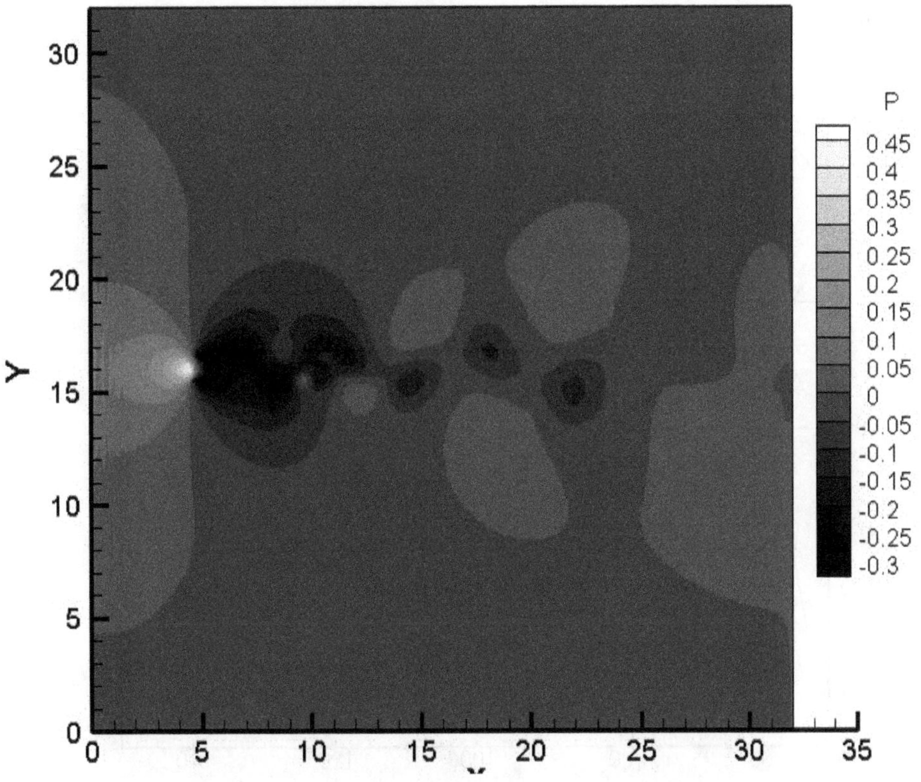


Figure 24. Pressure contours for Two Cylinders S/D=5

*Sciences and Engineering, 26*(3), 330-339.

Elias, R., Martins, M., & Coutinho, A. (2006). Parallel edge-based solution of viscoplastic flows with the supg/pspg formulation. *Computational Mechanics, 38*(4), 365-381. https:/doi.org/10.1007/s00466-005-0012-y

Ern, A., Guermond, J. L. (2004). Theory and practice of finite elements. *Applied mathematical sciences, 159.* https:/doi.org/10.1007/978-1-4757-4355-5

Gervasio, P., & Saleri, F. (1998). Stabilized spectral element approximation for the navier-stokes equations. *Numerical Methods for Partial Differential Equations, 14*(1), 115-141. https:/doi.org/10.1002/(SICI)1098-2426(199801)14:1<115::AID-NUM7>3.0.CO;2-T

Ghia, U., Ghia, K., & Shin, C. (1982). High-re solutions for incompressible flow using the navier-stokes equations and a multigrid method. *Journal of computational physics, 48*(3), 387-411. https:/doi.org/10.1016/0021-9991(82)90058-4

Gupta, M. M., & Kalita, J. C. (2005). A new paradigm for solving navier-stokes equations: streamfunction-velocity formulation. *Journal of Computational Physics, 207*(1), 52-68. https:/doi.org/10.1016/j.jcp.2005.01.002

Hughes, T. J., & Brooks, A. (1979). A multidimensional upwind scheme with no crosswind diffusion. *Finite element methods for convection dominated flows, AMD 34*, 19-35.

Hughes, T. J., & Brooks, A. (1982). A theoretical framework for petrov-galerkin methods with discontinuous weighting functions: Application to the streamline-upwind procedure. *Finite elements in fluids, 4*, 47-65.

Hughes, T. J., Franca, L. P., & Balestra, M. (1986). A new finite element formulation for computational fluid dynamics: V. circumventing the babuška-brezzi condition: A stable petrov-galerkin formulation of the stokes problem accommodating equal-order interpolations. *Computer Methods in Applied Mechanics and Engineering, 59*(1), 85-99.

Jansen, K. E., Whiting, C. H., & Hulbert, G. M. (2000). A generalized-alpha method for integrating the filtered navier-stokes equations with a stabilized finite element method.

Johnson, C. (2012). Numerical solution of partial differential equations by the finite element method. Courier Corporation.

Karniadakis, G., & Sherwin, S. (1999). Spectral/hp element methods for CFD. Oxford University Press, USA.

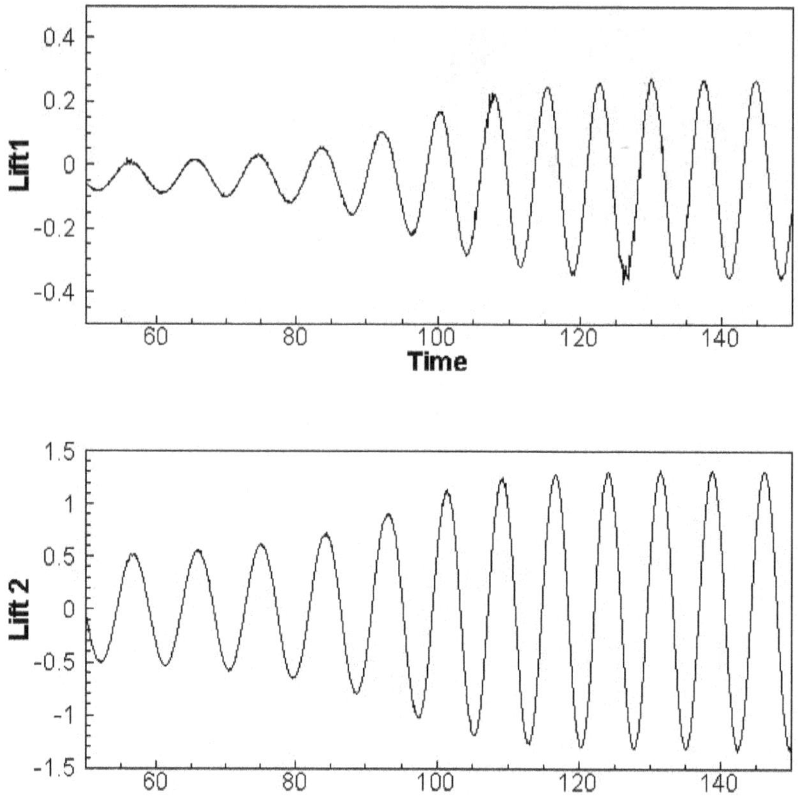


Figure 25. Lift on Two Cylinders in Tandem S/D=5

Kovasznay, L. (1948). Laminar flow behind a two-dimensional grid. In: *Proc. Camb. Philos. Soc,* Cambridge Univ Press, *44*, 58-62. https:/doi.org/10.1017/s0305004100023999

Laval, H., & Quartapelle, L. (1990). A fractional-step taylor-galerkin method for unsteady incompressible flows. *International Journal for Numerical Methods in Fluids, 11*(5), 501-513. https:/doi.org/10.1002/fld.1650110504

Lin, P. (1997). A sequential regularization method for time-dependent incompressible navier-stokes equations. *SIAM Journal on Numerical Analysis, 34*(3), 1051-1071. https:/doi.org/10.1137/S0036142994270521

Lin, P., Chen, X., & Ong, M. T. (2004). Finite element methods based on a new formulation for the non-stationary incompressible navierstokes equations. *International journal for numerical methods in fluids, 46*(12), 1169- 1180. https:/doi.org/10.1002/fld.794

Mittal, S., Kumar, V., & Raghuvanshi, A. (1997). Unsteady incompressible flows past two cylinders in tandem and staggered arrangements. *International Journal for Numerical Methods in Fluids, 25*(11), 1315-1344. https:/doi.org/10.1002/(SICI)1097-0363(19971215)25:11<1315::AID-FLD617>3.0.CO;2-P

Oden, J. T., & Reddy, J. N. (2012). An introduction to the mathematical theory of finite elements. Courier Corporation.

Oden, J. T., Kikuchi, N., & Song, Y. J. (1982). Penalty-finite element methods for the analysis of stokesian flows. *Computer Methods in Applied Mechanics and Engineering, 31*(3), 297-329. https:/doi.org/10.1016/0045-7825(82)90010-X

Pavlov, A., Sazhin, S., Fedorenko, R., & Heikal, M. (2000). A conservative finite difference method and its application for the analysis of a transient flow around a square prism. *International Journal of Numerical Methods for Heat & Fluid Flow, 10*(1), 6-47. https:/doi.org/10.1108/09615530010306894

Pontaza, J., & Reddy, J. (2003). Spectral hp least-squares finite element formulation for the navier-stokes equations. *Journal of Computational Physics, 190*(2), 523-549. https:/doi.org/10.1016/S0021-9991(03)00296-1

Ranjan, R. (2010). Hp-spectral methods for structural mechanics and fluid dynamics problems. Texas A&M University.

Ranjan, R. (2011). Nonlinear finite element analysis of bending of straight beams using hp-spectral approximations.

*Journal of Solid Mechanics, 3*(1), 96-113.

Ranjan, R., & Reddy, J. (2009). Hp-spectral finite element analysis of shear deformable beams and plates. *Journal of Solid Mechanics, 1*(3), 245-259.

Ranjan, R., Chronopoulos, A. T., & Feng, Y. (2016a). Computational algorithms for solving spectral/hp stabilized incompressible flow problems. *Journal of Mathematics Research, 8*(4), 21.

Ranjan, R., Feng, Y., & Chronopoulos, A. (2016b). Procedures for solving spectral/hp stabilized incompressible flow problems. University of Texas, San Antonio, Department of Computer Science, Technical Report.

Ranjan, R., Feng, Y., & Chronopoulos, A. (2016c). Stabilized and galerkin least squares formulations. University of Texas, San Antonio, Department of Computer Science, Technical Report.

Reddy, J. (1982). On penalty function methods in the finite-element analysis of flow problems. *International Journal for Numerical Methods in Fluids, 2*(2), 151-171. https:/doi.org/10.1002/fld.1650020204

Shakib, F. (1989). Finite element analysis of the compressible euler and navier-stokes equations.

Shakib, F., Hughes, T. J., & Johan, Z. (1991). A new finite element formulation for computational fluid dynamics: X. the compressible euler and navier-stokes equations. *Computer Methods in Applied Mechanics and Engineering, 89*(1), 141-219. https:/doi.org/10.1016/0045-7825(91)90041-4

Sohankar, A., Norberg, C., & Davidson, L. (1998). Low-reynolds-number flow around a square cylinder at incidence: study of blockage, onset of vortex shedding and outlet boundary condition. *International Journal for Numerical Methods in Fluids, 26*(1), 39-56. https:/doi.org/10.1002/(SICI)1097-0363(19980115)26:1<39::AID-FLD623>3.0.CO;2-P

Stenberg, R. (1990). Error analysis of some finite element methods for the stokes problem. *Mathematics of Computation, 54*(190), 495-508. https:/doi.org/10.1090/S0025-5718-1990-1010601-X

Tezduyar, T. (1992). Stabilized finite element formulations for incompressible flow computations. *Advances in applied mechanics, 28*(1), 1-44.

Tezduyar, T., & Sathe, S. (2003). Stabilization parameters in supg and pspg formulations. *Journal of computational and applied mechanics, 4*(1), 71-88.

Williamson, C. (1985). Evolution of a single wake behind a pair of bluff bodies. *Journal of Fluid Mechanics, 159*, 1-18. https:/doi.org/10.1017/S002211208500307X

Xiao-liang, C., & Shaikh, A. W. (2006). Analysis of the iterative penalty method for the stokes equations. *Applied mathematics letters, 19*(10), 1024-1028. https:/doi.org/10.1016/j.aml.2005.10.021

9

# Existence of Positive Solutions of Fisher-KPP Equations in Locally Spatially Variational Habitat with Hybrid Dispersal

Liang Kong[1]

[1] Department of Mathematical Sciences, University of Illinois at Springfield, Springfield, Illinois

Correspondence: Liang Kong, Department of Mathematical Sciences, University of Illinois at Springfield, Springfield, Illinois, United States. E-mail: lkong9@uis.edu

**Abstract**

The current paper investigate the persistence of positive solutions of KPP type evolution equations with random/nonlocal dispersal in locally spatially inhomogeneous habitat. By the constructions of super/sub solutions and comparison principle, we prove that such an equation has a unique globally stable positive stationary solution.

**Keywords:** KPP equations, random dispersal, nonlocal dispersal, localized spatial inhomogeneity, positive stationary solution, principal eigenvalue, sub-solution, super-solution, comparison principle

**Mathematics subject classification.** 35K57, 45G10, 58D20, 92D25.

**1. Introduction**

The current paper is concerned with persistence of species in locally spatially variational environments or habitat,

$$u_t(t,x) = \mathcal{A}u(t,x) + u(t,x)f(x,u(t,x)), \quad x \in \mathbb{R}^N \tag{1.1}$$

where $(\mathcal{A}u)(t,x) = \nu_1 \Delta u + \nu_2[\int_{\mathbb{R}^N} \kappa(y-x)u(t,y)dy - u(t,x)]$, $\nu_1, \nu_2 \geq 0$ and $\nu_1 + \nu_2 > 0$.

If $\nu_2 = 0$, then (1.1) is the classical reaction diffusion equation, so called random dispersal equation, (without loss of generality, let $\nu_1 = 1$)

$$u_t(t,x) = \Delta u(t,x) + u(t,x)f(x,u(t,x)), \quad x \in \mathbb{R}^N \tag{1.2}$$

which is broadly used to model the population dynamics of many species in unbounded environments, where $u(t,x)$ is the population density of the species at time $t$ and location $x$, $\Delta u$ characterizes the internal interaction of the organisms, and $f(x,u)$ represents the growth rate of the population, which satisfies that $f(x,u) < 0$ for $u \gg 1$ and $\partial_u f(x,u) < 0$ for $u \geq 0$ (see Aronson & Weinberger, 1957; Aronson & Weinberger, 1978; Cantrell & Cosner, 2003; Fife, 1979; Fife & Peletier, 1977; Fisher, 1937; Kolmogorov, Petrowsky, & Poscunov, 1937; Murray, 1989; Shigesada & Kawasaki, 1997; Skellam, 1951; Weinberger, 1982; Weinberger, 2002; Zhao, 2003; etc.).

If $\nu_1 = 0$, then (1.1) is so called nonlocal dispersal equation, (without loss of generality, let $\nu_2 = 1$)

$$u_t(t,x) = \int_{\mathbb{R}^N} \kappa(y-x)u(t,y)dy - u(t,x) + u(t,x)f(x,u(t,x)), \quad x \in \mathbb{R}^N, \tag{1.3}$$

where $\kappa(\cdot)$ is a smooth convolution kernel supported on a ball centered at the origin (that is, there is a $\delta_0 > 0$ such that $\kappa(z) > 0$ if $\|z\| < \delta_0$, $\kappa(z) = 0$ if $\|z\| \geq \delta_0$, where $\|\cdot\|$ denotes the norm in $\mathbb{R}^N$ and $\delta_0$ represents the nonlocal dispersal distance), $\int_{\mathbb{R}^N} \kappa(z)dz = 1$, and $f(\cdot,\cdot)$ is of the same property as $f$ in (1.2) (see Bates & Zhao, 2007; Chasseigne, Chaves, & Rossi, 2006; Cortazar, Coville, Elgueta, & Martinez, 2007; Cousens, Dytham, & Law, 2008; Fife, 2003; Grinfeld, Hines, Hutson, Mischaikow, & Vickers, 2005; Hutson, Martinez, Mischaikow & Vickers, 2003; Lee, Hoopes, Diehl, Gilliland, Huxel, Leaver, McCain, Umbanhowar, & Mogilner, 2001; Levin, Muller-Landau, Nathan, & Chave, 2003; etc.).

When using (1.2) to model the population dynamics of a species, it is assumed that the underlying environment is continuous and the dispersal of cells or organisms are based on the hypothesis that the movement of the dispersing species can be described as a random walk in which there is no correlation between steps. However, dispersal of large organisms often involves mechanisms that may introduce correlations in movements. To model the population dynamics of such species in the case that the underlying environment is continuous, the nonlocal dispersal equation (1.3) is often used. This paper propose to study a mixed dispersal strategy, that is, a hybrid of random and non-local dispersal. We assume that a fraction of individuals in the population adopt random dispersal, while the rest fraction assumes non-local dispersal. Some research has been done on the hybrid dispersal in the spatially periodic habitat (see Kao, Lou & Shen, 2010; Kao, Lou & Shen, 2012; and Zhang, 2013). Our main goal is to study how the hybrid dispersal affects the persistence of a single species and how the hybrid dispersal strategies will evolve in spatially locally inhomogeneous environment (see H1 and H2).

Since the seminal works by Fisher (Fisher, 1937) and Kolmogorov, Petrowsky, Piscunov (Kolmogorov, Petrowsky, & Poscunov, 1937) on the following special case of (1.2),

$$\frac{\partial}{\partial t}u(t,x) = \frac{\partial^2}{\partial x^2}u(t,x) + u(t,x)(1-u(t,x)), \qquad x \in \mathbb{R}. \tag{1.4}$$

A great deal of research has been carried out toward the spatial spreading dynamics of (1.2) and (1.3) with $f(\cdot,\cdot)$ being independent of the space variable or periodic in the space variable, which reflects the spatial periodicity of the media. We refer to (Aronson & Weinberger, 1957; Aronson & Weinberger, 1978; Berestycki, Hamel, & Nadirashvili, 2010; Kametaka, 1976; Liang & Zhao, 2007; Liang, Yi, & Zhao, 2006; Sattinger, 1976; Uchiyama, 1978; Weinberger, 1982; etc). for the study of (1.2) in the case that $f(x,u)$ is independent of $x$ and refer to (Berestycki, Hamel, & Nadirashvili, 2005; Berestycki, Hamel, & Roques, 2005; Freidlin & Gärtner, 1979; Hamel, 2008; Hudson & Zinner, 1995; Nadin, 2009; Nolen, Rudd, & Xin, 2005; Nolen & Xin, 2005; Weinberger, 2002; etc). for the study of (1.2) in the case that $f(x,u)$ is periodic in $x$; refer to (Coville & Dupaigne, 2005; Coville, Dávila, & Martínez, 2008; Li, Sun, & Wang, 2010; etc). for the study of (1.3) in the case that $f(x,u)$ is independent of $x$ and refer to (Hetzer, Shen, & Zhang, 2013; Shen & Zhang, 2012; etc.) for the study of (1.3) in the case that $f(x,u)$ is periodic in $x$ and refer to (Berestycki & Nadin, 2016; Berestycki, Jin, & Silvestre, 2016; Kong & Shen, 2011; Kong & Shen, 2014; Nolen, Roquejoffre, & Ryzhik; Shen, 2011; etc). for the study of (1.2) and/or (1.3) in the case that $f(t,x,u)$ is temporally and/or spatially heterogeneous.

For example, consider (1.2) and assume that $f(x + p_i\mathbf{e_i}, u) = f(x,u)$ for $i = 1,2,\cdots,N$, where $p_i$ $(i = 1,2,\cdots,N)$ are positive constants and

$$\mathbf{e_i} = (\delta_{i1}, \delta_{i2}, \cdots, \delta_{iN}), \ \delta_{ij} = 1 \text{ if } i = j \text{ and } 0 \text{ if } i \neq j.$$

If the principal eigenvalue of the following eigenvalue problem associated to the linearized equation of (1.2) at $u = 0$,

$$\begin{cases} \Delta u(x) + f(x,0)u(x) = \lambda u(x), & x \in \mathbb{R}^N \\ u(x + p_i\mathbf{e_i}) = u(x), & x \in \mathbb{R}^N, \end{cases} \tag{1.5}$$

is positive, then (1.2) has a unique positive stationary solution $u^*(\cdot)$ with $u^*(\cdot + p_i\mathbf{e_i}) = u^*(\cdot)$.

In this paper, we consider (1.1) in the case that the growth rates depend on the space variable, but only when it is in some bounded subset of the underlying habitat, which reflects the localized spatial inhomogeneity of the media. More precisely, let $\mathcal{H} = \mathbb{R}^N$, we assume

**(H1)** *$f : \mathcal{H} \times \mathbb{R} \to \mathbb{R}$ is a $C^2$ function, $f(x,u) < 0$ for all $(x,u) \in \mathcal{H} \times \mathbb{R}^+$ with $u \geq \beta_0$ for some $\beta_0 > 0$, and $\partial_u f(x,u) < 0$ for all $(x,u) \in \mathcal{H} \times \mathbb{R}^+$.*

**(H2)** *$f(x,u) = f^0(u)$ for some $C^2$ function $f^0 : \mathbb{R} \to \mathbb{R}$ and all $(x,u) \in \mathcal{H} \times \mathbb{R}$ with $\|x\| \geq L_0$ for some $L_0 > 0$, and $f^0(0) > 0$.*

Assume (H1) and (H2). Then (1.1) has the following limit equations as $\|x\| \to \infty$,

$$u_t(t,x) = \mathcal{A}u(t,x) + u(t,x)f^0(u(t,x)), \quad x \in \mathbb{R}^N. \tag{1.6}$$

Equations (1.6) will play an important role in the study of (1.1). Equations (1.6) has a unique positive constant stationary solution $u^0$. We introduce some standing notations and then state the main results of the paper.

Let $p = (p_1, p_2, \cdots, p_N)$ with $p_i > 0$ for $i = 1,2,\cdots,N$. We define the Banach spaces $X_p$ by

$$X_p = \{u \in C(\mathbb{R}^N, \mathbb{R}) \,|\, u(\cdot + p_i\mathbf{e_i}) = u(\cdot), \quad i = 1, ..., N\} \tag{1.7}$$

with norm $\|u\|_{X_p} = \max_{x\in\mathbb{R}^N} |u(x)|$.

Let

$$X_p^+ = \{u \in X_p \,|\, u(x) \geq 0 \ \forall x \in \mathcal{H}\} \tag{1.8}$$

and

$$X_p^{++} = \{u \in X_p \,|\, u(x) > 0 \ \forall x \in \mathcal{H}\}. \tag{1.9}$$

We define $X$ by

$$X = \{u \in C(\mathbb{R}^N, \mathbb{R}) \,|\, u \text{ is uniformly continuous and bounded}\} \tag{1.10}$$

with norm $\|u\|_X = \sup_{x\in\mathbb{R}^N} |u(x)|$.

Let

$$X^+ = \{u \in X_i \,|\, u(x) \geq 0 \ \forall x \in \mathcal{H}\} \tag{1.11}$$

and

$$X^{++} = \{u \in X_i^+ \mid \inf_{x\in\mathcal{H}} u(x) > 0\}. \tag{1.12}$$

Without occurring confusion, we may write $\|\cdot\|_{X_p}$ and $\|\cdot\|_X$ as $\|\cdot\|$.

Assume (H1). By general semigroup theory (see Henry, 1981; Pazy, 1983), for any $u_0 \in X$, (1.2) has a unique local solution $u(t,\cdot;u_0)$ with $u(0,\cdot;u_0) = u_0(\cdot)$. Moreover, if $u_0 \in X^+$, then $u(t,\cdot;u_0)$ exist and $u(t,\cdot;u_0) \in X^+$ for all $t \geq 0$ (see Proposition 2.2).

Our objective is to explore the spatial spreading dynamics of (1.1) with localized spatial inhomogeneity. The main results of this paper are stated in the following two theorems.

**Theorem 1.1** (Positive stationary solutions). *Assume* (H1) *and* (H2). *Equation* (1.2) *has a unique stationary solution* $u = u^*(\cdot) \in X^{++}$. *Moreover,*

$$\lim_{r\to\infty} \sup_{x\in\mathcal{H},\|x\|\geq r} |u^*(x) - u^0| = 0,$$

*where* $u^0 > 0$ *is such that* $f^0(u^0) = 0$.

**Theorem 1.2** (Stability). *Assume* (H1) *and* (H2). *For any* $u_0 \in X^{++}$,

$$\lim_{t\to\infty} \|u(t,\cdot;u_0) - u^*(\cdot)\|_X = 0.$$

The rest of the paper is organized as follows. In section 2, we present some preliminary materials to be used in later sections. Section 3 is devoted to the study of positive stationary solutions of (1.1). Theorem 1.1 and Theorem 1.2 are proved in this section.

## 2. Preliminary

In this section, we present some preliminary materials to be used in later sections, including some basic properties of solutions of (1.1); principal eigenvalue theories for spatially periodic dispersal operators with random, and nonlocal; and spatial spreading dynamics of KPP equations in spatially periodic media.

### *2.1 Basic Properties of KPP Equations*

In this subsection, we present some basic properties of solutions of (1.1), including comparison principle, global existence, convergence in open compact topology, and decreasing of the so called part metric along the solutions. Throughout this subsection, we assume (H1).

Let $X$ be as in (1.10). For given $u_0 \in X$, let $u(t,\cdot;u_0)$ be the (local) solution of (1.2) with $u(0,\cdot;u_0) = u_0(\cdot)$.

Let $X^+$ and $X^{++}$ be as in (1.11) and (1.12). For given $u, v \in X$, we define

$$u \leq v \ (u \geq v) \quad \text{if } v - u \in X^+ \ (u - v \in X^+) \tag{2.1}$$

and

$$u \ll v \ (u \gg v) \quad \text{if } v - u \in X^{++} \ (u - v \in X^{++}). \tag{2.2}$$

For given continuous and bounded function $u : [0,T) \times \mathbb{R}^N \to \mathbb{R}$, it is called a *super-solution* (*sub-solution*) of (1.1) on $[0,T)$ if

$$u_t(t,x) \geq (\leq)\mathcal{A}u(t,x) + u(t,x)f(x,u(t,x)) \quad \forall (t,x) \in (0,T)\times\mathbb{R}^N.$$

**Proposition 2.1** (Comparison principle). *Assume* (H1).

*(1) Suppose that* $u^1(t,x)$ *and* $u^2(t,x)$ *are sub- and super-solutions of* (1.1) *on* $[0,T)$ *with* $u^1(0,\cdot) \leq u^2(0,\cdot)$. *Then* $u^1(t,\cdot) \leq u^2(t,\cdot)$ *for* $t \in (0,T)$. *Moreover, if* $u^1(0,\cdot) \neq u^2(0,\cdot)$, *then* $u^1(t,x) < u^2(t,x)$ *for* $x \in \mathcal{H}$, *and* $t \in (0,T)$.

*(2) If* $u_{01}, u_{02} \in X$ *and* $u_{01} \leq u_{02}$, $u_{01} \neq u_{02}$ , *then* $u(t,x;u_{01}) < u(t,x;u_{02})$ *for all* $x \in \mathcal{H}$ *and* $t > 0$ *at which both* $u(t,\cdot;u_{01})$ *and* $u(t,\cdot;u_{02})$ *exist.*

*(3) If* $u_{01}, u_{02} \in X$ *and* $u_{01} \ll u_{02}$, *then* $u(t,\cdot;u_{01}) \ll u(t,\cdot;u_{02})$ *for* $t > 0$ *at which both* $u(t,\cdot;u_{01})$ *and* $u(t,\cdot;u_{02})$ *exist.*

*Proof.* (1) The case $v_2 = 0$ follows from comparison principle for parabolic equations. The case $v_1 = 0$ follows from (Shen & Zhang, 2010) [Propositions 2.1 and 2.2].

(2) follows from (1).

(3) We provide a proof for the case $v_1 = 0$. Other cases can be proved similarly. Take any $T > 0$ such that both $u(t,\cdot;u_{01})$ and $u(t,\cdot;u_{02})$ exist on $[0,T]$. It suffices to prove that $u(t,\cdot;u_{02}) \gg u(t,\cdot;u_{01})$ for $t \in [0,T]$. To this end, let $w(t,x) = u(t,x;u_{02}) - u(t,x;u_{01})$. Then $w(t,x)$ satisfies the following equation,

$$w_t(t,x) = \int_{\mathbb{R}^N} \kappa(y-x)w(t,y)dy - w(t,x) + a(t,x)w(t,x),$$

where

$$a(t,x) = f(x, u(t,x;u_{02})) + u(t,x;u_{01}) \int_0^1 \partial_u f(x, su(t,x;u_{02}) + (1-s)u(t,x;u_{01}))ds.$$

Let $M > 0$ be such that $M \geq \sup_{x\in\mathbb{R}^N, t\in[0,T]}(1 - a(t,x))$ and $\tilde{w}(t,x) = e^{Mt}w(t,x)$. Then $\tilde{w}(t,x)$ satisfies

$$\tilde{w}_t(t,x) = \int_{\mathbb{R}^N} \kappa(y-x)\tilde{w}(t,y)dy + [M - 1 + a(t,x)]\tilde{w}(t,x).$$

Let $\mathcal{K} : X \to X$ be defined by

$$(\mathcal{K}u)(x) = \int_{\mathbb{R}^N} \kappa(y-x)u(y)dy \quad \text{for} \quad u \in X. \tag{2.3}$$

Then $\mathcal{K}$ generates an analytic semigroup on $X$ and

$$\tilde{w}(t,\cdot) = e^{\mathcal{K}t}(u_{02} - u_{01}) + \int_0^t e^{\mathcal{K}(t-\tau)}(M - 1 + a(\tau,\cdot))\tilde{w}(\tau,\cdot)d\tau.$$

Observe that $e^{\mathcal{K}t}u_0 \geq 0$ for any $u_0 \in X^+$ and $t \geq 0$ and $e^{\mathcal{K}t}u_0 \gg 0$ for any $u_0 \in X^{++}$ and $t \geq 0$. Observe also that $u_{02} - u_{01} \in X^{++}$. By (2), $\tilde{w}(\tau,\cdot) \geq 0$ and hence $(M - 1 + a(\tau,\cdot))\tilde{w}(\tau,\cdot) \geq 0$ for $\tau \in [0,T]$. It then follows that $\tilde{w}(t,\cdot) \gg 0$ and then $w(t,\cdot) \gg 0$ (i.e. $u(t,\cdot;u_{02}) \gg u(t,\cdot;u_{01})$) for $t \in [0,T]$. □

**Proposition 2.2.** *Assume* (H1). *For any given $u(t,\cdot;u_0)$ exists for all $t \geq 0$.*

*Proof.* Let $u_0 \in X^+$ be given. There is $M \gg 1$ such that $0 \leq u_0(x) \leq M$ and $f(x,M) < 0$ for all $x \in \mathcal{H}$. Then by Proposition 2.1,

$$0 \leq u(t,\cdot;u_0) \leq M$$

for any $t > 0$ at which $u(t,\cdot;u_0)$ exists. It is then not difficult to prove that for any $T > 0$ such that $u(t,\cdot;u_0)$ exists on $(0,T)$, $\lim_{t\to T} u(t,\cdot;u_0)$ exists in $X$. This implies that $u(t,\cdot;u_0)$ exists and $u(t,\cdot;u_0) \geq 0$ for all $t \geq 0$. □

For given $u, v \in X^{++}$, define

$$\rho(u,v) = \inf\{\ln\alpha \mid \frac{1}{\alpha}u \leq v \leq \alpha u,\ \alpha \geq 1\}.$$

Observe that $\rho(u,v)$ is well defined and there is $\alpha \geq 1$ such that $\rho(u,v) = \ln\alpha$. Moreover, $\rho(u,v) = \rho(v,u)$ and $\rho(u,v) = 0$ iff $u \equiv v$. In literature, $\rho(u,v)$ is called the *part metric* between $u$ and $v$.

**Proposition 2.3.** *For given $u_0, v_0 \in X^{++}$ with $u_0 \neq v_0$, $\rho(u(t,\cdot;u_0), u(t,\cdot;v_0))$ is non-increasing in $t \in (0,\infty)$.*

*Proof.* It can be proved by similar argument in (Kong & Shen, 2011) [Proposition 3.3]. For completeness, we provide a proof here.

First, note that there is $\alpha^* > 1$ such that $\rho(u_0,v_0) = \ln\alpha^*$ and $\frac{1}{\alpha^*}u_0 \leq v_0 \leq \alpha^* u_0$. By Proposition 2.1,

$$u(t,\cdot;v_0) \leq u(t,\cdot;\alpha^* u_0) \quad \text{for} \quad t > 0.$$

Let $v(t,x) = \alpha^* u(t,x;u_0)$. Then

$$\begin{aligned} v_t(t,x) &= \mathcal{A}v(t,x) + v(t,x)f(x,u(t,x;u_0)) \\ &= \mathcal{A}v(t,x) + v(t,x)f(x,v(t,x)) + v(t,x)f(x,u(t,x;u_0)) - v(t,x)f(x,v(t,x)) \\ &> \mathcal{A}v(t,x) + v(t,x)f(x,v(t,x)). \end{aligned}$$

This together with Proposition 2.1 implies that

$$u(t,\cdot;\alpha^* u_0) \leq \alpha^* u(t,\cdot;u_0) \quad \text{for} \quad t > 0$$

and then

$$u(t,\cdot;v_0) \leq \alpha^* u(t,\cdot;u_0) \quad \text{for} \quad t > 0.$$

Similarly, it can be proved that

$$\frac{1}{\alpha^*}u(t,\cdot;u_0) \leq u(t,\cdot;v_0) \quad \text{for} \quad t > 0.$$

It then follows that

$$\rho(u(t,\cdot;u_0), u(t,\cdot;v_0)) \leq \rho(u_0,v_0) \quad \forall t > 0$$

and hence

$$\rho(u(t_2,\cdot;u_0), u(t_2,\cdot;v_0)) \leq \rho(u(t_1,\cdot;u_0), u(t_1,\cdot;v_0)) \quad \forall 0 \leq t_1 < t_2.$$

□

*2.2 Principal Eigenvalues of Spatially Periodic Dispersal Operators*

In this subsection, we present some principal eigenvalue theories for spatially periodic dispersal operators with hybrid dispersals.

Let $p = (p_1, p_2, \ldots, p_N)$ with $p_i > 0$ for $i = 1, 2, \cdots, N$ and $X_p$ be as in (1.7). We will denote $I$ as an identity map on the Banach space under consideration. For given $\xi \in S^{N-1}$, $\mu \in \mathbb{R}$, $a \in X_p$, consider the following eigenvalue problems,

$$\begin{cases} Ou(x) = \lambda u(x), & x \in \mathbb{R}^N \\ u(x + p_i \mathbf{e}_i) = u(x), & x \in \mathbb{R}^N, \end{cases} \tag{2.4}$$

where

$$Ou(x) := \nu_1 \Delta u(x) + \nu_2 \int_{\mathbb{R}^N} e^{-\mu(y-x)\cdot\xi} \kappa(y - x)u(y)dy - 2\mu\nu_2\xi \cdot \nabla u(x) + (a(x) + \nu_1\mu^2 - \nu_2)u(x) \tag{2.5}$$

and $O : \mathcal{D}(O) \subset X_p \to X_p$.

Observe that if $\nu_1 = 1$, and $\nu_2 = 0$,

$$(Ou)(x) = \Delta u(x) - 2\mu\xi \cdot \nabla u(x) + (a(x) + \mu^2)u(x) \quad \forall\, u \in \mathcal{D}(O) \subset X_p, \tag{2.6}$$

If $\nu_1 = 0$, and $\nu_2 = 1$,

$$(Ou)(x) = \int_{\mathbb{R}^N} e^{-\mu(y-x)\cdot\xi} \kappa(y - x)u(y)dy - u(x) + a(x)u(x) \quad \forall\, u \in \mathcal{D}(O) \subset X_p \tag{2.7}$$

Let $\sigma(O)$ be the spectrum of $O$.

**Definition 2.1.** *Let $\mu \in \mathbb{R}$, and $\xi \in S^{N-1}$ be given. A real number $\lambda(\mu, \xi, a) \in \mathbb{R}$ is called the* principal eigenvalue *of $O$ if it is an isolated algebraic simple eigenvalue of $O$ with a positive eigenfunction and for any $\lambda \in \sigma(O) \setminus \{\lambda(\mu, \xi, a)\}$,* $\mathrm{Re}\lambda < \lambda(\mu, \xi, a)$.

For given $\mu \in \mathbb{R}$, and $\xi \in S^{N-1}$, let

$$\lambda^0(\mu, \xi, a) = \sup\{\mathrm{Re}\mu \,|\, \mu \in \sigma(O)\}. \tag{2.8}$$

Observe that for any $\mu \in \mathbb{R}$ and $\xi \in S^{N-1}$, $O$ generates an analytic semigroup $\{T(t)\}_{t\geq 0}$ in $X_p$ and moreover, $T(t)$ is strongly positive (that is, $T(t)u_0 \geq 0$ for any $t \geq 0$ and $u_0 \in X_p^+$ and $T(t)u_0 \gg 0$ for any $t > 0$ and $u_0 \in X_p^+ \setminus \{0\}$). Then by (Meyer-Nieberg, 1991) [Proposition 4.1.1], $r(T(t)) \in \sigma(T(t))$ for any $t > 0$, where $r(T(t))$ is the spectral radius of $T(t)$. Hence by the spectral mapping theorem (see Chicone & Latushkin, 1999; [Theorem 2.7]), $\lambda^0(\mu, \xi, a) \in \sigma(O)$. Observe also that $\lambda^0(0, \xi, a)$ are independent of $\xi \in S^{N-1}$. We may then put

$$\lambda^0(a) = \lambda^0(0, \xi, a).$$

It is well known that the principal eigenvalue $\lambda(\mu, \xi, a)$ in (2.6) exist for all $\mu \in \mathbb{R}$ and $\xi \in S^{N-1}$ and

$$\lambda(\mu, \xi, a) = \lambda^0(\mu, \xi, a).$$

The principal eigenvalue of $O$ in (2.7) may not exist (see Shen & Zhang, 2010 for examples). If the principal eigenvalue $\lambda(\mu, \xi, a)$ exists in (2.7), then

$$\lambda(\mu, \xi, a) = \lambda^0(\mu, \xi, a).$$

Regarding the existence of principal eigenvalue of $O$ in (2.7), the following proposition is proved in (Shen & Zhang, 2010; Shen & Zhang, 2012).

**Proposition 2.4** (Existence of principal eigenvalue). *(1) If $a \in C^N(\mathbb{R}^N, \mathbb{R}) \cap X_p$ and the partial derivatives of $a(x)$ up to order $N - 1$ are zero at some $x_0$ satisfying that $a(x_0) = \max_{x\in\mathbb{R}^N} a(x)$, then the principal eigenvalue $\lambda(\mu, \xi, a)$ of $O$ exists for all $\mu \in \mathbb{R}$ and $\xi \in S^{N-1}$.*

*(2) If $a(x)$ satisfies that $\max_{x\in\mathbb{R}^N} a(x) - \min_{x\in\mathbb{R}^N} a(x) < \inf_{\xi\in S^{N-1}} \int_{z\cdot\xi\leq 0} k(z)dz$, then the principal eigenvalue $\lambda(\mu, \xi, a)$ of $O$ exists for all $\mu \in \mathbb{R}$ and $\xi \in S^{N-1}$.*

*Proof.* (1) It follows from (Shen & Zhang, 2010) [Theorem B].

(2) It follows from (Shen & Zhang, 2012) [Theorem B$'$]. □

Let $\hat{a}$ be the average of $a(\cdot)$, that is,

$$\hat{a} = \frac{1}{|D|}\int_D a(x)dx \quad \text{for} \tag{2.9}$$

where

$$D = [0, p_1] \times [0, p_2] \times \cdots \times [0, p_N] \cap \mathcal{H} \tag{2.10}$$

and

$$|D| = p_1 \times p_2 \times \cdots \times p_N \ \text{for} \tag{2.11}$$

By Proposition 2.4 (2), $\lambda(\mu, \xi, \hat{a})$ exists for all $\mu \in \mathbb{R}$ and $\xi \in S^{N-1}$. The following proposition shows a relation between $\lambda^0(\mu, \xi)$ and $\lambda^0(\mu, \xi, \hat{a})$.

**Proposition 2.5** (Influence of spatial variation). *For given $\mu \in \mathbb{R}$, and $\xi \in S^{N-1}$, there holds*

$$\lambda^0(\mu, \xi) \geq \lambda^0(\mu, \xi, \hat{a}).$$

*Proof.* It follow from (Hetzer, Shen, & Zhang, 2013) [Theorem 2.1]. □

We remark that $\lambda(\mu, \xi, \hat{a})(= \lambda^0(\mu, \xi, \hat{a}))$ have the following explicit expressions,

$$\lambda(\mu, \xi, \hat{a}) = \nu_1\mu^2 + \nu_2\Big(\int_{\mathbb{R}^N} e^{-\mu z\cdot\xi}\kappa(z)dz - 1\Big) + \hat{a} \tag{2.12}$$

*2.3 KPP Equations in Spatially Periodic Media*

In this subsection, we recall some spatial spreading dynamics of KPP equations in spatially periodic media.

Consider

$$u_t(t, x) = \mathcal{A}u(t, x) + u(t, x)g(x, u(t, x)), \quad x \in \mathbb{R}^N, \tag{2.13}$$

where $g(\cdot, \cdot)$ are periodic in the first variable and monostable in the second variable. More precisely, we assume

**(P1)** *$g : \mathcal{H} \times \mathbb{R} \to \mathbb{R}$ is a $C^2$ function, $g(x + p_l\mathbf{e}_l, u) = g(x, u)$, where $p_l > 0$ and $g(x, u) < 0$ for all $(x, u) \in \mathcal{H} \times \mathbb{R}^+$ with $u \geq \alpha_0$ for some $\alpha_0 > 0$ and $\partial_u g(x, u) < 0$ for all $(x, u) \in \mathcal{H} \times \mathbb{R}^+$.*

**(P2)** $\lambda^0(g(\cdot, 0)) > 0$.

Assume (P1). Similarly, by general semigroup theory, for any $u_0 \in X$, (2.13) has a unique (local) solution $u(t, \cdot; u_0, g(\cdot, \cdot))(\in X)$ with initial data $u_0(\cdot)$. Moreover, if $u_0 \in X_p$, then $u(t, \cdot; u_0, g(\cdot, \cdot)) \in X_p$ for any $t > 0$ at which $u(t, \cdot; u_0, g(\cdot, \cdot))$ exists. By Proposition 2.1, if $u_0 \in X^+$, then $u(t, \cdot; u_0, g(\cdot, \cdot))$ exists and $u(t, \cdot; u_0, g(\cdot, \cdot)) \in X^+$ for all $t > 0$.

**Proposition 2.6** (Spatially periodic positive stationary solution). *Assume (P1) and (P2). Then* (2.13) *has a unique spatially periodic stationary solution $u^*(\cdot; g(\cdot, \cdot)) \in X_p^{++}$ which is globally asymptotically stable with respect to perturbations in $X_p^+ \setminus \{0\}$.*

*Proof.* It follows from (Zhao, 1996) [Theorem 2.3] and (Shen, & Zhang, 2012) [Theorem C]. □

Let $\hat{g}(u)$ be the spatial average of $g(x, u)$, that is,

$$\hat{g}(u) = \frac{1}{|D|}\int_D g(x, u)dx \quad \text{for} \tag{2.14}$$

where $D$, $|D|$ is as in (2.10).

Assume

**(P3)** $\hat{g}(0) > 0$.

Observe that $\lambda(\hat{g}(0)) = \hat{g}(0)$. Then by Proposition 2.5, (P3) implies (P2).

### 3. Positive Stationary Solutions

In this section, we explore the existence of positive stationary solutions of (1.1), and prove Theorem 1.1 and 1.2. Throughout this section, we assume (H1) and (H2). We first prove some lemmas.

**Lemma 3.1.** *For any $\epsilon > 0$, there are $p = (p_1, p_2, \cdots, p_N) \in \mathbb{N}^N$ and $h \in X_p \cap C^N(\mathcal{H}, \mathbb{R})$ such that*

$$f(x,0) \geq h(x) \quad \text{for} \quad x \in \mathcal{H},$$

$$\hat{h} \geq f^0(0) - \epsilon \quad (\text{hence} \quad \lambda^0(h(\cdot)) \geq f^0(0) - \epsilon),$$

*and the partial derivatives of $h(x)$ up to order $N-1$ are zero at some $x_0 \in \mathcal{H}$ with $h(x_0) = \max_{x\in\mathcal{H}} h(x)$, where $\hat{h}$ is the average of $h(\cdot)$ (see (2.9) for the definition).*

*Proof.* By (H2), there is $L_0 > 0$ such that $f(x,0) = f^0(0)$ for $x \in \mathcal{H}$ with $\|x\| \geq L_0$. Let $M_0 = \inf_{x\in\mathcal{H}} f(x,0)$. Let $h_0 : \mathbb{R} \to [0,1]$ be a smooth function such that $h_0(s) = 1$ for $|s| \leq 1$ and $h_0(s) = 0$ for $|s| \geq 2$. For any $p = (p_1, p_2, \cdots, p_N) \in \mathbb{N}^N$ with $p_j > 4L_0$, let $h \in X_p \cap C^N(\mathcal{H}, \mathbb{R})$ be such that

$$h(x) = f^0(0) - h_0\Big(\frac{\|x\|^2}{L_0^2}\Big)(f^0(0) - M_0) \quad \text{for} \quad x \in \Big([-\frac{p_1}{2}, \frac{p_1}{2}] \times [-\frac{p_2}{2}, \frac{p_2}{2}] \times \cdots \times [-\frac{p_N}{2}, \frac{p_N}{2}]\Big) \cap \mathcal{H}.$$

Then

$$f(x,0) \geq h(x) \quad \forall x \in \mathcal{H}.$$

It is clear that the partial derivatives of $h(x)$ up to order $N-1$ are zero at some $x_0 \in \mathcal{H}$ with $h(x_0) = \max_{x\in\mathcal{H}} h(x) (= f^0(0))$. For given $\epsilon > 0$, choosing $p_j \gg 1$, we have

$$\hat{h} > f^0(0) - \epsilon.$$

By Proposition 2.5, $\lambda^0(h(\cdot)) \geq \lambda^0(\hat{h}) = \hat{h}$ and hence

$$\lambda^0(h(\cdot)) \geq f^0(0) - \epsilon.$$

The lemma is thus proved. □

**Lemma 3.2.** *Suppose that $\tilde{u}^* : \mathbb{R}^N \to [\sigma_0, M_0]$ is Lebesgue measurable, where $\sigma_0$ and $M_0$ are two positive constants. If*

$$\Delta\tilde{u}^*(x) + \int_{\mathbb{R}^N} \kappa(y-x)\tilde{u}^*(y)dy - \tilde{u}^*(x) + \tilde{u}^*(x)\tilde{f}(x, \tilde{u}^*(x)) = 0 \quad \forall x \in \mathbb{R}^N,$$

*where $\tilde{f}(x,u) = f(x,u)$ or $f^0(u)$ for all $x \in \mathbb{R}^N$ and $u \in \mathbb{R}$, then $\tilde{u}^*(\cdot) \in X^{++}$.*

*Proof.* We prove the case that $\tilde{f}(x,u) = f(x,u)$. The case that $\tilde{f}(x,u) = f^0(u)$ can be proved similarly.

Let $h^*(x) = \Delta\tilde{u}^*(x) + \int_{\mathbb{R}^N} \kappa(y-x)\tilde{u}^*(y)dy$ for $x \in \mathbb{R}^N$. Then $h^*(\cdot)$ is $C^1$ and has bounded first order partial derivatives. Let

$$F(x,\alpha) = h^*(x) - \alpha + \alpha f(x,\alpha) \quad \forall x \in \mathbb{R}^N, \ \alpha \in \mathbb{R}.$$

Then $F : \mathbb{R}^N \times \mathbb{R} \to \mathbb{R}$ is $C^1$ and $F(x, \tilde{u}^*(x)) = 0$ for each $x \in \mathbb{R}^N$. If $\alpha^* > 0$ is such that $F(x,\alpha^*) = 0$, then

$$-1 + f(x,\alpha^*) = -\frac{h^*(x)}{\alpha^*} < 0$$

and hence

$$\partial_\alpha F(x,\alpha^*) = -1 + f(x,\alpha^*) + \alpha^*\partial_u f(x,\alpha^*) < 0.$$

By Implicit Function Theorem, $\tilde{u}^*(x)$ is $C^1$ in $x$. Moreover,

$$\frac{\partial\tilde{u}^*(x)}{\partial x_j} = \frac{\frac{\partial h^*(x)}{\partial x_j}}{-1 + f(x,\tilde{u}^*(x)) + \partial_u f(x,\tilde{u}^*(x))\tilde{u}^*(x)} \quad \forall x \in \mathbb{R}^N, \ 1 \leq j \leq N.$$

Therefore, $\tilde{u}^*$ has bounded first order partial derivatives. It then follows that $\tilde{u}^*(x)$ is uniformly continuous in $x \in \mathbb{R}^N$ and then $\tilde{u}^* \in X^{++}$. □

**Lemma 3.3.** *Suppose that $u^*(\cdot) \in X^{++}$ and $u = u^*(\cdot)$ is a stationary solution of (1). Then*

$$u^*(x) \to u^0 \quad \text{as} \quad \|x\| \to \infty.$$

*Proof.* Assume that $u^*(x) \not\to u^0$ as $\|x\| \to \infty$. Then there are $\epsilon_0 > 0$ and $x_n \in \mathbb{R}^N$ such that $\|x_n\| \to \infty$ and

$$|u^*(x_n) - u^0| \geq \epsilon_0 \quad \text{for} \quad n = 1, 2, \cdots .$$

By the uniform continuity of $u^*(x)$ in $x \in \mathbb{R}^N$, without loss of generality, we may assume that there is a continuous function $\tilde{u}^* : \mathbb{R}^N \to [\sigma_0, M_0]$ for some $\sigma_0, M_0 > 0$ such that

$$u(x + x_n) \to \tilde{u}^*(x)$$

as $n \to \infty$ uniformly in $x$ on bounded sets. By the Lebesgue Dominated Convergence Theorem, we have

$$\Delta\tilde{u}^*(x) + \int_{\mathbb{R}^N} \kappa(y - x)\tilde{u}^*(y)dy - \tilde{u}^*(x) + \tilde{u}^*(x)f^0(\tilde{u}^*(x)) = 0 \quad \forall x \in \mathbb{R}^N.$$

By Lemma 3.2, $\tilde{u}^* \in X^{++}$. By Proposition 2.6 again, we have $\tilde{u}^*(x) \equiv u^0$ and then $u^*(x_n) \to u^0$ as $n \to \infty$. This is a contradiction. Therefore $u^*(x) \to u^0$ as $\|x\| \to \infty$. □

**Lemma 3.4.** *There is $u^- \in X^{++}$ such that for any $\delta > 0$ sufficiently small, $u(t, x; \delta u^-)$ is increasing in $t > 0$ and $u^{-,*,\delta} \in X^{++}$, where $u^{-,*,\delta}(x) = \lim_{t\to\infty} u(t, x; \delta u^-)$, and hence $u = u^{-,*,\delta}(\cdot)$ is a stationary solution of (1.1) in $X^{++}$.*

*Proof.* Let $M^* > 0$ be such that $f(x, M^*) < 0$. Let $\epsilon > 0$ be such that

$$f^0(0) - \epsilon > 0.$$

By Lemma 3.1, there are $p \in \mathbb{N}^N$ and $h(\cdot) \in X_p \cap C^N(\mathcal{H}, \mathbb{R})$ such that

$$f(x, 0) \geq h(x), \quad \text{and} \quad \hat{h} \geq f^0(0) - \epsilon (> 0).$$

Moreover, the partial derivatives of $h(x)$ up to order $N - 1$ are zero at some $x_0 \in \mathcal{H}$ with $h(x_0) = \max_{x\in\mathcal{H}} h(x)$. Let $u^-$ be the positive principal eigenfunction of $O$ with $a(\cdot) = h(\cdot)$ and $\|u^-\| = 1$. It is not difficult to verify that $u = \delta u^-$ is a sub-solution of (1.1) for any $\delta > 0$ sufficiently small. It then follows that for any $\delta > 0$ sufficiently small,

$$\delta u^-(\cdot) \leq u(t_1, \cdot; \delta u^-) \leq u(t, \cdot; \delta u^-) \quad \forall 0 < t_1 < t.$$

This implies that there is a Lebesgue measurable function $u^{-,*,\delta} : \mathcal{H} \to [\sigma_0, M_0]$ for some $\sigma_0, M_0 > 0$ such that

$$\lim_{t\to\infty} u(t, x; \delta u^-) = u^{-,*,\delta}(x) \quad \forall x \in \mathcal{H}.$$

By Lemma 3.2, $u^{-,*,\delta} \in X^{++}$. Therefore $u^{-,*,\delta} \in X^{++}$ and $u = u^{-,*,\delta}(\cdot)$ is a stationary solution of (1.1) in $X^{++}$. □

**Lemma 3.5.** *Let $M \gg 1$ be such that $f(x, M) < 0$ for $x \in \mathcal{H}$. Then $\lim_{t\to\infty} u(t, x; u_0)$ exists for every $x \in \mathcal{H}$, where $u_0(x) \equiv M$. Moreover, $u^{+,*,M}(\cdot) \in X^{++}$, where $u^{+,*,M}(x) := \lim_{t\to\infty} u(t, x; u_0)$, and hence $u = u^{+,*,M}(\cdot)$ is a stationary solution of (1.1) in $X^{++}$.*

*Proof.* For any $M > 1$ with $f(x, M) < 0$ for all $x \in \mathcal{H}$, $u = M$ is a super-solution of (1). Hence

$$u(t, \cdot; M) \leq u(t_1, \cdot; M) \leq M \quad \forall 0 \leq t_1 < t.$$

It then follows that $\lim_{t\to\infty} u(t, x; M)$ exists for all $x \in \mathbb{R}^N$. Let $u^{+,*,M}(x) = \lim_{t\to\infty} u(t, x; M)$. We have $u^{+,*,M}(x) \geq u^{-,*,\delta}(x)$ for $0 < \delta \ll 1$. By the similar arguments as in Lemma 3.4, $u^{+,*,M} \in X^{++}$ and $u = u^{+,*,M}(\cdot)$ is a stationary solution of (1.1) in $X^{++}$. □

*Proof of Theorem 1.1.* (1) First, by Lemmas 3.4 and 3.5, (1) has stationary solutions in $X^{++}$. We claim that stationary solution of (1) in $X^{++}$ is unique. In fact, suppose that $u^{1,*}$ and $u^{2,*}$ are two stationary solutions of (1) in $X^{++}$. Assume that $u^{1,*} \neq u^{2,*}$. Then there is $\alpha^* > 1$ such that $\rho(u^{1,*}, u^{2,*}) = \ln\alpha^* > 0$. Note that

$$\frac{1}{\alpha^*}u^{1,*} \leq u^{2,*} \leq \alpha^* u^{1,*}.$$

By Lemma 3.3, $\lim_{\|x\|\to\infty} u^{1,*}(x) = u^0$ and $\lim_{\|x\|\to\infty} u^{2,*}(x) = u^0$. This implies that there is $\epsilon > 0$ such that

$$\frac{1}{\alpha^* - \epsilon}u^{1,*}(x) \leq u^{2,*}(x) \leq (\alpha^* - \epsilon)u^{1,*}(x) \quad \text{for} \quad \|x\| \gg 1.$$

By Proposition 2.1 and the arguments in Proposition 2.3,

$$\frac{1}{\alpha^*}u^{1,*}(x) < u^{2,*}(x) < \alpha^* u^{1,*}(x) \quad \forall x \in \mathbb{R}^N.$$

It then follows that for $0 < \epsilon \ll 1$,

$$\frac{1}{\alpha^* - \epsilon}u^{1,*}(x) \leq u^{2,*}(x) \leq (\alpha^* - \epsilon)u^{1,*}(x) \quad \forall x \in \mathbb{R}^N$$

and then $\rho(u^{1,*}, u^{2,*}) \leq \ln(\alpha^* - \epsilon)$, this is a contradiction. Therefore $u^{1,*} = u^{2,*}$ and (1.1) has a unique stationary solution $u^*$ in $X^{++}$. □

*Proof of Theorem 1.2.* For any $u_0 \in X^{++}$, there is $\delta > 0$ sufficiently small and $M > 0$ sufficiently large such that $\delta u^- \leq u_0 \leq M$ and $u = \delta u^-$ is a sub-solution of (1) ($u^-$ is as in Lemma 3.4) and $u = M$ is a super-solution of (1.1). Then

$$\delta u^- \leq u(t, \cdot; \delta u^-) \leq u(t, \cdot; u_0) \leq u(t, \cdot; M) \leq M \quad \forall t \geq 0.$$

By Theorem 1.1, Lemmas 3.4 and 3.5, and Dini's Theorem,

$$u(t, x; \delta u^-) < u^*(x) < u(t, x; M) \quad \forall t > 0, \ x \in \mathcal{H}$$

and

$$\lim_{t\to\infty} u(t, x; \delta u^-) = \lim_{t\to\infty} u(t, x; M) = u^*(x)$$

uniformly in $x$ on bounded sets. It then follows that

$$\lim_{t\to\infty} u(t, x; u_0) = u^*(x)$$

uniformly in $x$ on bounded sets.

We claim that $\|u(t, \cdot; u_0) - u^*(\cdot)\| \to 0$ as $t \to \infty$. Assume the claim is not true. Then there are $\epsilon_0 > 0$, $t_n \to \infty$, and $x_n$ with $\|x_n\| \to \infty$ such that

$$|u(t_n, x_n; u_0) - u^*(x_n)| \geq \epsilon_0 \quad \forall n \in \mathbb{N}.$$

Then by Lemma 3.3,

$$|u(t_n, x_n; u_0) - u^0| \geq \frac{\epsilon_0}{2} \quad \forall n \gg 1.$$

Let $\tilde{\delta} > 0$ and $\tilde{M} > 0$ be such that

$$\tilde{\delta} \leq u(t, \cdot; u_0) \leq \tilde{M} \quad \forall t \geq 0.$$

For any $\epsilon > 0$, let $T > 0$ be such that

$$|u(T, \cdot; \tilde{\delta}, f^0(\cdot)) - u^0| < \epsilon, \quad |u(T, \cdot; \tilde{M}, f^0(\cdot)) - u^0| < \epsilon. \tag{3.1}$$

Observe that

$$\tilde{\delta} \leq u(t_n - T, x_n + x; u_0) \leq \tilde{M}$$

and

$$u(t_n, x_n + \cdot; u_0) = u(T, x_n + \cdot; u(t_n - T, \cdot; u_0)) = u(T, \cdot; u(t_n - T, \cdot + x_n; u_0), f(\cdot + x_n, \cdot))$$

for $n \gg 1$. Then

$$u(T, \cdot; \tilde{\delta}, f(\cdot + x_n)) \leq u(t_n, x_n + \cdot; u_0) \leq u(T, \cdot; \tilde{M}, f(\cdot + x_n, \cdot)). \tag{3.2}$$

Observe also that $f(x + x_n, u) \to f^0(u)$ as $n \to \infty$ uniformly in $(x, u)$ on bounded sets. Then

$$u(T, x; \tilde{\delta}, f(\cdot + x_n, \cdot)) \to u(T, x; \tilde{\delta}, f^0(\cdot))$$

and

$$u(T, x; \tilde{M}, f(\cdot + x_n, \cdot)) \to u(T, x; \tilde{M}, f^0(\cdot))$$

as $n \to \infty$ uniformly in $x$ on bounded sets. This together with (3.1) implies that

$$|u(T, 0; \tilde{\delta}, f(\cdot + x_n, \cdot)) - u^0| < 2\epsilon, \quad |u(T, 0; \tilde{M}, f(\cdot + x_n, \cdot)) - u^0| < 2\epsilon \quad \text{for} \quad n \gg 1$$

and then by (3.2),

$$|u(t_n, x_n; u_0) - u^0| < 2\epsilon \quad \text{for} \quad n \gg 1.$$

Hence $\lim_{n\to\infty} u(t_n, x_n; u_0) = u^0$, which is a contradiction. Therefore $\|u(t, \cdot; u_0) - u^*(\cdot)\| \to 0$ as $t \to \infty$.

□

**Acknowledgements**

The author thanks the referees for valuable comments and suggestions which improved the presentation considerably.

**References**

Aronson, D. G., & Weinberger, H. F. (1957). Nonlinear diffusion in population genetics, combustion, and nerve pulse propagation, in "Partail Differential Equations and Related Topics" (J. Goldstein, Ed.), *Lecture Notes in Math., 466*, Springer-Verlag, New York, 5-49. http://dx.doi.org/10.1007/BFb0070595

Aronson, D. G. & Weinberger, H. F. (1978). Multidimensional nonlinear diffusions arising in population genetics, *Adv. Math., 30*, 33-76. http://dx.doi.org/10.1016/0001-8708(78)90130-5

Bates, P. & Zhao, G. (2007). Existence, uniqueness and stability of the stationary solution to a nonlocal evolution equation arising in population dispersal, *J. Math. Anal. Appl., 332*, 428-440. http://dx.doi.org/10.1016/j.jmaa.2006.09.007

Berestycki, H., Hamel, F., & Nadirashvili, N. (2005). The speed of propagation for KPP type problems, I - Periodic framework, *J. Eur. Math. Soc., 7*, 172-213. http://dx.doi.org/10.1007/s00285-004-0313-3

Berestycki, H., Hamel, F., & Nadirashvili, N. (2010). The speed of propagation for KPP type problems, II - General domains, *J. Amer. Math. Soc., 23*, 1-34. http://dx.doi.org/10.1090/S0894-0347-09-00633-X

Berestycki, H., Hamel, F., & Roques, L. (2005). Analysis of periodically fragmented environment model: II - Biological invasions and pulsating traveling fronts, *J. Math. Pures Appl., 84* , 1101-1146.

Berestycki, H., & Nadin, G. (2016). Asymptotic spreading for general heterogeneous Fisher-KPP type equations, preprint. http://lodel.ehess.fr/cams/docannexe.php?id=1344

Berestycki, H., Jin, T., & Silvestre, L. (2016). Propagation in a nonlocal reaction diffusion equation with spatial and genetic trait structure, *Nonlinearity, 29*(4), 1434-1466. http://dx.doi.org/10.1088/0951-7715/29/4/1434

Cantrell, R. S., & Cosner, C. (2003). *Spatial Ecology via Reactiond-Diffusion Equations*, Series in Mathematical and Computational Biology, John Wiley and Sons, Chichester, UK. http://dx.doi.org/10.1002/0470871296.ch1

Chasseigne, E., Chaves, M., & Rossi, J. D. (2006). Asymptotic behavior for nonlocal diffusion equations, *J. Math. Pures Appl. 86*, 271-291. http://dx.doi.org/10.1016/j.matpur.2006.04.005

Cortazar, C., Coville, J., Elgueta, M. & Martinez, S. (2007). A nonlocal inhomogeneous dispersal process, *J. Differential Equations, 241*, 332-358. http://dx.doi.org/10.1016/j.jde.2007.06.002

Cousens, R., Dytham, C. & Law, R (2008). *Dispersal in plants: a population perspective*, Oxford University Press, Oxford. http://dx.doi.org/10.1093/aob/mcn243

Chicone, C. & Latushkin, Y. (1999). *Evolution Semigroups in Dynamical Systems and Differential Equations*, Mathematical Surveys and Monographs, *70*, American Mathematical Society. http://dx.doi.org/10.1090/surv/070

Coville, J. & Dupaigne, L. (2005). Propagation speed of travelling fronts in non local reaction-diffusion equations, *Nonlinear Analysis, 60*, 797-819. http://dx.doi.org/10.1016/j.na.2003.10.030

Coville, J., Dávila, J., & Martínez, S. (2008). Existence and uniqueness of solutions to a nonlocal equation with monostable nonlinearity, *SIAM J. Math. Anal., 39*, 1693-1709. http://dx.doi.org/10.1137/060676854

Fife, P. C. (1979). *Mathematical Aspects of Reacting and Diffusing Systems*, Lecture Notes in Biomathematics, *28*, Springer-Verlag, Berlin-New York. http://dx.doi.org/10.1007/978-3-642-93111-6

Fife, P. C. (2003). Some nonclassical trends in parabolic and parabolic-like evolutions. *Trends in nonlinear analysis*, 153-191, Springer, Berlin.

Fife, P. C., & Peletier, L. A. (1977). Nonlinear diffusion in population genetics. *Arch. Rational Mech Anal., 64*, 93-109.

Fisher, R. (1937). The wave of advance of advantageous genes. *Ann. of Eugenics, 7*, 335-369. http://dx.doi.org/10.1111/j.1469-1809.1937.tb02153.x

Freidlin, M., & Gärtner, J. (1979). On the propagation of concentration waves in periodic and random media. *Soviet Math. Dokl., 20*, 1282-1286.

Grinfeld, M., Hines, G., Hutson, V., Mischaikow, K., & Vickers, G. T. (2005). Non-local dispersal. *Differential Integral Equations, 18*, 1299-1320.

Hamel, F. (2008). Qualitative properties of monostable pulsating fronts: exponential decay and monotonicity, *J. Math. Pures Appl., 89*, 355-399. http://dx.doi.org/10.1016/j.matpur.2007.12.005

Henry, D. (1981). Geometric Theory of Semilinear Parabolic Equations. *Lecture Notes in Math., 840*, Springer-Verlag, Berlin. http://dx.doi.org/10.1007/BFb0089647

Hetzer, G., Shen, W., & Zhang, A. (2013). Effects of spatial variations and dispersal strategies on principal eigenvalues of dispersal operators and spreading speeds of monostable equations. *Rocky Mountain Journal of Mathematics, 43*, 489-513. http://dx.doi.org/10.1216/RMJ-2013-43-2-489

Hudson, W., & Zinner, B. (1995). Existence of traveling waves for reaction diffusion equations of Fisher type in periodic media, *Boundary value problems for functional-differential equations*, 187-199, World Sci. Publ., River Edge, NJ, 1995.
http://dx.doi.org/10.1006/jmaa.1996.0137

Hutson, V., Martinez, S. , Mischaikow, K., & Vickers, G.T. (2003). The evolution of dispersal. *J. Math. Biol., 47*,483-517. http://dx.doi.org/10.1007/s00285-003-0210-1

Kametaka, Y. (1976). On the nonlinear diffusion equation of Kolmogorov-Petrovskii- Piskunov type. *Osaka J. Math., 13*,

11-66.

Kao, C.-Y., Lou, Y., & Shen, W. (2010). Evolution of Mixed Dispersal in Periodic Environments. *Discrete and Continuous Dynamical Systems*, *26*, 551-596. http://dx.doi.org/10.3934/dcdsb.2012.17.2047

Kao, C.-Y., Lou, Y., & Shen, W. (2012). Random dispersal vs non-Local dispersal. *Discrete and Continuous Dynamical Systems, Series B*, *17*, 2047-2072. http://dx.doi.org/10.3934/dcds.2010.26.551

Kolmogorov, A., Petrowsky, I. , & Piscunov, N. (1937). A study of the equation of diffusion with increase in the quantity of matter, and its application to a biological problem. *Bjul. Moskovskogo Gos. Univ.*, *1* 1-26.

Kong, L., & Shen, W. (2011). Positive Stationary Solutions and Spreading Speeds of KPP Equations in Locally Spatially Inhomogeneous Media. *Methods and Applications of Analysis*, *18*, 427-456. http://dx.doi.org/10.4310/MAA.2011.v18.n4.a5

Kong, L., & Shen, W. (2014). Liouville Type Property and Spreading Speeds of KPP Equations in Periodic Media with Localized Spatial Inhomogeneity. *Journal of Dynamics and Differential Equations*, *26*, Issue 1, pp 181-215. http://dx.doi.org/10.1007/s10884-014-9351-8

Lee, C. T., Hoopes, M. F., Diehl, J., Gilliland, W., Huxel, G., Leaver, E. V., McCain, K., Umbanhowar, J., & Mogilner, A. (2001). Non-local concepts and models in biology. *J. theor. Biol.*, *210* , 201-219. http://dx.doi.org/10.1006/jtbi.2000.2287

Levin, S.A., Muller-Landau, H.C., Nathan, R., & Chave, J. (2003). The ecology and evolution of seed dispersal: a theoretical perspective. *Annu. Rev. Eco. Evol. Syst.*, *34*, 575-604. http://dx.doi.org/10.1146/annurev.ecolsys.34.011802.132428

Li, W.-T., Sun, Y.-J., & Wang, Z.-C. (2010). Entire solutions in the Fisher-KPP equation with nonlocal dispersal. *Nonlinear Anal. Real World Appl.*, *11*, 2302-2313. http://dx.doi.org/10.1016/j.nonrwa.2009.07.005

Liang, X., & Zhao, X.-Q. (2007). Asymptotic speeds of spread and traveling waves for monotone semiflows with applications. *Comm. Pure Appl. Math.*, *60*, 1-40. http://dx.doi.org/10.1002/cpa.20154

Liang, X., Yi, Y., & Zhao, X.-Q. (2006). Spreading speeds and traveling waves for periodic evolution systems. *J. Diff. Eq.*, *231*, 57-77. http://dx.doi.org/10.1016/j.jde.2006.04.010

Murray, J. D. (1989). *Mathematical Biology*, Springer-Verlag, New York. http://dx.doi.org/10.1007/b98868

Nadin, G. (2009). Traveling fronts in space-time periodic media. *J. Math. Pures Appl.*, *9* (92), 232-262. http://dx.doi.org/10.1016/j.matpur.2009.04.002

Meyer-Nieberg, P. (1991). *Banach Lattices*, Springer-Verlag. http://dx.doi.org/10.1007/978-3-642-76724-1

Nolen, J., Rudd, M., & Xin, J. (2005). Existence of KPP fronts in spatially-temporally periodic adevction and variational principle for propagation speeds. *Dynamics of PDE*, *2*, 1-24. http://dx.doi.org/10.4310/DPDE.2005.v2.n1.a1

Nolen, J., & Xin, J. (2005). Existence of KPP type fronts in space-time periodic shear flows and a study of minimal speeds based on variational principle. *Discrete and Continuous Dynamical Systems*, *13*, 1217-1234. http://dx.doi.org/10.3934/dcds.2005.13.1217

Nolen, J., Roquejoffre, J-M, & Ryzhik, L. (2015). Power-Like Delay in Time Inhomogeneous Fisher-KPP Equations. *Communications in Partial Differential Equations*, *40* (3), 475-505. http://dx.doi.org/10.1080/03605302.2014.972744

Pazy, A. (1983). *Semigroups of Linear Operators and Applications to Partial Differential Equations*, Springer-Verlag New York Berlin Heidelberg Tokyo. http://dx.doi.org/10.1007/978-1-4612-5561-1

Sattinger, D. H. (1976). On the stability of waves of nonlinear parabolic systems. *Advances in Math.*, *22*, 312-355. http://dx.doi.org/10.1016/0001-8708(76)90098-0

Shen, W. (2011). Existence, uniqueness, and stability of generalized traveling waves in time dependent monostable equations. *Journal of Dynamics and Differential Equations*, *23*, 1-44. http://dx.doi.org/10.1007/s10884-010-9200-3

Shen, W., & Zhang, A. (2010). Spreading speeds for monostable equations with nonlocal dispersal in space periodic habitats. *Journal of Differential Equations*, *249*, 749-795. http://dx.doi.org/10.1016/j.jde.2010.04.012

Shen, W., & Zhang, A. (2012). Stationary solutions and spreading speeds of nonlocal monostable equations in space periodic habitats. *Proceedings of the American Mathematical Society* *140*, 1681-1696. http://dx.doi.org/10.1090/S0002-9939-2011-11011-6

Shigesada, N., & Kawasaki, K. (1997). *Biological Invasions: Theory and Practice*, Oxford University Press.

Skellam, J. G. (1951). Random dispersal in theoretical populations. *Biometrika, 38*, 196-218.

Uchiyama, K. (1978). The behavior of solutions of some nonlinear diffusion equations for large time. *J. Math. Kyoto Univ., 183*, 453-508.

Weinberger, H. F. (1982). Long-time behavior of a class of biology models. *SIAM J. Math. Anal., 13*, 353-396. http://dx.doi.org/10.1137/0513028

Weinberger, H. F. (2002). On spreading speeds and traveling waves for growth and migration models in a periodic habitat. *J. Math. Biol., 45*, 511-548. http://dx.doi.org/10.1007/s00285-002-0169-3

Zhang, A. (2013). Traveling wave solutions with mixed dispersal for spatially periodic Fisher-KPP equations. *Discrete Contin. Dyn. Syst. Suppl.*, 815-824.

Zhao, X. Q. (1996). Global attractivity and stability in some monotone discrete dynamical systems, *Bull. Austral. Math. Soc., 53*, 305-324. https://dx.doi.org/10.1017/S0004972700017032

Zhao, X. Q. (2003). *Dynamical Systems in Population Biology*, CMS Books in Mathematics *16*, *Springer-Verlag*, New York. http://dx.doi.org/10.1007/978-0-387-21761-1

# Extremal Dependence Modeling with Spatial and Survival Distributions

Diakarya Barro[1]

[1] Université Ouaga II BP: 417 Ouagadougou 12, Burkina Faso

Correspondence: Diakarya Barro, Université Ouaga II BP: 417 Ouagadougou 12, Burkina Faso. E-mail: dbarro2@gmail.com

**Abstract**

This paper investigates some properties of dependence of extreme values distributions both in survival and spatial context. Specifically, we prospose a spatial Extremal dependence coefficient for survival distributions. Madogram is characterized in bivariate case and multivariate survival function and the underlying hazard distributions are given in a risky context.

**Keywords:** survival distribution, extreme values distributions, variogram, spatial process, hazard function

**2010 Math. Subject Classification:** 62H20, 62H11, 60G15

## 1. Introduction

Extreme values (EV) analysis finds wide applications in many areas including climatology, environment sciences (Beirlant, J., et al., 2005), risk management (Balkema, G. & Paul, E., 2007; Degen, M. & Embrechts, P., 2008) and survival analysis (Hougaard, P., 2000). The distributions of this domain can be obtained as limiting distributions of properly normalized maxima of independent and identically distributed random variables. In particular if $Z = \left\{Z_x; x \in \mathbb{R}^2\right\}$ is a max stable random field defined on a set $X = \{x_1, ..., x_k\}$, then the spatial EV analysis shows that Z results from observations of a stochasltic process such as

$$Z(s) = \lim_{n\to\infty}\left\{\max_{1\leq i\leq n}\left[\frac{x_i(s) - b_n(s)}{a_n(s)}\right]\right\} \text{ with } s \in D; \tag{1.1}$$

provided the limit exists, where $\{a_n(.) > 0; n \geq 1\}$ and $\{b_n(.), ; n \geq 1\}$ are sequences of real constants, $s$ being a spatial location of a domain $D \subset \mathbb{R}^d$ and $Z(s)$, a random quantity (Padoan, S. A., et al., 2010).

Survival analysis is a subdomain of statistics which deals with failure or death time or natural catastroph. It is a important topic in many areas including biomedical, biostatistics, environment, etc (Padoan, S. A., et al., 2010; Resnick, S. I., 2008). One may distinguish three kind of models in survival analysis: the non parametric models, the semi-parametric models and the parametric ones.

Let $T = (T_1, ..., T_n)$ be a vector of lifetimes of n individuals in a given population with distribution $F_T$. If in particular T describes the life long time, the fraction of the population which will survive past a given vector of times $t = (t_1, ..., t_n)$ is provided by the survival distribution, conventionally denoted $S_T$, such as

$$S_T(t_1, ..., t_n) = \bar{F}_T(t_1, ..., t_n) = P(T_1 \geq t_1, ..., T_n \geq t_n). \tag{1.2}$$

The hazard function $h_T$ of T specifies the instantaneous rate of failure (risk or mortality rate) at a given date $t$ given that the individual survived up to time $t$. If the margins are absolutly continuous the cumulative density function (cdf) is also related to $S_T$ such as

$$h_T(t_1, ..., t_n) = \frac{f_T(t_1, ..., t_n)}{S_T(t_1, ..., t_n)} \text{ and } f_T(t_1, ..., t_n) = (-1)^n \frac{\partial^n S(t_1, ..., t_n)}{\partial t_1 ... \partial t_n}. \tag{1.3}$$

Spatial analysis is a key component of statistic involving a collected from different locations. In particular, while studying in biostatistics, epidemiology, environment sciences, data have a common, that they are collected from different spatial locations and they are nether independent nor identically distributed. So, that in spatial framework, when survival times are spatially referenced, some of clusters of high or low times might be apparent on a visual inspection of the data. The question which naturally arises as to whether these observed spatial survival patterns can be explained by incorporating appropriate covariates into the model or whether, in order the unexplained spatial variation.

The main contribution of this paper is to investigate some asymptotic properties of multivariate dependence models both in survival and spatial context. Section 2 deals with spatial measures of extremal dependence. In particular the extremal dependance spatial coefficient is modeled and survival madogram is characterized in bivariate case. In Section 3, the survival and hazard distributions are given in a risky context.

## 2. Survival Framework for Modeling Spatial Extremal Dependance

### 2.1 Spatial Extremal Dependence Coefficient

In multivariate extreme values (EV) analysis, many related measures have been proposed for quantifying the magnitude of the extremal dependence when the random vector exhibits asymptotic dependence. In particular, in univariate EV study, even in spatial and survival framework, three types of distributions can summary the asymptotic behavior of conveniently normalized maximum of distributions (Beirlant, J., et al., 2005).

For a fixed k in $\mathbb{N}^*$ let $Y = (Y_{k,1}; ...; Y_{k,s})$ denote independent copies of stochastic process observed at given ereas s of a domain S. Assume that the process $\{Y(s), s \in S\}$ is parametric max-stable distribution. Then the asymptotic distribution modeling the stochastic behavior is the same type like one of the three extremal spatial distributions

$$F(y_i(s_i)) = \begin{cases} \exp\{-\exp(-y_i(s_i))\} & = \Lambda(y_i(s_i));\ y_i(s_i) \in \mathbb{R}, \text{(Gumbel)} \\ \exp\left\{-(y_i(s_i))^{-\theta}\right\} & = \Phi_\theta(y_i(s_i));\ y_i(s_i) > 0;\ \text{(Fréchet)} \\ \exp\left(-(-y_i(s_i))^{\theta}\right) & = \Psi_\theta(y_i(s_i));\ y_i(s_i) \le 0, \text{(Weibull)} \end{cases}. \tag{2.1}$$

Let $T_s = (T_1(s), ..., T_n(s))$ be a vector of lifetimes of n individuals in a given population observed at a given site s of spatial domain $S = \left\{(s_1, ...., s_m), s_j \in \mathbb{R}^2\right\}$. The process $T_s$ is the survival and stochastic random vector which with joint distribution $F_s = (F_{s,1}; ...; F_{s,n})$. Therefore, for all realization y,

$$F_s(y) = (F_{1,s}(y_1), ..., F_{n,s}(y_n)) = (F_1(y_1(s_1)), ..., F_n(y_n(s_n))) = F(y(s)).$$

In all this study, our key assumption is that the process $T_s$ is continuous, stationary and is max-stable with generalized Fréchet margins. So, for a given site s in $S_\xi$

$$S_\xi = \{s_i \in S; \sigma_i(s_i) + \xi_i(s_i)(y_i(s_i) - \mu_i(s_i)) > 0\} \subset S.$$

where $u_+ = \max(u, 0)$ and $\left\{\mu_X(x) \in \mathbb{R}\right\}$, $\left\{\sigma_X(x) > 0\right\}$ and $\xi_X(x) \in \mathbb{R}$,

$$F_\theta(y_i(s_i)) = \Phi_\theta(y_i(s_i)) = \exp\left\{-\left(\tfrac{y_i(s_i)-\mu_i(s_i)}{\sigma_i(s_i)}\right)_+^{-\theta}\right\}; \theta > 0.$$

Notice that such an assumption implies no loss of generality since even in survival and space-varying context, every one-dimensional EV distribution can be obtained by a functional transformation of another. In particular, if for a given site s,

$$Y(x_i) \sim \Phi_\theta(y_i(x_i)) \implies Z(x_i) = \mu_{x_i} + \frac{\sigma_{x_i}}{\xi_{x_i}}\left[Y(x_i)^{\xi_{x_i}} - 1\right]. \tag{2.2}$$

Among measures of extremal dependence there are the extremal coefficient (Hougaard, P., 2000) or the madogram and its nested model the link between two sets of $\mathbb{R}^d$ (Cooley, D., et al., 2006). Moreover and for simplicity reason let's denote like in (Barro, D., et al., 2016) that: $\tilde{F}_i^{s_j}(x_i) = F_i\left(x_i\left(s_j\right)\right)$ and $\tilde{x}_i^{s_j} = x_i\left(s_j\right)$. Under the restriction to the simplest case where $F_i\left(x_i\left(s_j\right)\right) = 0$ if $i \neq j$, the following result allows us to provide a characterisation of the spatial extremal dependence (SED) in a survival field.

**Theorem 1** Let $T_s$ be a vector of lifetimes of n individuals in a given population with distribution $\tilde{F}^s$ satisfying the key assumption.

i) The one-dimensional marginal law $\left\{\tilde{F}_i^{s_j}; 1 \le i \le n\right\}$ of $\tilde{F}^s$ is a max-stable process, that is there exists survival parametric normalizing sequences $\{\sigma_i(s_i) > 0\}$ and $\{\mu_i(s_i) \in \mathbb{R}\}$ and $\{\xi_i(x_t) \in \mathbb{R}\}$such that, for all i, $1 \le i \le n$

$$\left[\tilde{F}_i^{s_j}\left(\sigma_n^i\left(\tilde{x}_i^{s_j}\right) + \mu_n^i(s_i)\right)\right]^n \underset{n\to+\infty}{\rightarrow} \begin{cases} \left[1 + \xi_i(s_i)\left(\frac{\tilde{x}_i^{s_i} - \mu_i(s_i)}{\sigma_i(s_i)}\right)\right]_+^{\frac{-1}{\xi_i(s_i)}} & \text{if } \xi_i(s_i) \neq 0 \\ \exp\left\{-\left(\dfrac{\tilde{x}_i^{s_i} - \mu_i(s_i)}{\sigma_i(s_i)}\right)\right\} & \text{if } \xi_i(s_i) = 0 \end{cases} \tag{2.3}$$

on $D_\xi(s_i) = \left\{s_i \in S, \tilde{x}_i^{s_i} - \mu_i(s_i) > 0\right\}$

ii) *There exists spatio-survival parametric measure of probability $P_{\xi(s)}$ defined on $\mathbb{R}^2 \times \{s\}$ and a non-decreasing function $g_s$ defined on $D_\xi$ such that the survival and spatial extremal coefficent $\theta_s\left(h_{ij}\right)$ of the process is given by*

$$P_{s_{ij},\xi_s} = g\left[s_i, \theta_s\left(h_{ij}\right)\right]; \tag{2.4}$$

where $h_{ij} = \left|s_i - s_j\right|$ is the separating distance between these sites $s_i$ and $s_j$.

**Proof.** By assumption the distribution function $T_s$ of the process is max-stable. So, for all site $s$, there exist vectors of constants $\{\sigma_{n,s} > 0\}$ and $\{\mu_{n,s} \in \mathbb{R}\}$ such that,

$$\lim_{n \longrightarrow +\infty} P\left(\bigcap_{i=1}^{n}\left\{\frac{M_i\left(\tilde{x}_i^{s_i}\right) - \mu_i\left(s_i\right)}{\sigma_i\left(s_i\right)} \leq y_i\left(s_i\right)\right\}\right) = G_i\left(y_i\left(\tilde{x}_i^{s_i}\right)\right) \tag{2.5}$$

where $G_i$ is an EV distribution and $M_i\left(x_i\right)$ is the spatial survival vector of the maximum

$$M_i\left(x_i^{s_j}\right) = \left(\max_{1 \leq j \leq m}\left(x_i^{s_j}\right)\right)^T.$$

But since $F_i$ is the marginal distribution of $T_s$ then, it lies on the max-domain of attraction of $G_i$. Thus, for all site $s_i$ , the relation (2.5) is equivalent to the generalized EV model, given by

$$\lim_{n\rightarrow+\infty}\left[\tilde{F}_i^{s_j}\left(\sigma_n^i\left(\tilde{x}_i^{s_j}\right) + \mu_n^i\left(s_i\right)\right)\right]^n = \begin{cases} \left[1 + \xi_i\left(s_i\right)\left(\frac{\tilde{x}_i^{s_i} - \mu_i(s_i)}{\sigma_i(s_i)}\right)\right]_+^{\frac{-1}{\xi_i(s_i)}} & \text{if } \xi_i\left(s_i\right) \neq 0 \\ \exp\left\{-\left(\frac{\tilde{x}_i^{s_i} - \mu_i(s_i)}{\sigma_i(s_i)}\right)\right\} & \text{if } \xi_i\left(s_i\right) = 0 \end{cases}.$$

where $\{\sigma_i\left(s_i\right) > 0\}$ and $\{\mu_i\left(s_i\right) \in \mathbb{R}\}$ and $\{\xi_i\left(x_t\right) \in \mathbb{R}\}$ such that, for all i,$1 \leq i \leq n$ are respectively the spatio-survival parameters of location, scale and shape of the observation at the parametric site $s_i$.

ii) In bivariate case and for all pair of sites $s_i$ and $s_j$ the extremal dependence parametric coefficient $\theta\left(s_i, s_j\right) = \theta_{ij} = \theta\left(h_{ij}\right)$ depends on the of separating distance $h_{ij}$.

It follows that

$$P\left[\tilde{F}^s\left(\tilde{x}_i^{s_i}\right) \leq y; \tilde{F}^s\left(\tilde{x}_j^{s_j}\right) \leq y\right] = \exp\left[\frac{-\theta\left(h_{ij}\right)}{y}\right]. \tag{2.6}$$

Moreover, using in the relation (2.6) the general form of a univariate EV model with normalizing coefficients $\sigma > 0$, $\mu \in \mathbb{R}$, $\xi_i\left(s_i\right) \in \mathbb{R}$, it comes, in the particular bivariate context, that

$$P\left(\tilde{F}^s\left(\tilde{x}_i^{s_j}\right) \leq y, \tilde{F}^s\left(\tilde{x}_i^{s_j}\right) \leq y\right) = \exp\left[-\theta\left(h_{ij}\right)\left(\left[1 + \xi\left(\frac{\tilde{x}_i^{s_i} - \mu_i(s_i)}{\sigma_i(s_i)}\right)\right]_+^{\frac{1}{\xi_i(s_i)}}\right)\right]. \tag{2.7}$$

Then, by introducing the concept of probability measure the relation (2.7) is equivalent to

$$P\left(\tilde{F}^s\left(\tilde{x}_i^{s_j}\right) \leq y, \tilde{F}^s\left(\tilde{x}_i^{s_j}\right) \leq y\right) = \exp\left[-\theta\left(h_{ij}\right)\left(\left[1 + \xi\left(\frac{\tilde{x}_i^{s_i} - \mu_i(s_i)}{\sigma_i(s_i)}\right)\right]_+^{\frac{1}{\xi_i(s_i)}}\right)\right] = P_{s_{ij},\xi_s}$$

and finally

$$P_{s_{ij},\xi_s} = g\left[s_i, \theta_s\left(h_{ij}\right)\right]$$

Thus we obtain (2.4) as asserted ∎

The following proposition provides a consequence of theorem 1

**Corollary 2** Let *$\{T^s; s \in S\}$ a spatial process satisfying the key assumption. Then, f*or all site $s_i \in S$

i) the marginal survival parametric extremal density $f_{\xi_i}$ is given by

$$f_{\xi_i}\left(t_i\left(\tilde{x}_i^{s_j}\right)\right) = \begin{cases} \dfrac{t_i\left(\tilde{x}_i^{s_i}\right)\left(1 + \xi\log\left(t_i\left(\tilde{x}_i^{s_i}\right)\right)\right)^{1+\frac{1}{\xi_i(s_i)}}}{\exp\left[-\left(1 + \xi\log\left(t_i\left(\tilde{x}_i^{s_i}\right)\right)\right)^{\frac{1}{\xi_i(s_i)}}\right]} & \text{if } \xi_{s_j} \neq 0 \\ \dfrac{1}{t_i^2\left(\tilde{x}_i^{s_i}\right)}\exp\left(\dfrac{-1}{t_i\left(\tilde{x}_i^{s_i}\right)}\right) & \text{if } \xi_{s_j} = 0 \end{cases}. \tag{2.8}$$

ii) the parametric hazard functions $h_\xi(t)$ are given

$$h_{\xi_i}(t) = \begin{cases} \dfrac{\left\{\exp\left(\left(1+\xi log\left(t_i\left(\tilde{x}_i^{s_i}\right)\right)\right)^{-\frac{1}{\xi_i(s_i)}}\right)-1\right\}}{t_i\left(\tilde{x}_i^{s_i}\right)\left(1+\xi log\left(t_i\left(\tilde{x}_i^{s_i}\right)\right)\right)^{\frac{1}{\xi_i(s_i)}+1}} & if\ \xi_{s_j} \neq 0 \\ \dfrac{1}{t_i(x_i(s_i))^2\left|\exp\left(\dfrac{1}{t_i\left(\tilde{x}_i^{s_i}\right)}\right)-1\right|} & if\ \xi_{s_j} = 0 \end{cases}. \tag{2.9}$$

**Proof.** In such a case, the parametric survival function

$$S_\xi(t_1, ..., t_n) = P_\xi(T_1 \geq t_1, ..., T_n \geq t_n)$$

is given marginally by

$$S_\xi(t_i) = \begin{cases} 1 - \exp\left[-(1+\xi logt_i)^{-\frac{1}{\xi_i(s_i)}}\right] & if\ \xi_i(s_i) \neq 0 \\ 1 - \exp\left(-\frac{1}{t_i}\right) & if\ \xi_i(s_i) = 0 \end{cases}.$$

Hence, the hazard function $h_\xi(t) = \dfrac{f_\xi(t)}{S_t(t)}$ is given by

$$h_\xi(t) = \begin{cases} \dfrac{1}{t(1+\xi logt)^{\frac{1}{\xi}+1}\left\{exp\left((1+\xi logt)^{-\frac{1}{\xi}}\right)-1\right\}} & if\ \xi_i(s_i) \neq 0 \\ \dfrac{1}{t^2\left(exp\left(\frac{1}{t}\right)-1\right)} & if\ \xi_i(s_i) = 0 \end{cases} \tag{2.10}$$

■

*2.2 Distortional Function of Spatial Extremal Model*

The following result characterizes a multivariate survival distribution via a spatial and distortional measure of dependence.

**Proposition 3** *Let* $\{T^s; s \in S\}$ *a spatial process satisfying the key assumption.* Then there exists a spatial conditional dependence measure such as $D_s$ defined on the spatial unit simplex, for all $\tilde{x}^s = \left(\tilde{x}_1^{s_1}, ..., \tilde{x}_n^{s_n}\right) \in \bar{\mathbb{R}}^n$;

$$\widetilde{S}_n^s = \left\{(t_1(s) ..., t_n(s)) \in [0,1]^n\ ; \sum_{i=1}^n t_i(s) = 1\right\} \tag{2.11}$$

such that,

$$\tilde{F}^s(\tilde{x}^s) = 1 - \exp\left\{\sum\nolimits_{i=1}^n \frac{t_i(\tilde{x}_i^{s_i})}{\left(1+\xi\log(t_i(\tilde{x}_i^{s_i}))\right)^{1+\frac{1}{\xi}}} D_s\left(\frac{\tilde{x}_i^{s_i}}{\sum_{i=1}^n \tilde{x}_i^{s_i}}, \ldots, \frac{\tilde{x}_{m-1}^{s_{m-1}}}{\sum_{i=1}^n \tilde{x}_i^{s_i}}\right)\right\}. \tag{2.12}$$

**Proof.** The EV analysis results from asymptotic normalized vector of maxima of a random vector which converges to a non degenerated multivariate EV model G. One of extremal study approach is the Peacks-over threshold (POT). Then the vector of exceedances of the same sample have a generalized Pareto model H.

Particularly if the extremal function $\tilde{F}^s$ underlying the survival process $T_S$. It follows that its spatio-survival associated POT model $\tilde{\text{H}}^s$ satisfies, for all $\tilde{x}^s = \left(\tilde{x}_1^{s_1}, ..., \tilde{x}_n^{s_n}\right) \in \bar{\mathbb{R}}^n$, the relationship by

$$H(x^s) = 1 + \left(\sum_{i=1}^n \tilde{x}_i^{s_i}\right)\tilde{A}_s\left(\frac{\tilde{x}_1^{s1}}{\sum_{i=1}^n \tilde{x}_i^{s_i}}, \ldots, \frac{\tilde{x}_{m-1}^{s_{m-1}}}{\sum_{i=1}^n \tilde{x}_i^{s_i}}\right) = 1 + \log F(\tilde{x}^s); \tag{2.12}$$

where $\tilde{A}_F$ is a spatio-survival dependence function of Pickands associated to $\tilde{F}$.

Furthermore, for a given $1 < N < n$ let consider the N-partition of the spatial domain $S$ proposed in [9]

$$S = \left\{(s_1, ...., s_m), s_j \in \mathbb{R}^2\right\} = S_N \cup S_{\bar{N}}.$$

Then, it follows that the corresponding distorsional probability $\tilde{\delta}_s$ is such that;

$$\tilde{\delta}_s(x) = \tilde{\delta}\left(\tilde{x}_1^{s1}, ..\tilde{x}_n^{s_i}\right) = 1 - \frac{P\left(T_j \leq \tilde{x}_j^{s_j}; N \leq j \leq n\right)}{P(T_i \leq x_i ; 1 \leq \text{i} \leq \text{N-1})}.$$

Moreover let $\tilde{F}^s_N$ and $\tilde{F}^s_{\bar{N}}$ be the corresponding partitional distributions functions. So, it comes that

$$\tilde{\delta}\left(\tilde{x}_1^{s1},..\tilde{x}_n^{s_i}\right) = 1 - \frac{\tilde{F}^s_{\bar{N}}\left(\tilde{x}_N^{s_N};...;\tilde{x}_n^{s_i}\right)}{\tilde{F}^s_N\left(\tilde{x}_1^{s1};...;\tilde{x}_{N-1}^{s_i}\right)}.$$

Furthermore, from results due to Dossou et al. (Dossou,-G. S., 2009) both the partitional distributions functions $\tilde{F}^s_N$ and $\tilde{F}^s_{\bar{N}}$ lie also in the max-domain of attraction of two multivariate EV distribtutions. And by noting $\tilde{A}_N$ and $\tilde{A}_{\bar{N}}$ the Pickands dependence functions of t $\tilde{F}^s_N$ and $\tilde{F}^s_{\bar{N}}$ respectively, it come that, even in a spatio-survival context

$$\begin{aligned}\tilde{\delta}\left(\tilde{x}_1^{s1},...,\tilde{x}_n^{s_i}\right) &= \exp\left\{-\left(\sum_{i=1}^n \tilde{x}_i^{s_i}\right)\tilde{A}_s\left(\frac{\tilde{x}_1^{s1}}{\sum_{i=1}^n \tilde{x}_i^{s_i}},\ldots,\frac{\tilde{x}_{m-1}^{s_{m-1}}}{\sum_{i=1}^n \tilde{x}_i^{s_i}}\right)+\right.\\ &\left.\left(\sum_{i=1}^n \tilde{x}_i^{s_i}\right)\tilde{A}_s\left(\frac{\tilde{x}_1^{s1}}{\sum_{i=1}^n \tilde{x}_i^{s_i}},\ldots,\frac{\tilde{x}_{m-1}^{s_{m-1}}}{\sum_{i=1}^n \tilde{x}_i^{s_i}}\right)\right\}\end{aligned}$$

Which can be written equivalently,

$$\tilde{\delta}\left(\tilde{x}_1^{s1},..\tilde{x}_n^{s_i}\right) = \exp\left\{-\left(\sum_{i=1}^n \tilde{x}_i^{s_i}\right)\tilde{D}_s\left(\frac{\tilde{x}_1^{s1}}{\sum_{i=1}^n \tilde{x}_i^{s_i}},\ldots,\frac{\tilde{x}_{m-1}^{s_{m-1}}}{\sum_{i=1}^n \tilde{x}_i^{s_i}}\right)\right\}$$

where $\tilde{D}_s$ being a distortional spatial and survival dependence function

$$D(t(\tilde{x}^s)) = A\left(t_1\left(\tilde{x}_1^{s_1}\right),\ldots,t_{m-1}\left(\tilde{x}_{m-1}^{s_{m-1}}\right)\right)+\left(1-t_1\left(\tilde{x}_1^{s_1}\right)\right)A_{\bar{N}_1}\left(\frac{t_2(\tilde{x}_2^{s1})}{1-t_2(\tilde{x}_2^{s1})},\ldots,\frac{t_{m-1}}{1-t_2(\tilde{x}_2^{s1})}\right)$$

Particularly in bivariate case it is easy to show that $D(t(\tilde{x}^s))$ is defined from $\mathbb{R}^+$ to $\left[\frac{-1}{2},1\right]$ by

$$D(t(\tilde{x}^s)) = A\left(\frac{1}{1-t(\tilde{x}^s))}\right) - \frac{t}{1+t(\tilde{x}^s)}.$$

Specially, for the logistic model:

$$F_0^{\check{s}}\left(\tilde{x}_1^{s_i},\tilde{x}_2^{s_i}\right) = \exp\left\{-\left(\left(\tilde{x}_1^{s_i}\right)^\theta + \left(\tilde{x}_1^{s_i},\tilde{x}_2^{s_i}\right)^\theta\right)^{\frac{1}{\theta}}\right\}$$

the spatial conditional measure is

$$D^{\check{s}}_{\theta,t}(x) = \frac{x_t}{1+x_t}\left[\left(1+x_t^{-\theta}\right)^{\frac{1}{\theta}} - 1\right]$$

■

which is given graphically has follows

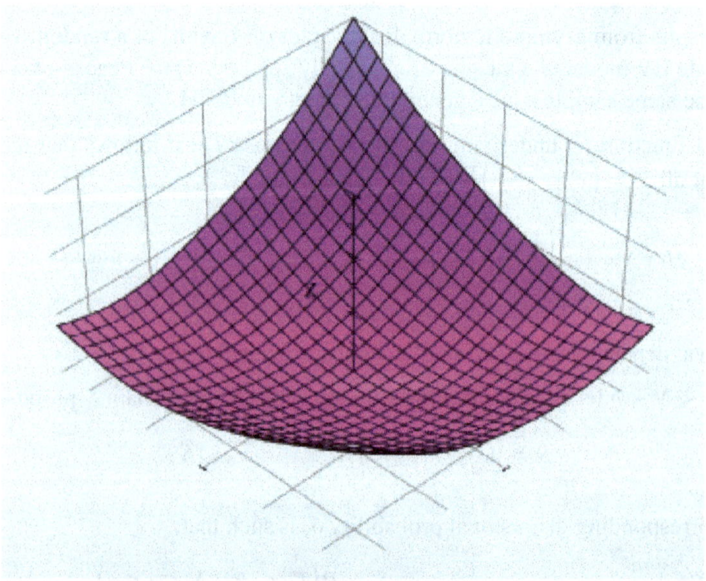

Figure 1. Bivariate logistic model for $\theta_1 = \theta_2 = 2$

*2.3 A New Charaterization of Survival Madogram*

Madogram is a measure of the full pairwise extremal dependence function to evaluate dependence among extreme regional observation. Some extensions of this tool have been proposed. Specifially while modeling spatial extreme variablity of an isotropic and max-stable field, Cooley (Cooley, D., et al., 2006) proposed the F-madogram $\gamma_F(h)$ which transforms the process via its marginal F. The following result provides a parametrization of characterizing of the madogram.

**Proposition 4** *Let* $\{T^s; s \in S\}$ *a spatial process satisfying the key assumption with distribution* $\tilde{F}^s$. Then, the survival $\lambda$-madogram associated to the bivariate margins of F is given by the ratio

$$\gamma_\lambda(h) = \frac{P(D_h, \lambda, s_i)}{Q\left(D_h(\lambda, 1-\lambda) + \lambda\right)}; \tag{2.13}$$

where P and Q are convenient polynoms and $D_h$ being a distortional spatial dependence measure.

**Proof.** In the previous proposition, the bivariate case implies that, particularly in bivariate case it is easy to shows that $D(t(\tilde{x}^s))$ is defined from $\mathbb{R}^+$ to $\left[\frac{-1}{2}, 1\right]$ by

$$D(t(\tilde{x}^s)) = A\left(\frac{1}{1 - t(\tilde{x}^s))}\right) - \frac{t}{1 + t(\tilde{x}^s)}.$$

Furthermore, in condional study, Proposition 6 of the paper (Barro, D. et al., 2012) provides that, under additional contraints, the $\lambda$–madogram can be expressed as

$$\gamma_\lambda(h) = \frac{1}{D_h(\lambda, 1-\lambda) + \lambda} - c(\lambda) \ with\ c(\lambda) = \frac{2\lambda(1-\lambda) + 1}{2(\lambda+1)(2-\lambda)} \tag{2.14}$$

where $D_h$ is a conditional spatial measure convex defined on the unit simplex of $\mathbb{R}^2$.

In our context by replacing $D(t(\tilde{x}^s))$ by $D(t(\tilde{x}^s))$ it follows easily that

$$\gamma_{\lambda_s}(h) = \frac{-[2\lambda(1-\lambda 1) + 1](D_h(\lambda, 1-\lambda) + \lambda)2(\lambda+1)(2-\lambda)}{(D_h(\lambda, 1-\lambda) + \lambda)[2(\lambda+1)(2-\lambda)]} = \frac{P(D_h, \lambda, s_i)}{Q\left(D_h(\lambda, 1-\lambda) + \lambda\right)} \tag{2.15}$$

where $D_h$ is a conditional spatial measure convex defined on the unit simplex of $\mathbb{R}2$

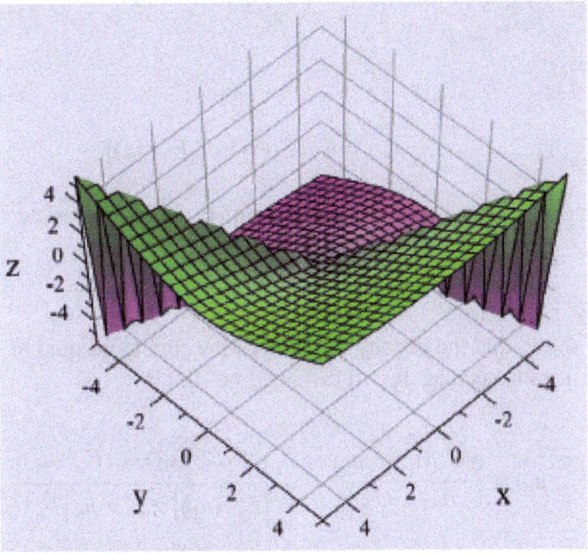


Figure 2. Bivariate distortional $\lambda - madogram$

### 3. Survival and Hazard Distributions in a Risky Context

In epidemiological studies, the intensity of contamination must change over time. For example, at begining of the epidemy, the intensity is high but it decreases when the sanitaries autorities take some dispositions epidemy.

Let $T_s = (T_1(s), ..., T_n(s))$ be the survival, continuous and stochastic random vector T satisfying the key assumption. Consider $H$ as a closed half space in $\mathbb{R}^n$ where $P(T_s \in H) > 0$. As in the high scenarios defined by Degen (see [1], [5] and [6]), the following results characterize the spatial and survival risk and some derivative properties.

**Definition** The spatial and survival high risk scenario $T_s^H$ associated to the process $T_s$ is defined as the vector T($s$) conditioned to lie in the half space $H$. The *spatial* probability distribution function ($pdf$) $F_{T_s}^H$ of $T_s^H$ on $\mathbb{R}^2$ is such that $F_{T_s}^H = P \circ T_s^{-1}$., then $Z_s^H$ has the high risk distribution $\pi^H$ given by

$$dF_{T_s}^H(z(s)) = \delta_H(z(s))\, d\pi(z(s))\, /\pi(H). \tag{3.1}$$

One obtains the following relation between T and its canical parametrisation in a risky context.

**Proposition** If $T_s$ has the distribution $\tilde{F}_s$, then $T_s^H$ has the high risk distribution $\tilde{F}_s^H$ given by, , for all $i = 1, \ldots, m$.

$$\tilde{F}_s^H(t_1, \ldots, t_d) = \tilde{f}_s^H(\tilde{t}_{s,1}, \ldots, \tilde{t}_{s,d}) \prod_{i=1}^{d} \delta_{H_i}(t_i)\tilde{F}_{1,s}^H(t_1)/\pi_i(F_i) \tag{3.2}$$

where, $\tilde{t}_{s,i} = \tilde{F}_{i,s}^H(t_i)$ are the marginals distributions functions of the distribution $T^H$ and given

$$\tilde{f}_s^H(u_1, ..., u_m) = \frac{f\left(F_1^{-1}(u_1), ..., F_m^{-1}(u_m)\right)}{f_1\left(F_1^{-1}(u_1), \right) ... f_m\left(F_m^{-1}(u_m)\right)} \tag{3.3}$$

for all $(u_1, ..., u_m) \in [0, 1]^m$.

**Proof.** Let $(X_1, ..., X_n), n \in \mathbb{N}$, be a vector of random *i.i.d* variables with a joint distribution F with continuous margins $F_i$. According to Sklar's theorem (see [15]), there exists a unique copula, $C_T$ providing a canonical parameterisation of F via its univariate marginal quantile functions $F_i^{-1}$ such that,

$$F_i^{-1}(u) = \inf\{x_i \in \mathbb{R}, F_i(x_i) \geq u\};$$

for all $x = (x_1, ..., x_m) \in (\mathbb{R} \cup \{\pm\infty\})^m$.

$$F(x_1, ..., x_m) = C_T[F_1(x_1), ..., F_m(x_m)]. \tag{3.4}$$

Or conversively, for $C_F$ providing a canonical parameterisation of F via its univariate marginal quantile functions $F_i^{-1}$ such that:

$$C_F(u_1, ..., u_n) = F\left(F_1^{-1}(u_1), ..., F_n^{-1}(u_n)\right). \tag{3.5}$$

Differentiating the formula (3.4) shows that the density function of the copula is equal to the ratio of the joint density h of H to the product of marginal densities $h_i$ such as, for all $(u_1, ..., u_n) \in [0, 1]^n$,

$$c(u_1, ..., u_n) = \frac{\partial^n C(u_1, ..., u_n)}{\partial u_1 ... \partial u_m} = \frac{f\left[F_1^{-1}(u_1), ..., F_n^{-1}(u_n)\right]}{f_1\left[F_1^{-1}(u_1)\right] \times ... \times h_n\left[F_n^{-1}(u_n)\right]}. \tag{3.6}$$

Let $Z$ be a random vector in $\mathbb{R}^m$ and $H = H_1 \times \ldots \times H_m \subset \mathbb{R}^m$ a closed half space with $P(Z \in H) > 0$. Let's suppose that $z_i \in \aleph = H_1 \cap \ldots \cap H_d$, for all $i = 1, \ldots, m$. If the marginals distributions of the high risk scenarios distribution $\pi^H$ are all continuous, then the density of the $\tilde{F}_s$ is given by

$$c(u_1, \ldots, u_n) = \frac{\delta_H d\pi\left(\tilde{u}_1^{H_1}, \ldots, \tilde{u}_m^{H_m}\right)\right) /\pi(H)}{d\pi_1\left(\tilde{u}_1^{H_1}\right) \times \ldots \times d\pi_m\left(\tilde{u}_m^{H_m}\right) / (\pi_1(H_1) \times \ldots \times \pi_m(H_m))} \tag{3.7}$$

where $\tilde{u}_i^{H_i} = (\pi_i^{H_i})^{(-1)}(u_i)$, the inverses of the marginals distributions $(\pi_i^{H_i})$. Suppose that the high risk scenarios margins $\pi_1^{H_1}, \ldots, \pi_d^{H_d}$ are continuous. Using the relation, it comes that

$$c(\pi_1^{H_1}(z_1), \ldots, \pi_d^{H_d}(z_d)) = \frac{d\pi^H(z_1, \ldots, z_d)}{\prod_{i=1}^d \delta_{H_i}(z_i) d\pi_i(z_i)/\pi_i(H_i)}$$

If $z_i \in \aleph$, $\forall\, i = 1, \ldots, d$ then we get this relation :

$$c(u_1, \ldots, u_d) = \frac{d\pi^H\left(\tilde{u}_i^{H_i}, \ldots, \tilde{u}_d^{H_d})\right)}{\prod_{i=1}^d \delta_{H_i}(z_i) d\pi_i(z_i)/\pi_i(H_i)}$$

where $\pi_i^{H_i}(z_i) = u_i$ or $z_i = (\pi_i^{H_i})^{(-1)}(u_i)$.

This gives

$$c(u_1, \ldots, u_d) = \frac{\delta_H d\pi\left(\tilde{u}_1^{H_1}, \ldots, \tilde{u}_m^{H_m}\right)/\pi(H)}{\delta_\aleph d\pi_1\left(\tilde{u}_1^{H_1}\right) \times \ldots \times d\pi_d\left(\tilde{u}_m^{H_m}\right)/(\pi_1(H_1) \times \ldots \times \pi_m(H_m))}$$

Since $\delta_\aleph = 1$, then it follows that $z_i \in \aleph = \bigcap_{i=1}^d H_i$.

Finally we obtain

$$\tilde{f}_s^H(u_1, ..., u_m) = \frac{f\left(F_1^{-1}(u_1), ..., F_m^{-1}(u_m)\right)}{f_1\left(F_1^{-1}(u_1), \right) ..., f_m\left(F_m^{-1}(u_m)\right)} \tag{3.8}$$

Thus, we obtain the relation (3.3) as asserted ■

## 4. Conclusion

The results of this study show that the survial and spatial framework are also convenient to model extremal dependence. Tools of dependence such as the extremal dependance coefficient, the multivariate dependence function, the madogram have been modeled both in spatial and survival context. The survival and hazard distributions are given in a risky context.

## References

Balkema, G., & Paul, E. (2007). *High risk scenarios and extreme: a geometric approach.*

Barro, D. (2009). Conditional Dependence of Trivariate Generalized Pareto Distributions, *Asian Journal of Mathematics and Statistics, 2*(3), 41-54. http://dx.doi.org/10.3923/ajms.2009.41.54

Barro, D., et al. (2012). Spatial Stochastic Framework For Sampling Time Parametric Max-stable Processes. *Inter.JSP, 1*(2), 203-210. http://dx.doi.org/10.5539/ijsp.v1n2p203

Barro, D., et al. (2016). Spatial Tail Dependence and Survival Stability in a Class of Archimedean Copulas.*International Journal of Mathematics and Mathematical Sciences*, (8). http://dx.doi.org/10.1155/2016/8927248

Benjamim, M. T. (2015). *Auxiliary Varables Markov Chain monte Carlo for spatial survival geostatistical Model,*

Beirlant, J., Goegebeur, Y., Segers, J., & Teugels, J. (2005). Statistics of Extremes : Theory and Applications. *John Wiley and Sons, Chichester.*

Cooley, D., Naveau, P., Jomelli, V., Rabatel, A., & Grancher, D., (2006). *A Bayesian hierarchical extreme value model for lichenometry. Environmetrics, 17*, 555–574. http://dx.doi.org/10.1002/env.764

Degen, M., E. (2008). EVT-based estimation of risk capital and convergence of high quantiles. *Advances in Applied Probability, 40*(3), 696-715.

Degen, M. (2006). *On Multivariate Generalised Pareto Distributions and High Risk Scenarios.* Department of Mathematics, ETH Zü rich, Diploma thesis. http://www.math.ethz.ch/~degen/thesis.pdf

Dossou, -G., S., (2009). Modelling the Dependence of Parametric Bivariate Extreme Value Copulas. *Asian Journal of Mathematics and Statistics, 2*(3), 41-54. http://dx.doi.org/10.3923/ajms.2009.41.54

Fonseca, et al., (2012). *Generalized madogram and pairwise dependence of maxima over two regions of a random field.*

Hougaard, P. (2000). *Analysis for Multivariate Survival Data, Statistics for Biology and Health.* Springer-Verlag, New York, Hall, London.

Nelsen, R. B. (1999). *An Introduction to copulas.* Lectures notes in Statistics 139, Springer-Verlag.

Padoan, S. A., Ribatet, M., & Sisson, S. A. (2010). Likelihood-Based Inference for Max-Stable Processes, *Journal of the American Statistical Association, 105*(489), Theory and Methods. http://dx.doi.org/10.1198/jasa.2009.tm08577

Resnick, S. I. (2008). *Extreme Values, Regular Variation, and Point Processes (reprint).* Springer, NY.

Tajvidi, N. (1996). *Characterisation and Some Statistical Aspects of Univariate and Multivariate Generalised Pareto Distributions.* Dissertation, Department of Mathematics, Chalmers.

# A New Heuristic Method for Transportation Network and Land Use Problem

Mouhamadou A.M.T. Baldé[1] & Babacar M. Ndiaye[1]

[1] Laboratory of Mathematics of Decision and Numerical Analysis, LMDAN-FASEG, University of Cheikh Anta Diop, BP 45087 Dakar-Fann, 10700 Dakar, Senegal

Correspondence: Babacar M. Ndiaye. E-mail: babacarm.ndiaye@ucad.edu.sn

**Abstract**

Our paper deals with the Transportation Network and Land Use (TNLU) problem. It consists in finding, simultaneously, the best location of urban area activities, as well as of the road network design that may minimize the moving cost in the network, and the network costs. We propose a new mixed integer programming formulation of the problem, and a new heuristic method for the resolution of TNLU. Then, we give a methodology to find locations or relocations of some Dakar region amenities (home, shop, work and leisure places), that may reduce travel time or travel distance. The proposed methodology mixes multi-agent simulation with combinatorial optimization techniques; that is individual agent strategies versus global optimization using Geographical Information System. Numerical results which show the effectiveness of the method, and simulations based on the scenario of Dakar city are given.

**Keywords**: transportation network, land use plan, quadratic assignment, heuristic, simulations.

## 1. Introduction

The Transportation Network and Land Use (TNLU) problem is a combination of transportation network optimization problem and land use planning or quadratic assignment problem. This model appear, for example, if we decided to find locations or relocations of some Dakar region amenities (home, shop, work, leisure places), that may reduce travel time or travel distance. By solving this problem we want to have the best layout of the city in the direction of the location of activities and roads linking these activities. A better configuration in the direction of the location of activities and transportation for a city would reduce congestion and pollution. We consider an activity ($i$) placed on a zone ($k$) and an important flow (cars) from other activities, in the direction of the activity ($i$). Suppose that there is traffic lights at intersections, thus flows will be subject to long waiting times before reaching activity ($i$), hence congestion. During this waiting, passengers of vehicles are subject to breathing harmful gases from cars. It would therefore make sense to ask whether the location of this activity is the best possible (reallocation of this activity or not). Since the 60s many authors are interested in these two problems separately. Indeed we can cite authors such as Scott (1969), Boyce, Farhi and Weischedel (1973), Hoang Hai Hoc (1973) and Billheimer (1970) for network design problems, Lawler(1963), Schlager (1965) and Gordon-McReynolds(1972), Maniezzo(1997), Xia and Yuan(2006), Huizhen Zhang(2010) and many others for the activities's location problems. Among the researchers who are interested in the combined problem we can cite Marc Los and more recently Lin and Feng (2003). They have considered a more complex model in the sense that their formulation allows the location of several activities in the same area. Marc Los gave an exact solving method and heuristic methods. The Los's model assumes the OD flow streams, once the activities have been fixed via the shortest paths in the sense of a fixed cost independent of the flow on the arcs (see Gueye, 2012).

In this paper, the heuristics that we propose are based on Shortest Path (SP), Location of Activity (LA) and Network Design (ND) problems. It gives results corresponding to the best affectation found, by solving a Quadratic Assignment Problem (QAP) which uses shortest path solutions for distance, and the network design (ND) problem associated to these affectations. We use heuristic methods (QAP-heur and ND-heur) for the resolution of QAP and ND problems.

After locating the $n$ activities into the $n$ locations, we develop a new methodology by which a very large set of relocation possibilities can be simulated, analyzed, and the "best" one can be found. The methodology was coded in a prototype software called Simulation-Driven Optimization (SDO), that originated from two projects DAMA (Balac et al., 2014). In this task, SDO uses the multi-agent simulator MATSim (see Balmer et al., 2009). In MATSim, the actors of the modelled system are the agents (i.e the city residents). The agents act according to given "realistic" rules. They try to perform some activities at different places and have learning capabilities. The overall traffic observed in the urban area emerges from the simulation as a consequence of individual agents behaviour, each pursuing his/her individual interests. MATsim basically needs three data to perform a simulation: the transportation network (network.xml), the amenities location (facilities.xml) and the initial agent plans (plans.xml). At the first MATSim iteration, each agent follows one or several possible initial plans contained in the agent plan file. Following a complete MATSim simulation, in SDO we adopt a global (or collective) view which contrasts with the individual behaviour of the agents in the simulation. Our problem is indeed slightly different: it aims at finding some suitable relocations to increase the global accessibility for a set of selected amenities. Let us remark that the MATSim simulations are operating on only one day, such as the global accessibility we seek to improve. So, to be pertinent, the simulated plans should be as representative as possible of what the agent do most frequently.

In the following Section 2, we give a mixed integer programming formulation of the problem. Then, in Section 3 we present one heuristic method, TNLU-heur, to solve it. The Section 4 is devoted to the computation of the lower bound for TNLU. We detail in Section 5 the simulation results of our algorithms. The Section 6 deals with SDO that calculates the optimal locations of activities based on the results of the simulations of MATSim. Finally, we conclude and give some perspectives in Section 7.

## 2. Formulation of the Model

We assume to have $n$ locations and $n$ activities to locate in these locations. Consider the graph $G(N, Ar)$ such as $N$ is the set of node consisting different areas ($n$) and $Ar$ is the set of links joining the n locations in pairs. Let:

$A = (a_{ik})$ : location cost matrix of the activity $i$ on the area $k$.

$Q = (q_{ij})$ : moving flow matrix per unit of times between the activities $i$ and $j$.

$C = (c_{ij})$ : construction cost matrix of link or road $(i, j)$.

$D = (d_{ij})$ : direct distance matrix between two locations $i$ and $j$ or distance of link $(i, j)$.

To determine the locations of activities, we consider the decision variables, called assignment, $x_{ij} = 1$ if the activity $i$ is affected on the location $j$ and 0 otherwise. To determine the configuration of the network, we need to introduce binary variables $y_{ij} = 1$ if the link $(i, j)$ is selected and 0 otherwise. Depending on the constructed network, the shortest path within the meaning of the fixed costs of building $d_{ij}$ associated with each arc $(i, j)$ may vary. Then it is necessary to introduce variables of shortest path $l_{kl}(y)$ between $k$ and $l$ and variables $z_a^{kl}$ which take value 1 if the shortest path between $k$ and $l$ goes by the link $a = (i, j)$ and take 0 otherwise. It is also estimated that if the activity $i$ is placed on the node $k$ and activity $j$ on the node $l$, then the cost of transportation of a vehicle from $k$ to $l$ is equal to the distance $l_{kl}(y)$. On the other hand, the locations of activities induce fixed costs $a_{ik}$ on the back of construction costs of roads $c_{ij}$ for the network construction (for more details see Gueye, 2012).

The sum of the location, transportation and construction costs is expressed mathematically by the following function:

$$\sum_{i,k} a_{ik}x_{ik} + \sum_{i,k,j,l} q_{ij}l_{kl}x_{ik}x_{jl} + \sum_{i,j} c_{ij}y_{ij}$$

This is a cubic integer model. The "Optimal" network and activity locations are obtained by minimizing the above function under a set of constraints described below.

$$Min \quad \sum_{i,k=1}^{n} a_{ik}x_{ik} + \sum_{\substack{i,k,j,l=1\\ i\neq j,k\neq l}}^{n} q_{ij}l_{kl}x_{ik}x_{jl} + \sum_{\substack{i,j=1\\ i\neq j}}^{n} c_{ij}y_{ij}$$

$s-t:$

$$\begin{array}{lll}
(1) & \sum_{i=1}^{n} x_{ik} = 1\,, & k = 1,\dots,n \\
(2) & \sum_{k=1}^{n} x_{ik} = 1\,, & i = 1,\dots,n \\
(3) & \sum_{a\in\delta^+(k)} z_a^{kl} = \sum_{a\in\delta^-(l)} z_a^{kl} = 1\,,\ k\neq l, & k,l = 1,\dots,n \\
(4) & \sum_{a\in\delta^+(m)} z_a^{kl} = \sum_{a\in\delta^-(m)} z_a^{kl},\ k\neq l, & k,l = 1,\dots,n \\
(5) & l_{kl} - \sum_{a\in Ar} d_a z_a^{kl} \geq 0\,,\ k\neq l, & k,l = 1,\dots,n \\
(6) & z_{a\in\delta^+(i)}^{kl} + z_{a\in\delta^-(i)}^{kl} \leq y_{a\in\delta^+(i)}\,,\ k\neq l, & k,l = 1,\dots,n \\
(7) & x_{ik},\ y_{ij} \in \{0,1\}\,,\quad i\neq j, & i,j,k = 1,\dots,n \\
(8) & z_{ij}^{kl} \in \{0,1\},\ l_{kl}\geq 0,\ k\neq l,\ i\neq j, & k,l,i,j = 1,\dots,n
\end{array}$$

where $m$ is a intermediate node different to $k$ and $l$ taken respectively as origin and destination of the considered path, $\delta^+(i) = \{(i, j) \in Ar / j \in N\}$, $\delta^-(i) = \{(j, i) \in Ar / j \in N\}$.

The constraints (1) and (2) represent assignment constraints ensuring that only one activity is localized in one area only. For the shortest paths, are used the flow variables $z_a^{kl}$ where are associated the constraints of flow conservation. Constraint (5) can give to variable $l_{kl}$ value of the shortest path from $k$ to $l$. The constraint (6) ensures that only are taken into account in the calculation of shortest paths, the link actually constructed.

Thus, in our model the variables are mainly $x_{ik}$, $y_{ij}$ and $l_{kl}$, involved in the objective function, and $z_a^{kl}$ which occurs only in the constraints.

This model is an extension of Koopmans-Beckmann's problem. In addition, our problem, which generalized quadratic assignment problem[1] is NP-complete (Garey et al., 1979; Sahni et al., 1976). Indeed, if we fix all the variables $y_{ij}$ then the problem becomes of finding activities assignments minimizing the sum of travel costs. This is equivalent to solve a quadratic assignment problem. In sequel, we can reformulate the model by replacing the arc $a$ by $(i, j)$.

$$Min \sum_{i,k=1}^{n} a_{ik}x_{ik} + \sum_{\substack{i,k,j,l=1 \\ i\neq j,k\neq l}}^{n} q_{ij}l_{kl}x_{ik}x_{jl} + \sum_{\substack{i,j=1 \\ i\neq j}}^{n} c_{ij}y_{ij}$$

$$s-t:$$

$$D = \begin{cases} (1)\ \sum_{i=1}^{n} x_{ik} = 1\ , & k = 1,\dots,n \\ (2)\ \sum_{k=1}^{n} x_{ik} = 1\ , & i = 1,\dots,n \\ (3)\ \sum_{\substack{j=1 \\ j\neq k}}^{n} z_{kj}^{kl} = \sum_{\substack{i=1 \\ i\neq l}}^{n} z_{il}^{kl} = 1\ ,\ k \neq l, & k,l = 1,\dots,n \\ (4)\ \sum_{\substack{j=1 \\ m\neq j}}^{n} z_{mj}^{kl} = \sum_{\substack{i=1 \\ i\neq m}}^{n} z_{im}^{kl}\ ,\ k\neq l, k\neq m, l\neq m, & m,k,l = 1,\dots,n \\ (5)\ \sum_{\substack{i,j=1 \\ i\neq j}}^{n} d_{ij}z_{ij}^{kl} \leq l_{kl}\ ,\quad k\neq l, & k,l = 1,\dots,n \\ (6)\ z_{ij}^{kl} + z_{ji}^{kl} \leq y_{ij}\ ,\quad i\neq j, k\neq l\ , & i,j,k,l = 1,\dots,n \\ (7)\ x_{ik},\ y_{ij} \in \{0,1\}\ ,\quad i\neq j, & i,j,k = 1,\dots,n \\ \quad z_{ij}^{kl} \in \{0,1\}, l_{kl} \geq 0,\ k\neq l,\ i\neq j, & k,l,i,j = 1,\dots,n \end{cases}$$

### 3. A Decomposition Method

We propose in this section to solve three sub problems for TNLU, which are:

- Shortest path problems;
- A location of activity problem;
- A network design problem.

It starts with Shortest Path problems (SP) which solutions are the shortest path between the different locations and denoted $\gamma_{kl}$. Then the solutions of shortest path problems are introduced into the Location of Activity problem (LA), which returns as solution $(x_{ik}^*)$ the allocation of different activities in different locations. Finally, $x_{ik}^*$ are used in the Network Design problem (ND), which returns as solution the shortest path and roads built, respectively denoted $l_{kl}^*$ and $y_{ij}^*$. So $(x_{ik}^*)$, $(l_{kl}^*)$ and $(y_{ij}^*)$ are solution of the decomposition method of TNLU. The TNLU-heur (Figure 1) is a decomposition method of TNLU which uses heuristic methods (QAP-heur and ND-heur in section 3.2 and 3.3) for QAP and ND problem.

[1] known to be NP-complete

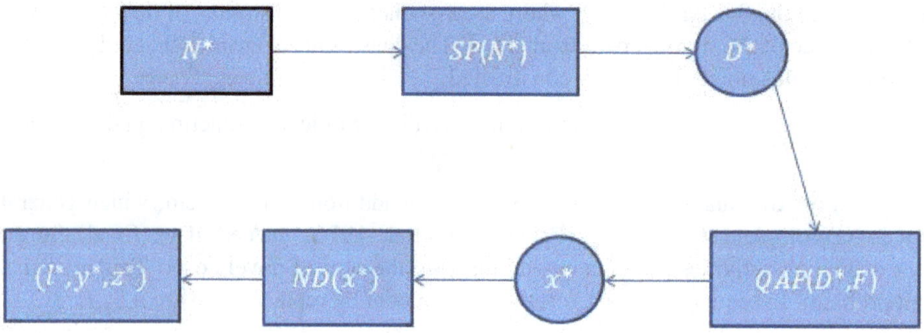


Figure 1. Diagram of TNLU-heur

*3.1 Shortest Path Problems (SP)*

This section contains the resolution of the SP problems which is formulated as follows:

$$Min \sum_{\substack{i,j=1,i\neq j\\ i\neq l,j\neq k}}^{n} d_{ij} z_{ij}^{kl}$$

$$s-t:$$

$$Z = \begin{cases} (1)\ \sum_{\substack{j=1\\ j\neq k}}^{n} z_{kj}^{kl} = \sum_{\substack{i=1\\ i\neq l}}^{n} z_{il}^{kl} = 1,\ k\neq l, & k,l = 1,\dots,n \\ (2)\ \sum_{\substack{j=1\\ j\neq m,j\neq k}}^{n} z_{mj}^{kl} = \sum_{\substack{i=1,\\ i\neq m,i\neq l}}^{n} z_{im}^{kl} \leq 1\ ,\ k\neq m, l\neq m & m,k,l = 1,\dots,n \\ (3)\ z_{ij}^{kl} + z_{ji}^{kl} \leq 1\ ,\ i\neq j,\ i\neq l,\ j\neq k, & i,j = 1,\dots,n \\ (4)\ z_{ij}^{kl} \in \{0,1\}\ ,\quad i\neq j,\ i\neq l,\ j\neq k & i,j,k,l = 1,\dots,n \end{cases}$$

$\forall\, k,l = 1,\dots,n,\ k\neq l.$

Thus we solve $n^2 - n$ SP problems using IBM CPLEX. Finally, at the end of the iterations (of these SP problems), we obtain $\gamma_{kl} = \min_{z\in Z} \sum_{\substack{i,j=1,i\neq j\\ i\neq l,j\neq k}}^{n} d_{ij} z_{ij}^{kl}\ \forall k,l = 1,\dots,n,\ k\neq l.$

*3.2 Activity Location Problem (AL)*

For the LA problem we use solutions of previous problems($\gamma_{kl}$), and we formulate the problem as following :

$$Min\quad g(x,\gamma) = \sum_{i,k=1}^{n} a_{ik}x_{ik} + \sum_{\substack{i,k,j,l=1\\ i\neq j,k\neq l}}^{n} q_{ij}\gamma_{kl}x_{ik}x_{jl}$$

$$s-t:$$

$$X = \begin{cases} (1)\ \sum_{i=1}^{n} x_{ik} = 1\ ,\ k = 1,\dots,n \\ (2)\ \sum_{k=1}^{n} x_{ik} = 1\ ,\ i = 1,\dots,n \\ (3)\ x_{ik} \in \{0,1\},\ i,k = 1,\dots,n \end{cases}$$

If we solve the Quadratic Assignment Problem (QAP), we find the solutions $x_{ik}^*$. Due to the fact that QAP is NP-hard, we propose a heuristic to solve it.

• **A heuristic for QAP (QAP-heur)**

We are given $n$ initial assignments in a random choice. The value of the objective function is evaluated, then two assignments are chosen according to a specific criterion. Then the value of the objective function is evaluated in the case of

swapping these two assignments. If there is a reduction of the cost then this permutation is preserved if not the previous location are kept. The process is repeated until a given number of iterations.
Thus for an initial allocation, we choose according to a criterion the best neighbour of the 2-opt permutation.
The selection criterion is the following (reducing the risk that the algorithm does not stop on a local solution during search):

- $i_0, k_0, j_0, l_0 \in \left\{i, k, j, l \in [1, \ldots, n] : x_{ik} = 1 = x_{jl}\right\}$ which maximizes the function :

$$a_{i_0k_0} + a_{j_0l_0} + \sum_{\substack{i,k=1 \\ i\neq j_0, k\neq l_0}}^{n} q_{ij_0} d_{kl_0} x_{ik} x_{j_0l_0} + \sum_{\substack{j,l=1 \\ j\neq i_0, l\neq k_0}}^{n} q_{i_0j} d_{k_0l} x_{i_0k_0} x_{jl}$$

- Otherwise, if it maximizes the value:

$$a_{i_0k_0} + a_{j_0l_0} + q_{i_0j_0} d_{k_0l_0} + q_{j_0i_0} d_{l_0k_0}$$

- Otherwise, if it maximizes :

$$a_{i_0k_0} + a_{j_0l_0}$$

We put a counter that gives the number of iterations since the last best solution, and if it exceeds a given threshold, then we consider that the algorithm is stuck on a local solution. The threshold is empirically and randomly chosen in the interval $[n^2/2, 2n^2]$.
The algorithm is a 2-opt neighbourhood search method associated with local search as tabou-search and stochastic search.

```
Data: n number of locations and activities, m number of iterations
Random initialization of x;
list ⟵ ∅;
i ⟵0;
while i < m do
    i ⟵i+1;
    Best-Solution ⟵ x;
    Choose i, k, j, l ∈ [1, ..., n] using the criterion;
    y ⟵ permute(x);
    if evaluate(x) ≤ evaluate(y) then
        UPDATE list;
        Choose i, k, j, l ∈ [1, ..., n] using the criterion;
        y ⟵ permute(x);
    else
        x ⟵ permute(x);
        UPDATE list;
    end
end
```

**Algorithm 1:** QAP-heur

*3.3 Network Design Problem (ND)*

We solve the following problem :

$$Min \sum_{i,k=1}^{n} a_{ik} x^*_{ik} + \sum_{\substack{i,k,j,l=1 \\ i\neq j, k\neq l}}^{n} q_{ij} l_{kl} x^*_{ik} x^*_{jl} + \sum_{\substack{i,j=1 \\ i\neq j}}^{n} c_{ij} y_{ij}$$

$$s-t:$$

$$R = \begin{cases} (1)\ \sum\limits_{\substack{j=1\\ j\neq k}}^{n} z_{kj}^{kl} = \sum\limits_{\substack{i=1\\ i\neq l}}^{n} z_{il}^{kl} = 1\ ,\ \ k\neq l, & k,l=1,\dots,n \\ (2)\ \sum\limits_{\substack{j=1\\ m\neq j}}^{n} z_{mj}^{kl} = \sum\limits_{\substack{i=1\\ i\neq m}}^{n} z_{im}^{kl}\ ,\ \ k\neq l, k\neq m, l\neq m, & m,k,l=1,\dots,n \\ (3)\ \sum\limits_{\substack{i,j=1\\ i\neq j}}^{n} d_{ij} z_{ij}^{kl} \leq l_{kl}\ ,\ \ \ k\neq l, & k,l=1,\dots,n \\ (4)\ \ z_{ij}^{kl} + z_{ji}^{kl} \leq y_{ij}\ ,\ \ \ i\neq j,\ k\neq l, & i,j,k,l=1,\dots,n \\ (5)\ \ y_{ij} \in \{0,1\}\ ,\ \ \ i\neq j, & i,j=1,\dots,n \\ \quad z_{ij}^{kl} \in \{0,1\}, l_{kl}\geq 0,\ k\neq l,\ i\neq j, & k,l,i,j=1,\dots,n \end{cases}$$

The resolution of ND give us the solutions $l_{kl}^*$, $y_{ij}^*$ and $z_{ij}^{kl*}$. We can simplify the objective function by $\sum\limits_{\substack{i,k,j,l=1\\ i\neq j,k\neq l}}^{n} q_{ij} l_{kl} x_{ik}^* x_{jl}^* + \sum\limits_{\substack{i,j=1\\ i\neq j}}^{n} c_{ij} y_{ij}$ without changing the problem as the sum $\sum\limits_{i,k=1}^{n} a_{ik} x_{ik}^*$ is constant.

- **A heuristic for Network Design (ND-heur)**

In this sub-section we propose a heuristic for the Network Design problem (ND). This heuristic is known in the literature as deletion method (see Gamvros). The ND-heur consists of:

1. The method start with all edges in accordance with the instance solved. This mean that $\forall i \neq j \in \{1,\dots,n\}$, if $y_{ij}^*$ indicates the constracted roads then :

$$y_{ij}^* = 1 \ \text{ if } \ (d_{ij} < \infty) \ \text{ and } \ (y_{ij}^* = 0 \ \text{ if } d_{ij} = \infty)$$

2. After that, choose $\bar{y}$ in the neighbourhood of $y^*$ such that :

$$\text{For } (k_0, l_0) \in \{1,\dots,n\}^2, k_0 \neq l_0 \ \ c_{k_0 l_0} = \max_{k,l} c_{kl} y_{kl}$$

$$\text{Set } \bar{y}_{k_0 l_0} = 0 \text{ and } \forall (k,l) \neq (k_0, l_0)\ \bar{y}_{kl} = y_{kl}^*$$

3. Then, solve the following problem :

$$Min \quad \sum_{\substack{k,l=1\\ k\neq l}}^{n} \Big( \sum_{\substack{i,j=1\\ i\neq j}}^{n} q_{ij} x_{ik}^* x_{jl}^* \Big) l_{kl}$$

$$s-t: \begin{array}{ll} (1)\ \sum\limits_{\substack{j=1\\ j\neq k}}^{n} z_{kj}^{kl} = \sum\limits_{\substack{i=1\\ i\neq l}}^{n} z_{il}^{kl} = 1\ ,\ \ k\neq l, & k,l=1,\dots,n \\ (2)\ \sum\limits_{\substack{j=1\\ m\neq j}}^{n} z_{mj}^{kl} = \sum\limits_{\substack{i=1\\ i\neq m}}^{n} z_{im}^{kl}\ ,\ \ k\neq l, k\neq m, l\neq m, & m,k,l=1,\dots,n \\ (3)\ \sum\limits_{\substack{i,j=1\\ i\neq j}}^{n} d_{ij} z_{ij}^{kl} \leq l_{kl}\ ,\ \ \ k\neq l, & k,l=1,\dots,n \\ (4)\ z_{ij}^{kl} + z_{ji}^{kl} \leq \bar{y}_{ij}\ ,\ \ \ i\neq j,\ k\neq l, & i,j,k,l=1,\dots,n \\ (5)\ z_{ij}^{kl} \in \{0,1\}, l_{kl}\geq 0,\ k\neq l,\ i\neq j, & k,l,i,j=1,\dots,n \end{array}$$

   where $x^*$ is a fixed activities allocation, to compute the shortest path associated to $\bar{y}$.
   So this problem is the network design problem associated with the activities assignment $x^*$. We solve this problem by using the Flyod-Warshall algorithm (Floyd, 1962).

4. After resolution of the problem, if there is a feasible solution and a cost reduction, update $y^* = \bar{y}$.

5. Finally, repeat all the process from phase 2 up to $n^2 - n$ times.

The step 2 specifies that a link which has the most contribution in the cost $\sum_{i,j=1,i\neq j}^{n} c_{ij}y_{ij}$ is a candidate for deletion and is effectively deleted in the step 3 $\left(\text{if there is a reduction of the global cost : } \sum_{\substack{k,l=1\\k\neq l}}^{n} (\sum_{\substack{i,j=1\\i\neq j}}^{n} q_{ij}x_{ik}^{*}x_{jl}^{*})l_{kl} + \sum_{\substack{i,j=1\\i\neq j}}^{n} c_{ij}y_{ij}\right)$.

## 4. Lower Bound Based on Distances

In this section, the proposed bound is based on a linearization of the cubic term of TNLU problem formulation (see section 2). We replace $\sum_{k=1}^{n}\sum_{l=1}^{n} l_{kl}x_{ik}x_{jl}$ by a new variable $D_{ij}$ $\forall i, j = 1, ..., n,\ i \neq j$, with additional constraints to keep the problem equivalent. So $D_{ij}$ have a structure of distance, which allow us to add triangular inequality to the constraints (equation (3)). We obtain the following linear problem:

$$\text{(TNLU-LIN)} \quad Min \sum_{i,k=1}^{n} a_{ik}x_{ik} + \sum_{i,j=1}^{n} q_{ij}D_{ij} + \sum_{\substack{i,j=1\\i\neq j}}^{n} c_{ij}y_{ij}$$

$$s-t:$$

$$D'_{\gamma} = \begin{cases}
(1)\ x_{ik} \in X & \\
(2)\ D_{ij} \geq \gamma_{kl}(x_{ik} + x_{jl} - 1)\ ,\ k \neq l,\ i \neq j, & k,l,i,j = 1,\ldots,n \\
(3)\ D_{ij} \leq D_{im} + D_{mj}\ ,\ i \neq j, i \neq m \neq j, & k,l,m = 1,\ldots,n \\
(4)\ \sum_{\substack{j=1\\j\neq k}}^{n} z_{kj}^{kl} = \sum_{\substack{i=1\\i\neq l}}^{n} z_{il}^{kl} = 1\ ,\ k \neq l, & k,l = 1,\ldots,n \\
(5)\ \sum_{\substack{j=1\\j\neq m}}^{n} z_{mj}^{kl} = \sum_{\substack{i=1\\i\neq m}}^{n} z_{im}^{kl}\ ,\ k \neq l, k \neq m, l \neq m, & m,k,l = 1,\ldots,n \\
(6)\ \sum_{\substack{i,j=1\\i\neq j}}^{n} d_{ij}z_{ij}^{kl} \geq \gamma_{kl}\ ,\quad k \neq l, & k,l = 1,\ldots,n \\
(7)\ z_{ij}^{kl} + z_{ji}^{kl} \leq y_{ij}\ ,\quad i \neq j, k \neq l, & k,l,i,j = 1,\ldots,n \\
(8)\ \sum_{j=1}^{n} D_{ij} \geq \sum_{k,l=1}^{n} \gamma_{kl}x_{ik}\ ,\quad i \neq j, & i,j = 1,\ldots,n \\
(9)\ \sum_{i=1}^{n} D_{ij} \geq \sum_{k,l=1}^{n} \gamma_{kl}x_{jl}\ ,\quad i \neq j, & i,j = 1,\ldots,n \\
(10)\ D_{ij} \geq 0\ ,\quad i \neq j, & i,j = 1,\ldots,n \\
(11)\ z_{ij}^{kl},\ y_{ij} \in \{0,1\}\ ,\quad i \neq j,\ i,j = 1,\ldots,n,\ \ k \neq l,\ k,l = 1,\ldots,n &
\end{cases}$$

with $X$ the domain of affectation variables presented in subsection 3.2, and $\gamma_{kl} = \min_{z\in Z} \sum_{\substack{i,j=1,i\neq j\\i\neq l,j\neq k}}^{n} d_{ij}z_{ij}^{kl}$.

The relaxation of TNLU-LIN is a lower bound of TNLU, and we solve the problem using IBM CPLEX. Depending on the structure of the instances, to have a tighter bound, we can replace the inequalities of the constraints (6), (8) and (9) of $(D'_{\gamma})$ by equalities.

## 5. Numerical Simulations on Test Problems

The behavior of TNLU-heur is evaluated using test problems on academics instances and Dakar city. For academics instances we use data from Los (1978) and QAPLIB (Burkard et al., 1996) benchmark. The QAPLIB problems are instances of Nugent, Vollman, and Ruml (denoted by Nug), Elshafei (Els), Wolkowicz (Had), Krarup and Pruznan (Kra). More precisely we construct the instances from flows and distances of "Nug", "Els", "Had" and "Kra", where there is no localization and construction costs. In addition, the costs of localization are null and construction costs are obtained by multiplying the distances by a mean cost 10 ie $c_{ij} = 10.0 \times d_{ij}\ \forall\ i, j = 1, \ldots, n$ for instances "Nug" and " Had", by 100 for instances "Els", and by 1000 for instances "Kra". The mean construction costs are fixed according to the order of size of the values of distances.

Recall the considered data:

$n$ : number of locations and activities.

$A = (a_{ik})$ : location cost matrix of the activity $i$ on the area $k$.

$Q = (q_{ij})$ : moving flow matrix per unit of times between the activities $i$ and $j$.

$C = (c_{ij})$ : construction cost matrix of link or road $(i, j)$.

$D = (d_{ij})$ : direct distance matrix between two locations $i$ and $j$ or distance of link $(i, j)$.

The lower bound has been solved with IBM Cplex Optimizer 12.5 solver interfaced with C++. The tests have been performed on a HP ProBook 450 Intel (R) Core (TM) i5 3230M CPU @2.40GHz.

For simulations on academic instances and Dakar city, we have executed ten times and have kept the best solution. This is due to the stochastic nature of the algorithm.

*5.1 Tests on Academic Instances*

In all tables, Los x means data of Los for n = x, Nug x means data of Nug for n = x and so on. In order to evaluate the result obtained by our heuristic we calculate the gap as follows:

$$Gap = \frac{opt - LB}{opt}$$

where $opt$ is the best value found (upper bound) and $LB$ is the lower bound based on distances. The following table 1 summarizes results for different step sizes. We see that the heuristic TNLU-heur provides good upper bounds for the TNLU problem. Indeed, for some instances of the academic test the gap is quite small. For instances Had and Nug, the CPU time grows at a linear rate with $n$. However, for instances Los the CPU time remains nearly constant, firstly for n=8, 10,12; and secondly for n=14,15.

*5.2 Tests on Dakar City*

For real life simulation of Dakar we could get from the DGTC (Direction des Travaux Géographiques et Cartographiques), a zonal division of the region into zones corresponding to the 2000 Census Report information with projection to 2015. And from this Census Report we could have information on the flow of movement $(q_{ij})$ between areas. For data of cost construction of roads $(c_{ij})$, we obtained information about mean costs updated in January 2008, and our choice fell on the road type monolayer whose width is $7m$ and the cost per km (all taxes included) is 33 246 500 FCFA ($\simeq 21.7764575e+09$ euro). So if we denote by $C_M$ the mean cost of road construction, then the cost of construction of the road is calculated by $c_{ij} = C_M * d_{ij}$.

To calculate the distance we used OSRM project (Open Source Routing Machine) which is a C++ implementation of a routing engine high performance for shortest paths in road networks. It combines sophisticated routing algorithms with the data of the road network and free from OpenStreetMap (OSM). OSRM is able to calculate and output a shortest path between any origin and destination in milliseconds.

For the cost of localization activities $(a_{ik})$, we did not find an institution that can provide us with this information. Therefore we decided to build ourselves these costs realistically. We start with a choice of activities to locate in zones. The tests were done for a maximum number of zones equal to $n = 20$. Table 2 illustrates the concerned activities and zones.

In addition, to give the cost of locating an activity on a zone, we evaluated the mean cost of land for this activity in this zone and the cost of construction of the building for the same activity. The sum of these two costs gives the cost of location. So we look for the mean cost of $150m^2$ and for a considered activity we estimate approximately a mean size of the area, then we can obtain a mean cost of land for the activity in this zone.

For example, the Plateau Sud price of $150m^2$ can be estimated to $11.0e + 07$ FCFA ($\simeq 7.205e + 10$ euro). Area can be estimated for a Ministry to $300m^2 = 2 \times 150m^2$, thus the cost of land for this ministry in this zone is $2 \times 11.0e + 07 = 2.2e + 08$ FCFA($\simeq 1.441e + 11$ euro). In addition, we can estimate the cost of construction of the building to $2.0e + 08$ FCFA($\simeq 1.310e + 11$ euro). So the cost of locating the Ministry in Plateau Sud is $2.2e + 08$ FCFA + $2.e + 08$ FCFA = $4.2e + 08$ FCFA ($\simeq 2.751e + 11$ euro).

The other data used for simulations on Dakar city are reported in the following tables 3, 4, 5, 6 . Each table contains a

Table 1. Results of TNLU-heur

| Instances | Best value found | Time CPU in second | Number of iterations QAP/ND | Lower bound | Gap |
|---|---|---|---|---|---|
| Els19 | 1.75309e+007 | 23 | 10000/342 | 8.18476+006 | 53.31% |
| Had12 | 1932 | 14 | 10000/132 | 1869.5 | 3.23% |
| Had16 | 4120 | 19 | 10000/240 | 3986.3 | 3.25% |
| Had20 | 7422 | 28 | 10000/380 | 7246 | 2.37% |
| Kra32 | 2.93893e+006 | 104 | 10000/992 | 1.55e+006 | 47.26% |
| Los8 | 6.83477e+006 | 15 | 10000/56 | 5.60862e+006 | 17.94% |
| Los10 | 1.24041e+007 | 15 | 10000/90 | 1.10511e+007 | 10.91% |
| Los12 | 2.40381e+007 | 15 | 10000/132 | 2.13929e+007 | 11.00% |
| Los14 | 4.03875e+007 | 25 | 10000/182 | 3.55418e+007 | 12.00% |
| Los15 | 4.92603e+007 | 25 | 10000/210 | 4.26734e+007 | 13.37% |
| Los16 | 5.89492e+007 | 32 | 10000/240 | 5.11031e+007 | 13.31% |
| Nug12 | 904 | 17 | 10000/132 | 824.222 | 8.83% |
| Nug14 | 1398 | 31 | 10000/182 | 1304.5 | 6.69% |
| Nug15 | 1570 | 20 | 10000/210 | 1453.78 | 7.40% |
| Nug16 | 2094 | 20 | 10000/240 | 1883.99 | 10.03% |
| Nug17 | 2232 | 20 | 10000/272 | 2053.28 | 8.01% |
| Nug18 | 2486 | 23 | 10000/306 | 2294.9 | 7.69% |
| Nug20 | 3196 | 26 | 10000/380 | 2918.45 | 8.68% |
| Nug22 | 10297 | 31 | 10000/462 | 8634.81 | 16.14% |
| Nug25 | 14353 | 43 | 10000/600 | 12041.4 | 16.11% |
| Nug30 | 7192 | 73 | 10000/870 | 6656.27 | 7.45% |

Description: For each instance, the Time CPU is the minimal execution time in the ten experiments.

Table 2. Activities and areas

| n | Activities | Zones |
|---|---|---|
| 1 | Ministry | Plateau sud / PL.SUD |
| 2 | Supermarket | Médina / MED |
| 3 | University | Grand dakar-usine / GD.US |
| 4 | Medical technology clusters | Castor / CAST |
| 5 | Medical Poles | Fann Point E - Amitié / F.P.E.A |
| 6 | Embassy | Ngor-Yoff / NG-YO |
| 7 | Police | Parcelles assainies / P.A |
| 8 | Fire Brigade | Patte d'oie-mariste / PO-MAR |
| 9 | Town Hall | Thiaroye-yembeul / TH.YEUM |
| 10 | Public Garden | Guediewaye-ouest / GUED-OUEST |
| 11 | Stadium | Guediewaye-centre / GUED-CENT |
| 12 | Residence | Guediewaye-est / GUED-EST |
| 13 | Telecom | Thiaroye sur Mer / TH-MER |
| 14 | Real Estate Agency | Malika-Keur Massar / MAL.KM |
| 15 | Post Office | Mbao / MB |
| 16 | Country-Police | Rufisque Quartiers traditionnels / RUF-TRADI |
| 17 | Secondary education | Diamniadio / DIAM |
| 18 | Olympic swimming pool | Hann mariste / H.M |
| 19 | Primary school | Sangalkam / SANG |
| 20 | Bar restaurant | Zone industrielle / Z.I |

17 × 17 matrix. The values of localization and construction costs are rounded to two decimal.

Table 3. Distance matrix in kilometer based on the road network

| | | | | | | | | | | | | | | | | |
|---|---|---|---|---|---|---|---|---|---|---|---|---|---|---|---|---|
| 0 | 4.05 | 6.82 | 9.13 | 7.33 | 19.90 | 12.90 | 10.40 | 20.20 | 18.90 | 17.00 | 15.40 | 16.20 | 27.00 | 27.30 | 29.80 | 40.20 |
| 3.71 | 0 | 8.44 | 7.55 | 4.57 | 11.20 | 11.90 | 8.68 | 18.20 | 17.80 | 15.80 | 14.10 | 15.00 | 18.16 | 25.70 | 27.90 | 38.70 |
| 6.14 | 8.38 | 0 | 2.87 | 3.78 | 11.10 | 7.09 | 5.65 | 15.80 | 14.90 | 11.60 | 10.30 | 12.00 | 22.80 | 23.00 | 25.30 | 35.90 |
| 7.97 | 7.37 | 2.79 | 0 | 4.93 | 11.80 | 5.11 | 3.79 | 13.60 | 12.20 | 9.82 | 8.40 | 9.91 | 19.80 | 21.10 | 23.60 | 34.00 |
| 7.18 | 4.59 | 3.81 | 4.86 | 0 | 7.37 | 8.91 | 8.04 | 18.10 | 15.40 | 12.90 | 11.50 | 14.40 | 24.60 | 25.90 | 27.80 | 38.40 |
| 12.74 | 12.30 | 14.60 | 11.40 | 7.50 | 0 | 10.20 | 13.90 | 20.80 | 18.30 | 15.90 | 14.40 | 16.80 | 27.10 | 28.00 | 30.50 | 40.90 |
| 11.29 | 11.40 | 8.40 | 5.23 | 8.63 | 9.42 | 0 | 5.47 | 10.90 | 9.29 | 6.85 | 5.40 | 8.88 | 17.00 | 10.82 | 22.50 | 32.90 |
| 9.43 | 9.85 | 8.31 | 6.86 | 10.80 | 13.30 | 5.76 | 0 | 10.50 | 9.88 | 7.51 | 6.05 | 6.89 | 16.60 | 18.00 | 20.40 | 30.80 |
| 15.82 | 17.20 | 15.40 | 13.70 | 17.80 | 20.60 | 10.90 | 10.70 | 0 | 3.86 | 4.84 | 5.58 | 4.81 | 6.59 | 14.70 | 16.90 | 27.50 |
| 15.54 | 16.60 | 14.40 | 13.00 | 15.40 | 18.30 | 9.30 | 9.97 | 3.86 | 0 | 2.82 | 4.07 | 8.72 | 10.80 | 15.50 | 17.70 | 28.40 |
| 16.00 | 15.90 | 12.70 | 9.92 | 12.90 | 15.80 | 6.80 | 8.36 | 4.84 | 2.82 | 0 | 1.59 | 7.08 | 10.80 | 15.50 | 17.80 | 28.30 |
| 15.20 | 14.10 | 10.70 | 8.50 | 11.50 | 14.30 | 5.35 | 6.16 | 5.58 | 4.01 | 1.59 | 0 | 8.21 | 11.80 | 16.60 | 18.80 | 29.40 |
| 13.58 | 15.90 | 12.40 | 10.90 | 15.10 | 17.70 | 10.20 | 8.00 | 4.35 | 6.99 | 5.25 | 6.23 | 0 | 10.60 | 11.90 | 14.40 | 24.80 |
| 27.10 | 18.16 | 22.40 | 19.90 | 24.30 | 26.90 | 17.00 | 16.80 | 6.59 | 10.80 | 10.80 | 11.80 | 11.00 | 0 | 10.60 | 12.80 | 23.40 |
| 23.40 | 21.70 | 18.70 | 17.30 | 21.70 | 23.90 | 16.30 | 14.30 | 10.60 | 11.50 | 11.50 | 12.50 | 8.44 | 6.61 | 0 | 8.47 | 19.60 |
| 30 | 27.90 | 25.00 | 23.70 | 27.60 | 30.40 | 22.80 | 20.70 | 16.90 | 17.70 | 17.70 | 18.80 | 14.90 | 12.80 | 12.00 | 0 | 24.80 |
| 40.40 | 38.40 | 35.50 | 34.10 | 38.20 | 40.80 | 33.10 | 31.10 | 27.50 | 28.30 | 28.30 | 29.40 | 25.30 | 23.40 | 17.30 | 25.30 | 0 |

Table 4. Matrix of moving flow(Origin-Destination matrix)

| | | | | | | | | | | | | | | | | |
|---|---|---|---|---|---|---|---|---|---|---|---|---|---|---|---|---|
| 0 | 13310 | 394 | 2266 | 7218 | 1579 | 9148 | 1583 | 3235 | 1139 | 5023 | 981 | 1410 | 400 | 1917 | 74 | 24 |
| 12428 | 0 | 759 | 1055 | 5197 | 1881 | 3565 | 425 | 762 | 123 | 1418 | 926 | 704 | 239 | 467 | 0 | 15 |
| 328 | 850 | 0 | 1433 | 3986 | 0 | 446 | 0 | 149 | 245 | 0 | 0 | 0 | 0 | 467 | 0 | 24 |
| 2737 | 1234 | 904 | 0 | 2625 | 1404 | 1937 | 465 | 0 | 167 | 186 | 0 | 744 | 330 | 0 | 0 | 0 |
| 7882 | 4712 | 4014 | 2836 | 0 | 1222 | 2626 | 545 | 677 | 24 | 837 | 346 | 110 | 0 | 749 | 0 | 0 |
| 2948 | 1467 | 105 | 1218 | 1210 | 0 | 2543 | 453 | 0 | 0 | 0 | 327 | 903 | 84 | 0 | 0 | 24 |
| 9635 | 3479 | 446 | 1397 | 3302 | 2621 | 0 | 2937 | 761 | 734 | 1928 | 291 | 185 | 18 | 0 | 0 | 0 |
| 1773 | 356 | 0 | 465 | 730 | 235 | 2820 | 0 | 0 | 249 | 59 | 0 | 153 | 0 | 12 | 0 | 0 |
| 3633 | 1346 | 149 | 11 | 492 | 283 | 832 | 174 | 0 | 559 | 7610 | 2564 | 3939 | 2550 | 2634 | 0 | 39 |
| 1405 | 123 | 245 | 167 | 24 | 0 | 902 | 119 | 694 | 0 | 4138 | 654 | 101 | 0 | 0 | 0 | 0 |
| 5245 | 1841 | 59 | 447 | 1416 | 0 | 1786 | 153 | 7536 | 4920 | 0 | 6292 | 268 | 91 | 167 | 128 | 68 |
| 1616 | 926 | 0 | 0 | 291 | 327 | 291 | 0 | 2582 | 945 | 7380 | 0 | 0 | 149 | 0 | 0 | 0 |
| 1973 | 397 | 0 | 310 | 165 | 903 | 185 | 306 | 3583 | 101 | 463 | 0 | 0 | 0 | 534 | 0 | 0 |
| 685 | 239 | 0 | 330 | 0 | 0 | 0 | 0 | 2316 | 0 | 91 | 0 | 0 | 0 | 0 | 0 | 49 |
| 1956 | 467 | 467 | 0 | 282 | 0 | 0 | 0 | 2499 | 0 | 302 | 0 | 534 | 12 | 0 | 467 | 39 |
| 69 | 0 | 0 | 0 | 0 | 0 | 0 | 0 | 0 | 0 | 138 | 0 | 0 | 0 | 560 | 0 | 0 |
| 39 | 15 | 24 | 0 | 0 | 24 | 0 | 0 | 39 | 0 | 53 | 49 | 0 | 49 | 278 | 0 | 0 |

Table 5. Matrix of localization cost ($e + 08$ FCFA($\simeq 655\ e + 08$ euro))

| | | | | | | | | | | | | | | | | |
|---|---|---|---|---|---|---|---|---|---|---|---|---|---|---|---|---|
| 4.2 | 3.1 | 2.5 | 2.8 | 3.3 | 2.6 | 2.4 | 2.6 | 2.1 | 2.3 | 2.4 | 2.5 | 2.3 | 2.1 | 2.2 | 2.1 | 2.1 |
| 74 | 71.8 | 70.6 | 71.1 | 72.2 | 70.8 | 70.4 | 70.7 | 69.7 | 70.2 | 70.4 | 70.5 | 70.2 | 69.8 | 69.9 | 69.7 | 69.7 |
| 1190 | 975 | 850 | 906 | 1016.5 | 871.6 | 838 | 860 | 767.1 | 816 | 830 | 848 | 810 | 778 | 784 | 764 | 762 |
| 0.4 | 0.3 | 0.2 | 0.2 | 0.3 | 0.2 | 0.1 | 0.2 | 0.1 | 0.1 | 0.1 | 0.2 | 0.1 | 0.1 | 0.1 | 0.1 | 0.1 |
| 970 | 862.5 | 800 | 828 | 883.2 | 810.8 | 794 | 805 | 758.5 | 783 | 790 | 799 | 780 | 764 | 767 | 757 | 756 |
| 7.6 | 5.5 | 4.2 | 4.8 | 5.9 | 4.4 | 4.1 | 4.3 | 3.4 | 3.9 | 4 | 4.2 | 3.8 | 3.5 | 3.5 | 3.3 | 3.3 |
| 41.6 | 21.5 | 9.8 | 15.1 | 25.4 | 11.9 | 8.7 | 10.8 | 2.1 | 6.7 | 8 | 9.6 | 6.1 | 3.1 | 3.7 | 1.8 | 1.6 |
| 20.7 | 13.5 | 9.3 | 11.2 | 14.9 | 10.1 | 8.9 | 9.7 | 6.6 | 8.2 | 8.7 | 9.3 | 8 | 6.9 | 7.1 | 6.5 | 6.4 |
| 46 | 24.5 | 12 | 17.6 | 28.6 | 14.2 | 10.8 | 13 | 3.7 | 8.6 | 10 | 11.8 | 8 | 4.8 | 5.4 | 3.4 | 3.2 |
| 37.7 | 19.8 | 9.3 | 14 | 23.2 | 11.1 | 8.3 | 10.2 | 2.4 | 6.5 | 7.7 | 9.2 | 6 | 3.3 | 3.8 | 2.2 | 2 |
| 307.7 | 284.4 | 270.8 | 276.9 | 288.9 | 273.2 | 269.5 | 271.9 | 261.8 | 267.2 | 268.7 | 270.6 | 266.5 | 263 | 263.7 | 261.5 | 261.3 |
| 246.7 | 175 | 133.3 | 152 | 188.8 | 140.5 | 129.3 | 136.7 | 105.7 | 122 | 126.7 | 132.7 | 120 | 109.3 | 111.3 | 104.7 | 104 |
| 6 | 4.4 | 3.5 | 3.9 | 4.7 | 3.6 | 3.4 | 3.5 | 2.8 | 3.2 | 3.3 | 3.4 | 3.2 | 2.9 | 3 | 2.8 | 2.8 |
| 0.7 | 0.4 | 0.3 | 0.3 | 0.5 | 0.3 | 0.2 | 0.3 | 0.1 | 0.2 | 0.2 | 0.2 | 0.2 | 0.2 | 0.2 | 0.1 | 0.1 |
| 8.1 | 4.9 | 3 | 3.8 | 5.5 | 3.3 | 2.8 | 3.2 | 1.8 | 2.5 | 2.7 | 3 | 2.4 | 1.9 | 2 | 1.7 | 1.7 |
| 52.3 | 27.3 | 12.7 | 19.2 | 32.1 | 15.2 | 11.3 | 13.8 | 3 | 8.7 | 10.3 | 12.4 | 8 | 4.3 | 5 | 2.6 | 2.4 |
| 290 | 236.3 | 205 | 219 | 246.6 | 210.4 | 202 | 207.5 | 184.3 | 196.5 | 200 | 204.5 | 195 | 187 | 188.5 | 183.5 | 183 |

Table 6. Matrix of construction cost ($e + 08$ FCFA($\simeq 655\ e + 08$ euro))

| | | | | | | | | | | | | | | | | |
|---|---|---|---|---|---|---|---|---|---|---|---|---|---|---|---|---|
| 0 | 1.35 | 2.27 | 3.04 | 2.44 | 6.62 | 4.29 | 3.46 | 6.72 | 6.28 | 5.65 | 5.12 | 5.39 | 8.98 | 9.08 | 9.91 | 13.37 |
| 1.23 | 0 | 2.81 | 2.51 | 1.52 | 3.72 | 3.96 | 2.89 | 6.05 | 5.92 | 5.25 | 4.69 | 4.99 | 6.04 | 8.54 | 9.28 | 12.87 |
| 2.04 | 2.79 | 0 | 0.95 | 1.26 | 3.69 | 2.36 | 1.88 | 5.25 | 4.95 | 3.86 | 3.42 | 3.99 | 7.58 | 7.65 | 8.41 | 11.94 |
| 2.65 | 2.45 | 0.93 | 0 | 1.64 | 3.92 | 1.70 | 1.26 | 4.52 | 4.06 | 3.26 | 2.79 | 3.29 | 6.58 | 7.02 | 7.85 | 11.30 |
| 2.39 | 1.53 | 1.27 | 1.62 | 0 | 2.45 | 2.96 | 2.67 | 6.02 | 5.12 | 4.29 | 3.82 | 4.79 | 8.18 | 8.61 | 9.24 | 12.77 |
| 4.24 | 4.09 | 4.85 | 3.79 | 2.49 | 0 | 3.39 | 4.62 | 6.92 | 6.08 | 5.29 | 4.79 | 5.59 | 9.01 | 9.31 | 10.14 | 13.60 |
| 3.75 | 3.79 | 2.79 | 1.74 | 2.87 | 3.13 | 0 | 1.82 | 3.62 | 3.09 | 2.28 | 1.80 | 2.95 | 5.65 | 3.60 | 7.48 | 10.94 |
| 3.14 | 3.27 | 2.76 | 2.28 | 3.59 | 4.42 | 1.91 | 0 | 3.49 | 3.28 | 2.50 | 2.01 | 2.29 | 5.52 | 5.98 | 6.78 | 10.24 |
| 5.26 | 5.72 | 5.12 | 4.55 | 5.92 | 6.85 | 3.62 | 3.56 | 0 | 1.28 | 1.61 | 1.86 | 1.60 | 2.19 | 4.89 | 5.62 | 9.14 |
| 5.17 | 5.52 | 4.79 | 4.32 | 5.12 | 6.08 | 3.09 | 3.31 | 1.28 | 0 | 0.94 | 1.35 | 2.90 | 3.59 | 5.15 | 5.88 | 9.44 |
| 5.32 | 5.29 | 4.22 | 3.30 | 4.29 | 5.25 | 2.26 | 2.78 | 1.61 | 0.94 | 0 | 0.53 | 2.35 | 3.59 | 5.15 | 5.92 | 9.41 |
| 5.05 | 4.69 | 3.56 | 2.83 | 3.82 | 4.75 | 1.78 | 2.05 | 1.86 | 1.33 | 0.53 | 0 | 2.73 | 3.92 | 5.52 | 6.25 | 9.77 |
| 4.51 | 5.29 | 4.12 | 3.62 | 5.02 | 5.88 | 3.39 | 2.66 | 1.45 | 2.32 | 1.75 | 2.07 | 0 | 3.52 | 3.96 | 4.79 | 8.25 |
| 9.01 | 6.04 | 7.45 | 6.62 | 8.08 | 8.94 | 5.65 | 5.59 | 2.19 | 3.59 | 3.59 | 3.92 | 3.66 | 0 | 3.52 | 4.26 | 7.78 |
| 7.78 | 7.21 | 6.22 | 5.75 | 7.21 | 7.95 | 5.42 | 4.75 | 3.52 | 3.82 | 3.82 | 4.16 | 2.81 | 2.20 | 0 | 2.82 | 6.52 |
| 9.97 | 9.28 | 8.31 | 7.88 | 9.18 | 10.11 | 7.58 | 6.88 | 5.62 | 5.88 | 5.88 | 6.25 | 4.95 | 4.26 | 3.99 | 0 | 8.25 |
| 13.43 | 12.77 | 11.80 | 11.34 | 12.70 | 13.56 | 11.00 | 10.34 | 9.14 | 9.41 | 9.41 | 9.77 | 8.41 | 7.78 | 5.75 | 8.41 | 0 |

The map in Figure 2 give zones we have considered for simulation for n=17. The results of simulations are given in the

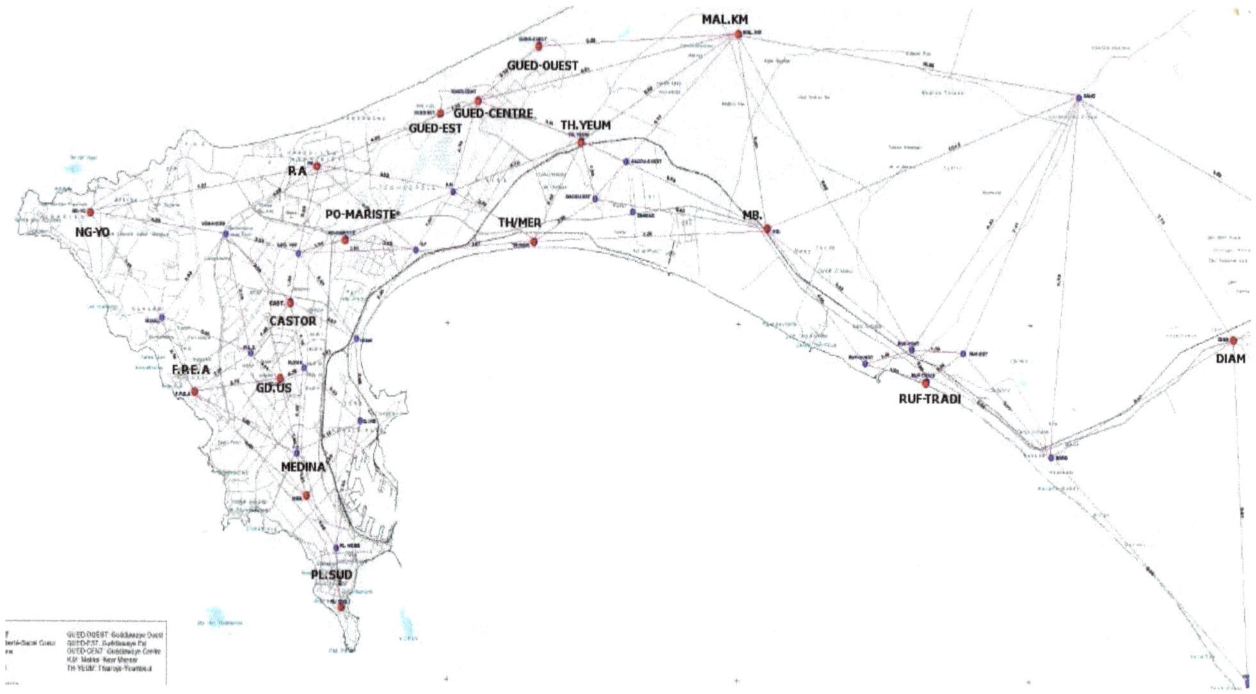


Figure 2. The zones for n=17.

Table 7, where DKx means data on Dakar city intances with size n=x. With this data (information) the simulations has been performed. Figure 3 shows the logical connections between the different areas, given by the optimal solution for

Table 7. Results of TNLU-heur

| Data size | Best value found | Time CPU(s) | Number of Iterations QAP/ND | Lower bound | Gap |
|---|---|---|---|---|---|
| DK6 | 1.75613e+011 | 17 | 10000/30 | 1.75339e+011 | 0.16% |
| DK10 | 1.69992e+011 | 18 | 10000/90 | 1.69259e+011 | 0.43% |
| DK12 | 2.09745e+011 | 19 | 10000/132 | 2.08995e+011 | 0.36% |
| DK17 | 2.26454e+011 | 25 | 10000/272 | 2.24796e+011 | 0.73% |
| DK20 | 2.43257e+011 | 30 | 10000/380 | 2.42865e+011 | 0.16% |

$n = 17$. We can see that Diamniadio must be connected to Rufisque-Traditionnel and itself joined to Medina and Plateau-Sud, through direct links or roads. That is, it would be interesting to join the suburbs to downtown of Dakar, through direct connections to make the road network more fluid. In order not to overload the road network, we can develop a network of maritime shuttle linking these suburb areas and the downtown. And this is in perfect harmony with the geometrical shape of the entire south-eastern part of the region of Dakar. We see that in Table 7 the good result for simulations, where the gap is quite small for all sizes of tested problem.

In addition to the analysis of the SP, AL and ND problems, a new analysis of the set of relocation possibilities is proposed in the next section.

## 6. Multi-agent Simulations

In this section, the behavior of SDO, only limited to the test of **localisation**, is evaluated using test problems on Dakar city. We show how the simulator has been used to give the SDO software that can be seen as an over-layer of MATSim. We use MATSim as a multi-agent simulator system, and need for that to generate agent plans. The necessary adaptation elements of the simulations for Dakar city are presented in sub-section 6.3.

Figure 3. Logical links between nodes

*6.1 MATSim*

A " multi-agent system (MAS)" is a system composed of a set of agents located in a certain environment and interacting according to certain relations. An agent is an entity characterized by the fact that it is, at least partially, autonomously. The implementation of this system on a computer by computer programs, translating these interactions is the multi-agent simulation. In our paper, the agents are individuals (flows of activities) of a city that interact during the displacement. MATSim is a simulation tool multi-agent especially in transport system simulation. The tool is designed since a decade thanks to the collaboration of volunteers and researchers. This is one of the biggest free systems in that category. With adequate computer resources, MATSim is able to simulate the behavior of millions of individuals on very large transportation networks. It is compatible with the data sets of various geographic information systems as OpenStreetMap. And being free, it is designed to be extensible and customizable.

Very schematically, the MATSim simulations are conducted in three steps (see Figure 4). Each agent (or individual) has

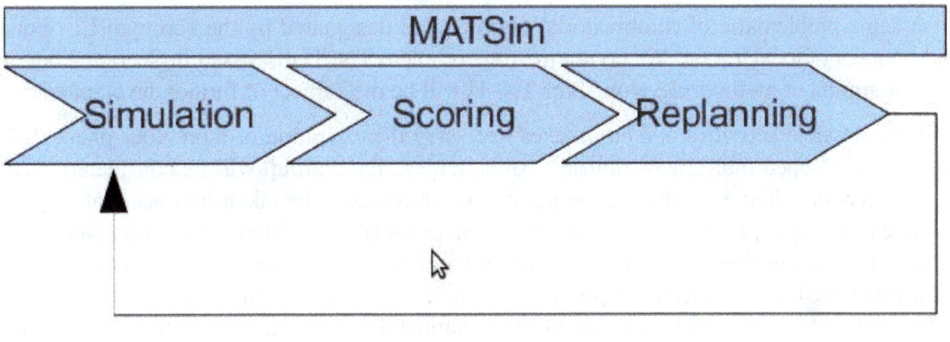


Figure 4. Diagram of MATSim simulations

at the beginning of the simulation a plan of activities that he wants to perform on the day. This can be for example: leave home at 8:00 am to bring his children to school and then go to work, take back their children at 4:30 pm to lead them in a sporting activity, retrieve them to 6:00 pm to do some shopping and then come back to home. Initially, the plan thus specifies the sequence of tasks to be done by individuals but does not mention where exactly and how these activities will be made. The execution of the plan will specifically consists in making choices on these specific places and displacement routes to perform these activities.

After the first iteration of the simulation, a phase called "scoring" is performed. It will be to evaluate, to put a note (or a score) on the selected locations and routes. This score is a function called **utility** in the mathematical sense (function) and economic (utility) of the term that the alleged agents thinking and all acting in the same way (which is restrictive,...) will seek to maximize.

In this objective of maximizing that a replanning phase follows that of "scoring". Replanning allows agents to change eventually their plans by deciding other execution venues of their activities, other routes, or of elimination of activities, helping to increase the utility of their displacements.

The system iterate through these three steps until a maximum number of iterations, or until observation of minimal changes of utility scores. Note, however, that from a mathematical point of view, there is no evidence, in this system, of convergence to a stable state. MATSim simulation requires data on the transport network, the plans of the agents and the location of all the activities they will perform.

*6.2 Simulation-Driven Optimization (SDO)*

The SDO is an overlay of MATSim as part of a similar project called DAMA (Balac et al., 2014). The tool calculates the optimal locations of activities based on the results of the simulations of MATSim. The general operation is described in the Figure 5 below. Generally, after a MATSim simulation, SDO takes into account the displacements generated by the

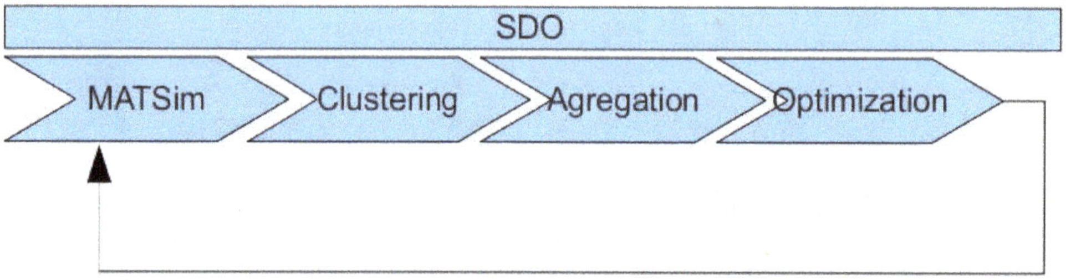


Figure 5. Simulation-Driven Optimization (SDO)

simulation and propose a relocation of some activities (possibly all) to increase the utility scores. The problem that solves SDO is whether where should be located the activities so that the sum of utility scores of the agents is the highest possible. Due to the complexity of the function used by MATSim, we substitute in SDO, to that function the sum of displacement time or traveled distances. Intuitively, more displacement will be shorter in time and distance, and more one can hoped that the utility for the agent will be strong.

Thus we have the same problematic of combinatorial optimization designated by the acronym LU. For the time, SDO only limited to the aspect of localization. We do not treat the problem TNLU discussed in Section 2 but only a part (AL) thereof. The full integration of methods developed for TNLU will be the subject of further developments.

The large quantity of possible activities in a city makes necessary the execution of a previous phase called "clustering" where activities will be grouped into groups (clusters) of activities. Each group will be considered as representative of a single entity. This has the effect of reducing the number of activities to be taken into account. The total amount of displacement between these groups, as well as the displacement times are calculated in the phase known as "aggregation". These data represent matrices of flow and distance of the problem AL. Due to them, the heuristics developed for solving the problem AL are executed, after which an "optimal" relocation is proposed. This relocation changes the geographical coordinates of the proposed activities. Then, a new MATsim simulation is restarted, with these new localizations, and so on until a fixed number of iterations. Analogously to the MATSim simulation, there is no mathematical proof that the loop which concatenates simulation and optimization of localizations converges to a stable state of the displacements.

*6.3 SDO computations on Dakar City*

The use of SDO tool to the cases of Dakar requires to acquire the necessary files for MATSim simulations. Three types of files are generally used: the network file, the file containing the original plans of the agents, and one containing the list of activities and localizations.

Files describing transport networks were first obtained by the OpenStreetMap website (2014). The OpenStreetMap (OSM) is an international project founded in 2004 with the aim of creating a free map. Volunteer contributors send GPS data to the computer system which is responsible to integrate them in the geographic database. The geographical information about Dakar city are refined but remains incomplete. So we used partial information. These OSM files were transformed into the format used by MATSim as shown by the following network diagrams (Figure 6) obtained with the company's software Senozon (2014). The second type of file contains agents' plans. In order to make sure first that the proposed methods

Figure 6. The Dakar network

were operating in computational point of view, we considered for Dakar, a sample of 102 agents (from n=17 zones) and their selected randomly activities. The third type is the file indicating the selected activities and their localizations, extracted from files of OSM networks.
With this information, the simulation and optimization loop explained above has been performed. Each iteration of the loop generates the outputs of the MATSim simulation which can be analyzed at the end of process. One can for example for each iteration perform the average of displacement times (or distances traveled, or scores) of all agents and see its evolution during the iterations. As in each loop an optimization phase is realized, the shape of the curve on a plane (horizontal axis: the iteration number, vertical axis: the average) allows to analyze the relevance of the approach. The decreasing displacement times indicate that the proposed relocations allow to facilitate access to activities. We give in Figures 7, 8 and 9 the evolution of averages obtained for the case of Dakar. The TravelTime is in seconde (s) and the TravelDistance in kilometer (km).

The graph on the utility (utility) of the Figure 7 shows that during the iterations, performed relocations made it possible to increase the average utility of users because the curve is increasing. In other words, people do displacements more and more considered as "useful" in the sense of the metric of the utility function of MATSim. Thus, the relocations increase the average "satisfaction" of users.

The curve of displacement times (TravelTime) in Figure 8 is generally decreasing, showing that times decrease with the proposed relocations.

On the other side, It is surprising that the distances (TravelDistance) traveled (Figure 9) increase slightly and decrease simultaneously. Everything goes as if the agents were doing, sometimes, a little more of distance, but in less time on average. This may be explained by a phenomenon called "congestion" that we have certainly confirmed by a more detailed analysis of the functioning of MATSim.

The congestion hypothesis consists in considering that the transport time between two locations not only depends on the distance between them, but also the number of people following the same displacement itinerary. If we place under this

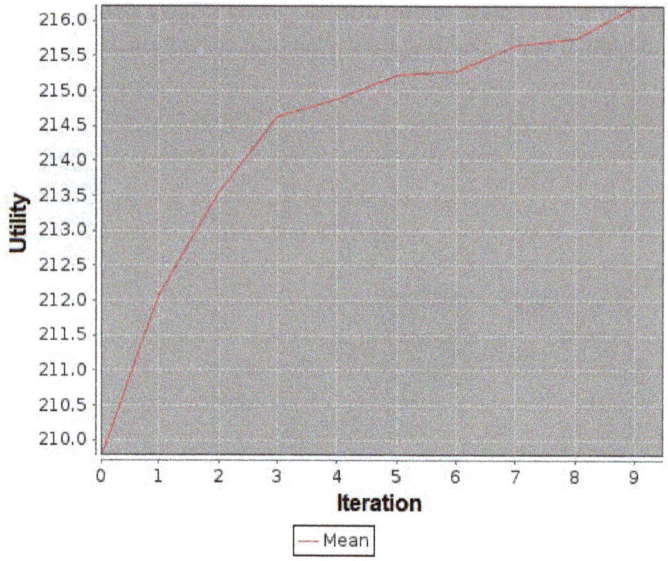


Figure 7. Average of the utilities

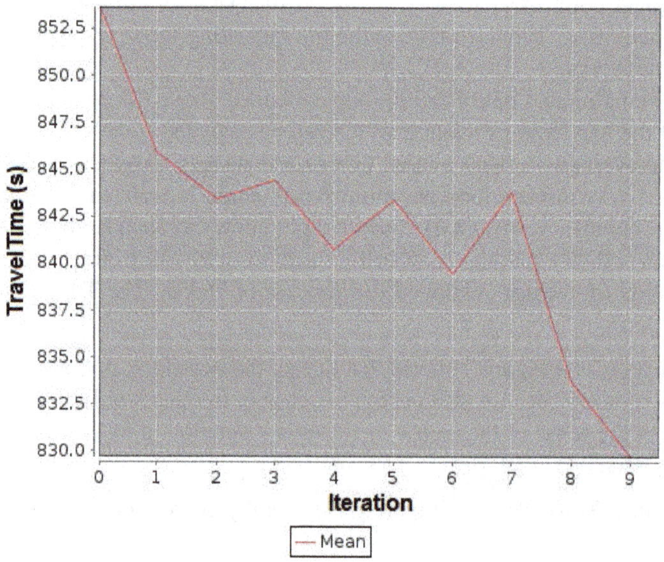


Figure 8. Average of displacement times

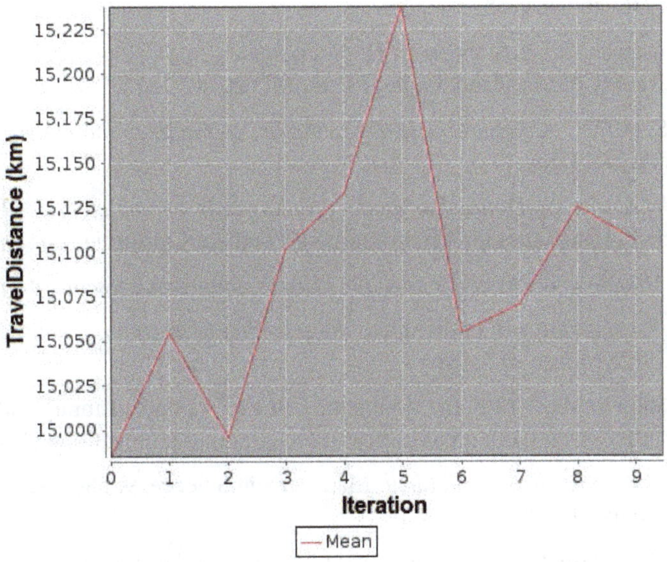


Figure 9. Average of the distances covered

hypothesis, a longer route, less frequented than another may still generate lower displacement times.
The variations (in reverse directions) of curves of distance and time appear to show that it is this phenomenon that has occurred for the case study in Dakar.

## 7. Conclusion

In this paper, we have proposed and analysed for the first time a new class of heuristic (TNLU-heur) for the TNLU problem. The new mixed integer formulation is a reformulated version of the Los's one in a more general sense. Unlike the model of Los we do not impose symmetry on the roads. Our new algorithm for TNLU was presented and applied to problems from the literature designed to highlight some intrinsic difficulties of TNLU, and from Dakar city. In addition to the convergence analysis, a new methodology of relocation possibilities is proposed. The SDO perfomance is evaluated by studying the quality of solution set in terms of optimal locations of activities. We investigated the average of displacement times, distances and scores; and established an interesting result characterizing the Utility, Travel Time and Travel Distance. In future work, we plan to study the combination between TNLU-heur and genetic or greedy algorithm. Another aspect that we wish to study will comprise the full integration of SP, AL and ND problems for further development.

**Acknowledgement**

The authors thank Serigne Gueye for his invaluable suggestions. They also thank the anonymous referees for useful comments and suggestions.

## References

Balac, M., Ciari, F., Genre-Grandpierre, C., Voituret, F., Gueye, S., & Michelon P. (2014). *Decoupling accessibility and automobile mobility in urban areas.* In Transport Research Arena, Paris.

Balmer, M., Rieser, M., Meister, K., Lefebvre, N., Charypar, D., & Nagel, K. (2009). *Matsimt: Architecture and simulation times.* Multi-Agent Systems for Trac and Transportation Engineering, pages 57-78. http://dx.doi.org/10.4018/978-1-60566-226-8.ch003.

Billheimer, J.W. (1970). *Optimal route configurations with fixed link construction costs*, Stanford Research Institut. SRI-Project 454531-309.

Boyce, D.E., Farhi, A. & Weischedel, R. (1973). *Optimal network problem: A branch-and-bound algorithm.* Environment and Planning, 5,519.-533. http://dx.doi.org/10.1068/a050519.

Burkard R. E., Karisch S. E., & Rendl F. (1996, September). *QAPLIB- A Quadratic Assignment Problem Library.* European Journal of Operational Research.

Floyd, R. W. (1962, June). *Algorithm 97: Shortest Path.* Communications of the ACM 5 (6): 345.

http://dx.doi.org/10.1145/367766.368168.

Gamvros, I., Golden, B., Raghavan, S., & Stanojeviś, D. *Heuristique search fo network design*. The Robert H. Smith School of Business. University of Maryland. College Park, MD 20742-1815.

Garey, M-R., & Johnson, D-S. (1979). *Computers and Intractability: A Guide to the Theory of NP-Completeness*. W. H. Freeman & Co. New York, NY, USA.

Gordon, P., & McRcynolds, W.K. (1972). *Optimal urban forms : Towards a complete modelling of urban interdependencies*. in : Research Papers in Economics no.4 (University of Southern California, Los Angeles, CA).

Gueye, S. (2012 January 8). *Modèle affectation quadratique / Dimensionnement réseaux projet ortrans*.

Hoang Hai Hoc. (1973 Jan). *A computational approach to the selection of an optimal network*, Management Science 19. no.5. http://dx.doi.org/10.1287/mnsc.19.5.488.

IBM ILOG CPLEX Optimization Studio V12.5, Inc. Using the CPLEX$^R$ Callable Library and CPLEX Barrier and Mixed Integer Solver Options (http://www-01.ibm.com/software/integration/optimization/cplex-optimization-studio).

Lawler, E.L. (1963). *The quadratic assignment problem*. Management Science, 9, 586-599. http://dx.doi.org/10.1287/mnsc.9.4.586.

Lin, J-J., & Feng, C-M. (2003). *A bi-level programming model for the land use-network design problem*, The Annals of Regional Science, 37:93-105. http://dx.doi.org/10.1007/s001680200112.

Los, M. (1978). *Simultaneous optimization of land use and transportation. A Synthesis of the Quadratic Assignment Problem and the Optimal Network Problem*, Regional Science and Urban Economics, 8, 21-42. http://dx.doi.org/10.1016/0166-0462(78)90010-8.

Maniezzo, V. (1997). *Exact and approximate nondeterministic tree-search procedures for the quadratic assignment problem*. Research Report CSR 97-1, Scienze dell'Informazione, Cesena site, University of Bologna.

MATSim. (2014). Matsim. URL www.matsim.org.

OpenStreetMap Development Team. (2014). Open Street Map. URL www.openstreetmap.org. The Free Wiki World Map.

Open Source Routing Machine. URL http://project-osrm.org

Sahni, S., & Gonzalez, T. (1976). *P-complete approximation problems*, Journal of the Association of Computing Machinery, 23, 555C565. http://dx.doi.org/10.1145/321958.321975.

Schlager, K.J. (1965). *A land use plan design model*, Journal of the American Institute of Planners. http://dx.doi.org/10.1080/01944366508978151.

Scott, A.J. (1969). *The optimal network problem: Some computational procedures*, Transportation Research, 3,201-210. http://dx.doi.org/10.1016/0041-1647(69)90152-X.

Senozon. (2014). Senozon. URL http://senozon.com/

Xia, Y. & Yuan, Y.X. (2006). *A new linearization method for quadratic assignment problem*. Optimization Methods and Software, 21(5):803C816. http://dx.doi.org/10.1080/10556780500273077.

Zhang, H., Beltran-Royo, C., & Ma, L. (2010). *Solving the quadratic assignment problem by means of general purpose mixed integer linear programming solvers*.

# 12

# A Viscosity Approximation Method for the Split Feasibility Problems in Hilbert Space

Li Yang[1]

[1] College of Mathematics and Information, China West Normal University, China

Correspondence: Li Yang, College of Mathematics and Information, China West Normal University, Nanchong, Sichuan, 637009, China. E-mail: yangli@cwnu.edu.cn

**Abstract**

In this paper, the most basic idea is to apply the viscosity approximation method to study the split feasibility problem (SFP), we will be in the infinite-dimensional Hilbert space to study the problem . We defined $x_0 \in C$ as arbitrary and $x_{n+1} = (1-\alpha_n)P_C(I-\lambda_n A^*(I-P_Q)A)x_n + \alpha_n f(x_n)$, for $n \geq 0$, where $\{\alpha_n\} \subset (0,1)$. Under the proper control conditions of some parameters, we show that the sequence $\{x_n\}$ converges strongly to a solution of SFP. The results in this paper extend and further improve the relevant conclusions in Deepho (Deepho, J. & Kumam, P., 2015).

**Keywords:** split feasibility problem, viscosity approximation method, strong convergence

**Mathematics Subject Classifications(2010): 47H05, 47H09, 47H20**

## 1. Introduction

In recent years, a large number of scholars have done a lot of meaningful research on the split feasibility problem (SFP), because the problem in signal processing and linear constrained optimization problems such as the feasible solution plays an important role (Censor, Y., et al, 2006; Byrne, C., 2002; Byrne, C., 2004.; Yang, Q., 2004.; Qu, B. & Xiu, N., 2005.; Xu, H. K., 2006.; Xu, H. K., 2010). In 1994,the SFP was first introduced by Censor and Elfving (Censor, Y. & Elfving, T., 1994), which is to find a point $x^*$ satisfying the property:

$$x^* \in C, Ax^* \in Q, \tag{1}$$

where $C$ and $Q$ be nonempty closed convex subsets of the real Hilbert spaces $H_1$ and $H_2$, $A : H_1 \to H_2$ be a bounded linear operator.

In order to find the solution of the problem SFP (1), many authors have proposed a variety of algorithms, it is worth noting that Byrne (Byrne, C., 2002) proposed the so-called CQ algorithm, the algorithm is this: take an initial point $x_0 \in H_1$ arbitrarily, and define the iterative step as

$$x_{n+1} = P_C(x_n - \lambda A^*(I - P_Q Ax_n), n \geq 0, \tag{2}$$

Where $0 < \lambda < 2/\rho(A^*A)$ and $P_C$ denotes the projector onto $C$ and $\rho(A^*A)$ is the spectral radius of the self-adjoint operator $A^*A$, $I$ denotes the identity operator.Then the sequence $\{x_n\}_{n\geq 0}$ generated by (2) converges strongly to a solution of SFP whenever $H_1$ is finite-dimensional and whenever there exists a solution to SFP(1).

By Byrne's CQ algorithm and Xu's viscosity approximatiom method (Xu, H. K., 2004), In 2015, Deepho and Kumam (Deepho, J. & Kumam, P., 2015) proposed the following algorithm:

$$x_{n+1} = (1-\alpha_n)P_C(I - \lambda A^*(I-P_Q)A)x_n + \alpha_n f(x_n), n \geq 1, \tag{3}$$

where $\{\alpha_n\} \in (0,1), 0 < \lambda < 2/||A||^2, f : C \to C$ is a contraction on C, and they proved that when the parameter $\{\alpha_n\}$ satisfied certain conditions ,then the algorithm (3) is strong converges to a solution of SFP(1). In this paper, we study the following more general algorithm which generates a sequence according to the recursive formula:

$$x_{n+1} = (1-\alpha_n)P_C(I - \lambda_n A^*(I-P_Q)A)x_n + \alpha_n f(x_n), n \geq 0, \tag{4}$$

And we will show that the sequence $\{x_n\}_{n\geq 0}$ defined by (4) strongly converges to a solution of SFP(1).

## 2. Preliminaries

Throughout this paper, we always assumes that $H_1$ and $H_2$ are two real Hilbert spaces with inner product $\langle \cdot \rangle$ and norm $|| \cdot ||$, we use $\Omega$ to denote the solution set of SFP(1), that is $\Omega = \{x \in C : Ax \in Q\} = C \cap A^{-1}Q$, The notation: $\rightharpoonup$ denotes weak convergence and $\rightarrow$ denotes strong convergence. Below we first list the definitions and theorems to be used in this paper.

**Definition 2.1.** Assume $H$ is a real Hilbert space. Let $T : H \rightarrow H$ be the nonlinear operators,

(i) ) $T$ is nonexpansive if $||Tx - Ty|| \leq ||x - y||$, for all $x, y \in H$ ;

(ii) $T$ is firmly nonexpansive if $\langle x - y, Tx - Ty \rangle \geq ||Tx - Ty||^2$, $x, y \in H$;

(iii)$T$ is $v-$ inverse strongly monotone ($v-$ ism), with $v > 0$ , if

$$\langle x - y, Tx - Ty \rangle \geq v||Tx - Ty||^2, \quad x, y \in H.$$

(iv) $T$ is averaged if $T = (1 - \alpha)I + \alpha S$ £where $\alpha \in (0, 1)$ £and $S : H \rightarrow H$ is nonexpansive. In this case ,we also say that $T$ is $\alpha-$ averaged. Thus firmly nonexpansive mappings (in particular ,the projections ) is $\frac{1}{2}-$ averaged mappings.

**Definition 2.2.** An operator $T : H \rightarrow H$ is called oriented operator if $Fix(T) \neq \Phi$, and

$$\langle z - Tx, x - Tx \rangle \leq 0, \quad x \in H.$$

In fact, we know that the oriented operator also contains firmly nonexpansive operator. The following is a useful characterization of projections.
**Proposition 2.1** Given $x \in H$ and $z \in C$. Then $z = P_C x$ if and only if

$$< x - z, y - z > \leq 0, \forall y \in C.$$

We collect some basic properties of averaged mappings and inverse strongly monotone operators in the following lemma.

**Lemma 2.1** (Qu, B. & Xiu, N., 2005; Xu, H. K., 2011) Let $T : H \rightarrow H$ be a given mapping.

(i) ) $T$ is nonexpansive if and only if the complement $I - T$ is $\frac{1}{2} - ism$;

(ii)If $T$ is $\upsilon - ism$,and $\gamma > 0$, then $\gamma T$ is $\frac{\upsilon}{\gamma} - ism$;

(iii) $T$ is averaged if and only if the complement $I - T$ is $\upsilon - ism$ for some $\upsilon > \frac{1}{2}$ . Indeed, for $\alpha \in (0, 1)$, $T$ is $\alpha-$ averaged if and only if $I - T$ is $\frac{1}{2\alpha} - ism$.

(iv) If $T_1$ is $\alpha_1-$ averaged and $T_2$ is $\alpha_2-$ is averaged, where $\alpha_1, \alpha_2 \in (0, 1)$, then the composite $T_1T_2$ is $\alpha-$ averaged, where $\alpha = \alpha_1 + \alpha_2 - \alpha_1\alpha_2$.

**Lemma 2.2** (Wang, F. H. & Xu., H. K., 2010) Suppose $C \bigcap A^{-1}Q \neq \Phi$. Let $U = I - \lambda A^*(I - P_Q)A$, where $0 < \lambda < 2/\rho(A^*A)$, and $\rho(A^*A)$ is the spectral radius of the self-adjoint operator $A^*A$.

(i) $U$ is an averaged mapping; namely, $U = (1-\beta)I + \beta V$, where $\beta \in (0, 1)$ is a constant and $V : H_1 \rightarrow H_1$ is nonexpansive;

(ii) $Fix(U) = A^{-1}Q$; consequently, $Fix(P_C U) = Fix(P_C) \bigcap Fix(U) = \Omega = C \cap A^{-1}Q$.

**Lemma 2.3** (Geobel, K. & Kirk, W. A., 1990) Let $H$ be a Hilbert space and let $C$ be a nonempty closed convex subset of $H$, let $T : C \rightarrow C$ is a nonexpansive mapping with $Fix(T) \neq \Phi$, Suppose that $\{x_n\} \subset C$ is such that $x_n \rightharpoonup z$ and $x_n - Tx_n \rightarrow 0$. Then $z \in F(T)$.

**Lemma 2.4** (Cui, H. H., Su, M. L. & Wang, F. H., 2013) Suppose $A : H_1 \rightarrow H_2$ be a bounded linear operator, and $T : H_2 \rightarrow H_2$ is an oriented operator, Let $V_\lambda = I - \lambda A^*(I - T)A$, where $0 < \lambda < \frac{2}{||A||^2}$. If $A^{-1}(Fix(T)) \neq \Phi$, then

$$||V_\lambda x - z||^2 \leq ||x - z||^2 - \frac{2 - \lambda||A||^2}{\lambda||A||^2}||V_\lambda x - x||^2$$

where $z \in A^{-1}(Fix(T))$ and $x \in H_1$.

**Lemma 2.5** (Mainge, P. E. & Maruster, S. 2011) Let $\{a_n\}$ be a nonnegative real sequence satisfying

$$a_{n+1} \leq (1 - \gamma_n)a_n + \delta_n$$

Where $\{\gamma_n\} \subset (0, 1)$, and $\{\delta_n\}$ is a sequences such that
(i) $\Sigma_{n=0}^{\infty}\gamma_n = \infty$;

(ii)$limsup_{n\rightarrow\infty}\delta_n/\gamma_n \leq 0$ or $\Sigma_{n=0}^{\infty}|\delta_n| < \infty$.

Then $\lim\limits_{n\rightarrow\infty} a_n = 0$.

## 3. NSTL Condition

Let $C$ be a nonempty closed convex subset of a real Hilbert space $H$. Motivated by Nakajo, Shimoji and Takahashi (Takahashi, W., 2009), we give the following definition: Let $T_n$ be families of nonexpansive mappings of $C$ into itself such that $\bigcap_{n=1}^{\infty} F(T_n) \neq \Phi$ ,where $F(T_n)$ is the set of all fixed points of $T_n$. Then $T_n$ is said to be satisfy NSTL-condition if for each bounded sequence $\{z_n\} \subset C$,

$$\lim_{n\to\infty} ||z_{n+1} - T_n z_n|| = 0, \tag{5}$$

implies that

$$\lim_{n\to\infty} ||z_n - T_n z_n|| = 0. \tag{6}$$

## 4. Main Results

**Theorem 3.1** Suppose the SFP(1) is consistent and $0 < \lambda' < \lambda_n < \lambda'' < \frac{2}{||A||^2}$. Let $C$ be a nonempty closed convex subset of a real Hilbert space $H_1$. Let $f : C \to C$ be a contraction with constant $\rho \in (0, 1)$. Take an initial guess $x_0 \in H_1$ arbitrarily, and we define the sequence $\{x_n\}$ by

$$x_{n+1} = (1 - \alpha_n)P_C(I - \lambda_n A^*(I - P_Q)A)x_n + \alpha_n f(x_n), n \geq 0, \tag{7}$$

where

$\{\alpha_n\} \subset (0, 1)$ such that

$$(C1) \lim_{n\to\infty} \alpha_n = 0;$$

$$(C2) \textstyle\sum_{n=0}^{\infty} \alpha_n = \infty;$$

$$(C3) \textstyle\sum_{n=0}^{\infty} |\alpha_{n+1} - \alpha_n| < \infty;$$

Then the sequence $\{x_n\}$ generated by algorithm(7) converges strongly to $\tilde{x} \in \Omega$, where $\tilde{x} = P_\Omega f(\tilde{x})$.

*Proof.* The proof of the process will be divided into four steps. First we show that the sequence $\{x_n\}$ is bounded. For our convenience, we take $T_n = P_C(I - \lambda_n A^*(I - P_Q)A)$. We assume $\bigcap_{n=1}^{\infty} F(T_n) \neq \Phi$ and $T_n$ satisfy NSTL condition. By lemma 2.2, we know that $\bigcap_{n=1}^{\infty} F(T_n)$ is a solution of SFP(1).Now, we note that the condition $0 < \lambda' < \lambda_n < \lambda'' < \frac{2}{||A||^2}$ implies that the operator $P_C(I - \lambda_n A^*(I - P_Q)A)$ is averaged. Since $I - P_Q$ is firmly nonexpansive mappings and so is $\frac{1}{2}$ -averaged, which is 1 -ism. Also observe that $A^*(I - P_Q)A$ is $\frac{1}{||A||^2}$-ism so that $\lambda_n A^*(I - P_Q)A$ is $\frac{1}{||A||^2}$ -ism. Further, from the fact that $I - \lambda_n A^*(I - P_Q)A$ is $\frac{1}{\lambda_n ||A||^2}$ averaged and $P_C$ is $\frac{1}{2}$ -averaged, by lemma 2.1, we may obtain that $P_C(I - \lambda_n A^*(I - P_Q)A)$ is $\mu_n$ - averaged, where

$$\mu_n = \frac{1}{2} + \frac{\lambda_n ||A||^2}{2} - \frac{1}{2}\frac{\lambda_n ||A||^2}{2} = \frac{2 + \lambda_n ||A||^2}{4} \in (0, 1),$$

This implies that $T_n = \mu_n I + (1 - \mu_n)S$, where $\mu_n = \frac{2+\lambda_n ||A||^2}{4} \in (0, 1)$ for some nonexpansive mappings S. Note that $T_n$ is also nonexpansive mappings, then for $p \in \bigcap_{n=1}^{\infty} F(T_n) \in \Omega$, we have $T_n p = p$, then

$$\begin{aligned} ||x_{n+1} - p|| &\leq (1 - \alpha_n)||T_n x_n - p|| + \alpha_n ||f(x_n - p|| \\ &\leq (1 - \alpha_n)||x_n - p|| + \alpha_n(||f(x_n) - f(p)|| + ||f(p) - p||) \\ &\leq (1 - \alpha_n)||x_n - p|| + \alpha_n(\rho||x_n - p|| + ||f(p) - p||) \\ &= (1 - (1 - \rho)\alpha_n)||x_n - p|| + \alpha_n ||f(p) - p|| \\ &\leq max\{||x_n - p||, \frac{1}{1 - \rho}||f(p) - p||\}, \end{aligned}$$

By induction

$$||x_n - p|| \leq max\{||x_0 - p||, \frac{1}{1 - \rho}||f(p) - p||\}, n \geq 0,$$

So $\{x_n\}$ is bounded, we also have that $\{T_n x_n\}$ and $\{f(x_n)\}$ are bounded.
Next, we claim that

$$\lim_{n\to\infty} \|x_n - T_n x_n\| = 0, \tag{8}$$

Indeed, by the definition of the (7),that is $x_{n+1} = (1-\alpha_n)T_n x_n + \alpha_n f(x_n)$, so $\lim_{n\to\infty} \|x_{n+1} - T_n x_n\| = \alpha_n \|T_n x_n - f(x_n)\|$, By the condition (C1), we have $\lim_{n\to\infty} \|x_{n+1} - T_n x_n\| = 0$. This together with the NSTL condition, Thus,(8) is clearly established.

Next, we will show that

$$\limsup_{n\to\infty} \langle \tilde{x} - x_n, \tilde{x} - f(\tilde{x})\rangle \le 0, \tag{9}$$

Indeed take a subquence $\{x_{n_k}\}$ of $\{x_n\}$ such that

$$\limsup_{n\to\infty} \langle \tilde{x} - x_n, \tilde{x} - f(\tilde{x})\rangle = \limsup_{n\to\infty} \langle \tilde{x} - x_{n_k}, \tilde{x} - f(\tilde{x})\rangle$$

We may assume that $x_{n_k} \rightharpoonup \bar{x}$. It follows from Lemma 2.3 and $\|T_n x_n - x_n\| \to 0$ that is $\bar{x} \in Fix(T_n) \in \Omega$. Hence from Lemma 2.3, we obtain

$$\limsup_{n\to\infty} \langle \tilde{x} - x_n, \tilde{x} - f(\tilde{x})\rangle = \langle \tilde{x} - \bar{x}, \tilde{x} - f(\tilde{x})\rangle \le 0$$

Finally, we will show that $x_n \to \tilde{x}$ in norm. It follows from Lemma 2.4, we obtain

$$\begin{aligned}\|T_n x_n - \tilde{x}\|^2 &\le \|x_n - \tilde{x}\|^2 - \frac{2 - \lambda_n\|A\|^2}{\lambda_n\|A\|^2}\|T_n x_n - x_n\|^2 \\ &\le \|x_n - \tilde{x}\|^2 - \frac{2 - \lambda''\|A\|^2}{\lambda''\|A\|^2}\|T_n x_n - x_n\|^2\end{aligned}$$

Thus, we have

$$\begin{aligned}\|x_{n+1} - \tilde{x}\|^2 &\doteq (1-\alpha_n)\|(T_n x_n - \tilde{x}) + \alpha_n(f(x_n) - \tilde{x})\|^2 \\ &\doteq (1-\alpha_n)^2\|T_n x_n - \tilde{x}\|^2 + \alpha_n^2\|f(x_n) - \tilde{x}\|^2 + 2\alpha_n(1-\alpha_n)\langle T_n x_n - \tilde{x}, f(x_n) - \tilde{x}\rangle \\ &\le (1-\alpha_n)^2[\|x_n - \tilde{x}\|^2 - \frac{2 - \lambda''\|A\|^2}{\lambda''\|A\|^2}\|T_n x_n - x_n\|^2] \\ &+ \alpha_n^2\|f(x_n) - \tilde{x}\|^2 + 2\alpha_n(1-\alpha_n)\langle T_n x_n - \tilde{x}, f(x_n) - \tilde{x}\rangle \\ &\le (1-\alpha_n)^2\|x_n - \tilde{x}\|^2 - \frac{(1-\alpha_n)^2(2 - \lambda''\|A\|^2)}{\lambda''\|A\|^2}\|T_n x_n - x_n\|^2 \\ &+ \alpha_n^2\|f(x_n) - \tilde{x}\|^2 + 2\alpha_n(1-\alpha_n)\langle T_n x_n - \tilde{x}, f(x_n) - f(\tilde{x})\rangle + 2\alpha_n(1-\alpha_n)\langle T_n x_n - \tilde{x}, f(\tilde{x}) - \tilde{x}\rangle \\ &\le (1-\alpha_n)^2\|x_n - \tilde{x}\|^2 - \frac{(1-\alpha_n)^2(2 - \lambda''\|A\|^2)}{\lambda''\|A\|^2}\|T_n x_n - x_n\|^2 \\ &+ \alpha_n^2\|f(x_n) - \tilde{x}\|^2 + 2\rho\alpha_n(1-\alpha_n)\|x_n - \tilde{x}\|^2 + 2\alpha_n(1-\alpha_n)\langle T_n x_n - \tilde{x}, f(\tilde{x}) - \tilde{x}\rangle \\ &\doteq [1 - (2\alpha_n - \alpha_n^2 - 2\rho\alpha_n(1-\alpha_n))]\|x_n - \tilde{x}\|^2 - \frac{(1-\alpha_n)^2(2 - \lambda''\|A\|^2)}{\lambda''\|A\|^2}\|T_n x_n - x_n\|^2 \\ &+ \alpha_n^2\|f(x_n) - \tilde{x}\|^2 + 2\alpha_n(1-\alpha_n)\langle T_n x_n - \tilde{x}, f(\tilde{x}) - \tilde{x}\rangle \\ &\doteq (1-\gamma_n)\|x_n - \tilde{x}\|^2 + \gamma_n\delta_n,\end{aligned}$$

That is

$$\|x_{n+1} - \tilde{x}\|^2 \le (1-\gamma_n)\|x_n - \tilde{x}\|^2 + \gamma_n\delta_n \tag{10}$$

where

$$\gamma_n \doteq 2\alpha_n - \alpha_n^2 - 2\rho\alpha_n(1-\alpha_n),$$

$$\delta_n \doteq -\frac{(1-\alpha_n)^2(2-\lambda''\|A\|^2)}{[2\alpha_n+\alpha_n^2-2\rho\alpha_n(1-\alpha_n)]\lambda''\|A\|^2}\|T_n x_n - x_n\|^2 + \frac{\alpha_n\|f(x_n)-z\|^2}{2-\alpha_n-2\rho(1-\alpha_n)} + \frac{2(1-\alpha_n)}{2-\alpha_n-2\rho(1-\alpha_n)}\langle T_n x_n - \tilde{x}, f(\tilde{x}) - \tilde{x}\rangle$$

It is easily seen from (C1),(C2),(8) and (9) that

$$\gamma_n \to 0, \sum_{n=1}^{\infty} \gamma_n = \infty, limsup_{n\to\infty}\delta_n \leq 0$$

Finally apply lemma 2.5 to (10), we conclude that $\|x_n - \tilde{x}\| \to 0$. □

**Corollary 3.1** (Deepho, J. & Kumam, P., 2015) Suppose the SFP(1.1) is consistent and $0 < \lambda < \frac{2}{\|A\|^2}$. Let $C$ be a nonempty closed convex subset of a real Hilbert space $H_1$. Let $f : C \to C$ be a contraction with constant $\rho \in (0,, 1)$.Take an initial guess $x_0 \in H_1$ arbitrarily,and we define the sequence $\{x_n\}$ by

$$x_{n+1} = (1-\alpha_n)P_C(I - \lambda A^*(I - P_Q)A)x_n + \alpha_n f(x_n), n \geq 0, \tag{11}$$

where $\{\alpha_n\} \subset (0, 1)$ such that
(1)$\lim_{n\to\infty} \alpha_n = 0$;
(2)$\sum_{n=0}^{\infty} \alpha_n = \infty$;
(3)$\sum_{n=0}^{\infty} |\alpha_{n+1} - \alpha_n| < \infty$;
Then the sequence $x_n$ generated by algorithm(11) converges strongly to $\tilde{x}$, where $\tilde{x}$ is the unique solution of the variational inequality

$$\langle (I - f)\tilde{x}, x - \tilde{x}\rangle \geq 0, \quad x \in \Omega.$$

**Remark1:** Let $\lambda_n = \lambda$ in algorithm(3.1), Thus it follows directly from Theorem 3.1 that the conclusion holds. The proof is complete. It is worth noting that our method of proof is different from the method of (Deepho, J. & Kumam, P., 2015).

**References**


Byrne, C. (2002). Iterative oblique projection onto convex sets and the split feasibility problem. *Inverse Problems., 18*, 441-453. http://dx.doi.org/10.1088/0266-5611/18/2/310

Byrne, C. (2004). A unified treatment of some iterative algorithms in signal processing and image reconstruction. *Inverse Problems., 20*, 103-120. http://dx.doi.org/10.1088/0266-5611/20/1/006

Censor, Y., Bortfeld, T., Martin, B., & Trofimov, A. (2006). A unified approach for inversion problems in intensity modulated radiation therapy. *Phys. Med. Biol., 51*,2253-2365. http://dx.doi.org/10.1088/0031-9155/51/10/001

Censor, Y., & Elfving, T. (1994). A multiprojection algorithm using Bregman projections in a product space. *Numer. Algorithms., 8*, 221-239. http://dx.doi.org/10.1007/BF02142692

Cui, H. H., Su, M. L., & Wang, F. H. (2013). Damped projection method for the split common fixed point problems. *Journal of Inequalities and Applications 2013*, 123. http://dx.doi.org/10.1186/1029-242x-2013-123

Deepho, J., & Kumam, P. (2015). A viscosity approximation method for the split feasibility problem. *Transactions on Engineering Technologies,* 69-77. http://dx.doi.org/10.1007/978-94-017-9588-3-6.

Geobel, K., & Kirk, W. A. (1990). *Topics in Metric Fixed Point Theory[M].* Cambridge: Cambridge University Press.

Mainge, P. E., & Maruster, S. (2011). Convergence in norm of modified Krasnoselski-Mann iterations for fixed points of demicontractive mappings. *Applied Mathematics and Computation, 217*, 9864-9874. http://dx.doi.org/10.1016/j.amc.2011.04.068

Qu, B., & Xiu, N. (2005). A note on the CQ algorithm for the split feasibility problem. *Inverse Problems., 21,* 1655-1665. http://dx.doi.org/10.1088/0266-5611/21/5/009

Takahashi, W. (2009). Viscosity approximation methods for countable families of nonexpansive mappings in Banach s-paces. *Nonlinear Analysis: Theory, Method and Applications, 70*,719-734. http://dx.doi.org/10.1016/j.na.2008.01.005

Wang, F. H., & Xu., H. K. (2010). Approximating curve and strong convergence of the Cqalgorithm for the split feasibility problem. *Journal of Inequalities and Applications. 2010,* 1-13. http://dx.doi.org/10.1155/2010/102085

Xu, H. K. (2006). A variable Krasnosel'skii-Mann algorithm and the multiple-set split feasibility problem. *Inverse Problems, 22*, 2021-2034. http://dx.doi.org/10.1088/0266-5611/22/6/007

Xu, H. K. (2010). Iterative methods for the split feasibility problem in infinite-dimensional Hilbert spaces. *Inverse Problems, 26*, 105018. http://dx.doi.org/10.1088/0266-5611/26/10/105018

Xu, H. K. (2004). Viscosity approximation methods for nonexpansive mapping. *J. Math. Anal. Appl., 298*, 279-291. http://dx.doi.org/10.1016/j.jmaa.2004.04.059

Xu, H. K. (2011). Averaged Mappings and the Gradient-Projection Algorithm. *J. Optim. Theory . Appl, 150*, 360-378. http://dx.doi.org/10.1007/s10957-011-9837-z

Yang, Q. (2004). The relaxed CQ algorithm solving the split feasibility problem. *Inverse Problems, 20*, 1261-1266. http://dx.doi.org/10.1088/0266-5611/20/4/014

# 13

# Some New Results on Super Heronian Mean

Labelling S. S. Sandhya[1], E. Ebin Raja Merly[2] & G. D. Jemi[3]

[1] Department of Mathematics, Sree Ayyappa College for Women, Chunkankadai-629 003, Tamilnadu, India

[2] Department of Mathematics, Nesamony Memorial Christian College, Marthandam-629 165, Tamilnadu, India

[3] Department of Mathematics, Narayanaguru College of Engineering, Manjalumoodu-629 151, Tamilnadu, India

Correspondence: G. D. Jemi. E-mail: gdjemi@gmail.com

**Abstract**

Here we discuss on Some new results on Super Heronian Mean Labelling of graphs. For the present investigation on Qn ⊙ K1, D(Qn) ⊙ K1, Middle graph, Total graph, Ln ⊙ K1.We discuss briefly about the summary and the other valid information with appropriate graphs and definitions.

**Keywords:** Middle graph, Super Heronian mean graph, Total graph

**1. Introduction**

The concept of graph Labelling was introduced by Rosa in 1967. A graph Labelling is an assignment of integers to the vertices or edges or both subject to certain conditions. If the domain of the mapping is the set of vertices (or edges), then the Labelling is called a vertex Labelling (or an edge Labelling). Here we consider simple, finite, undirected and connected graph G = (V, E). In this paper $Q_n$ and $D(Q_n)$ denotes Quadrilateral snake and Double Quadrilateral snake with n vertices. For all other general expressions and symbols we follow Harary.First we will provide some definitions useful for the present work.

*1.1 Definition*

Let f :V(G)→{1, 2, ------, p+q} be an injective function. For a vertex Labelling "f" the induced edge Labelling f*(e=uv) is defined by,

$$f^*(e)=\left\lfloor\frac{f(u)+\sqrt{f(u)f(v)}+f(v)}{3}\right\rfloor \quad [OR] \quad \left\lceil\frac{f(u)+\sqrt{f(u)f(v)}+f(v)}{3}\right\rceil.$$

Then "f" is called a Super Heronian Mean Labelling if {f(V(G)} U {f(e): e ϵ E(G)={1, 2, p+q}}. A graph which admits Super Heronian Mean Labelling is called Super Heronian Mean Graph.

*1.2 Definition*

The Total graph T(G) of a graph G is the graph whose vertex set is V(G)UE(G) and two vertices are adjacent whenever they are either adjacent or incident in G.

*1.3 Definition*

The Middle graph M(G) of a graph G is the graph whose vertex set is V(G)UE(G) and in which two vertices are adjacent iff either they are adjacent edges of G or one is a vertex of G and the other is an edge incident on it.

*1. 4 Theorem*

Quadrilateral snakes are Super Heronian mean graph.

*1.5 Theorem*

Double Quadrilateral snakes are Super Heronian mean graph.

**2. Main Results**

*2.1 Theorem*

Let G be a graph obtained by attaching pendant edges to each vertex of a Quadrilateral snake $Q_n$. Then G is a Super Heronian mean graph.

**Proof:** Consider a graph G which is obtained by attaching pendant edges to each vertex of a Quadrilateral snake $Q_n$. Let $Q_n$ be a Quadrilateral snake. Let $u_i$, $v_i$, $w_i$ be the vertices of Quadrilateral snake. Join $u_i v_i$, $u_{i+1} w_i$ and $v_i w_i$.Let $x_i$, $y_i$, $z_i$ be the pendant vertices.Join $u_i x_i$, $v_i y_i$ and $w_i z_i$.

Define a function $f: V(Q_n \odot K_1) \rightarrow \{1, 2, \text{-------}, p+q\}$ by

$$f(u_i)=13i-12;\ 1 \leq i \leq n$$
$$f(v_i)=13i-5;\ 1 \leq i \leq n-1$$
$$f(w_i)=13i-3;\ 1 \leq i \leq n-1$$
$$f(x_i)=13i-10;\ 1 \leq i \leq n$$
$$f(y_i)=13i-9;\ 1 \leq i \leq n-1$$
$$f(z_i)=13i;\ 1 \leq i \leq n$$

Edges are labeled with,

$$f(u_i u_{i+1})=13i-6;\ 1 \leq i \leq n-1$$
$$f(u_i v_i)=13i-8;\ 1 \leq i \leq n-1$$
$$f(v_i w_i)=13i-4;\ 1 \leq i \leq n-1$$
$$f(w_i u_{i+1})=13i-2;\ 1 \leq i \leq n-1$$
$$f(u_i x_i)=13i-11;\ 1 \leq i \leq n$$
$$f(v_i y_i)=13i-7;\ 1 \leq i \leq n-1$$
$$f(w_i z_i)=13i-1;\ 1 \leq i \leq n-1$$

Obviously f is a Super Heronian mean labelling and $Q_n \odot K_1$ is a Super Heronian meangraph.

*2.2. Example*

A Super Heronian mean labelling of $Q_4 \odot K_1$ is displayed below.

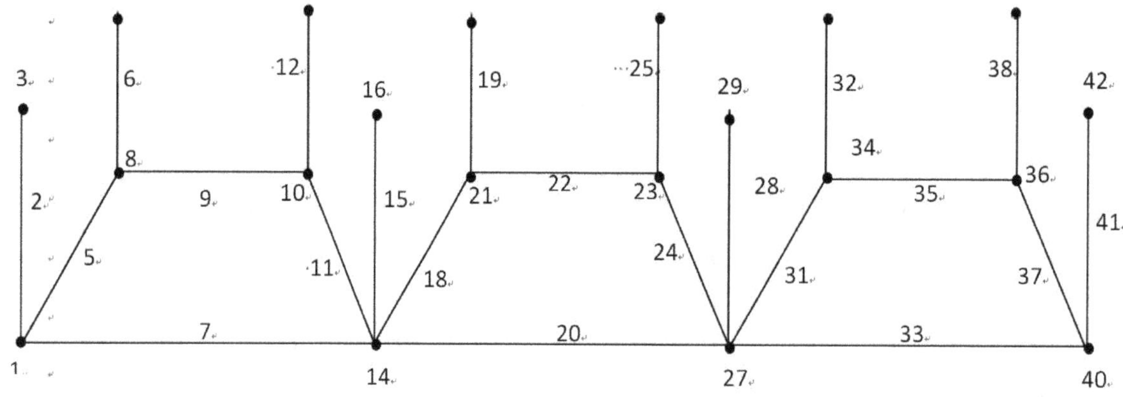


Figure 1

*2.3 Theorem*

**$D(Q_n) \odot K_1$** is a Super Heronian mean graph.

**Proof:** Let D(Qn) be a Double Quadrilateral snake.Let ui, vi, wi, xi, yi be the vertices of Double Quadrilateral snake. Join ui vi, ui+1 wi, vi wi and ui xi ui+1 yi, xi yi. Let ai, bi, ci, di, si, ti be the pendant vertices.Join vi ai, wi bi, xi ci, yi di, xi ti and ui si.

Define a function $f: V(D(Q_n) \odot K_1) \rightarrow \{1, 2, \text{-------}, p+q\}$ by,

$$f(u_i)=24i-19;\ 1 \leq i \leq n$$
$$f(v_i)=24i-16;\ 1 \leq i \leq n-1$$
$$f(w_i)=24i-11;\ 1 \leq i \leq n-1$$
$$f(x_i)=24i-12;\ 1 \leq i \leq n-1$$
$$f(y_i)=24i-4;\ 1 \leq i \leq n-1$$

$f(a_i)=24i-14;\ 1\le i\le n-1$

$f(b_i)=24i-6;\ 1\le i\le n-1$

$f(c_i)=24i-3;\ 1\le i\le n-1$

$f(d_i)=24i;\ 1\le i\le n-1$

$f(s_i)=24i-22;\ 1\le i\le n-1$

$f(t_i)=24i-21;\ 1\le i\le n-1$

Edges are labeled with,

$f(u_iu_{i+1})=24i-8;\ 1\le i\le n-1;\ f(u_iv_i)=24i-18;\ 1\le i\le n-1$

$f(v_iw_i)=24i-14;\ 1\le i\le n-1;\ f(w_iu_{i+1})=24i-4;\ 1\le i\le n-1$

$f(u_ix_i)=24i-17;\ 1\le i\le n-1;\ f(x_iy_i)=24i-10;\ 1\le i\le n-1$

$f(u_iu_{i+1})=24i-1;\ 1\le i\le n-1;\ f(x_ic_i)=24i-7;\ 1\le i\le n-1$

$f(y_id_i)=24i-2;\ 1\le i\le n-1;\ f(v_ia_i)=24i-15;\ 1\le i\le n-1$

$f(w_ib_i)=24i-9;\ 1\le i\le n-1;\ f(s_iu_i)=24i-22;\ 1\le i\le n$

$f(u_it_i)=24i-20;\ 1\le i\le n$

Clearly f is a Super Heronian mean Labelling and $D(Q_n)\odot K_1$ is a Super Heronian mean graph.

*2.4 Example*

A Super Heronian mean Labelling of $D(Q_4)\odot K_1$is displayed below.

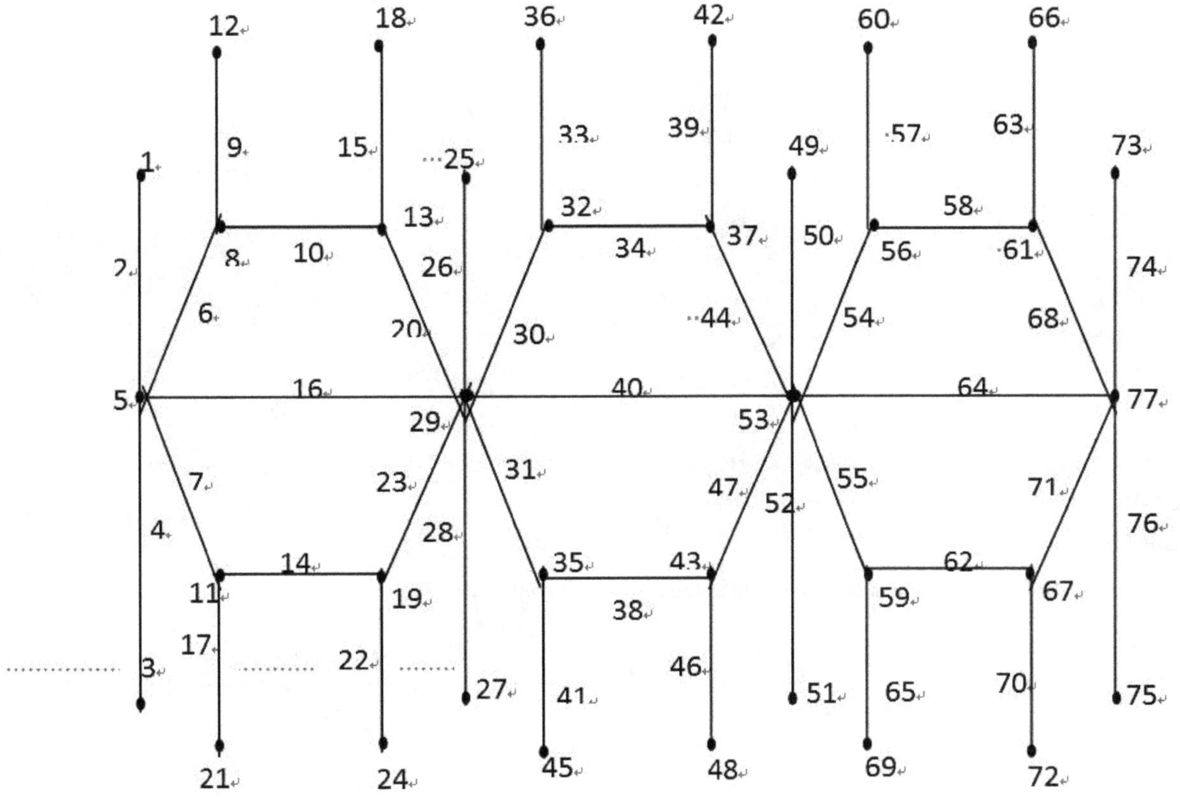


Figure 2

*2.5 Theorem*

Middle graph of path $P_n$ is a Super Heronian mean graph.

**Proof:** Let $u_1, u_2,$ --------, $u_n$ be the vertices and $v_1, v_2,$ ------, $v_{n-1}$ be the edges of path $P_n$ and $G=M(P_n)$ be the middle graph of path $P_n$.

Define a function $f: V(G)\rightarrow\{1, 2, ----, p+q\}$ by

$$f(u_i)=5i-4;\ 1\le i\le n$$
$$f(v_i)=5i-2;\ 1\le i\le n-1$$

Edges are labeled with,

$$f(u_iv_i)=5i-3;1\le i\le n-1$$
$$f(u_iv_{i-1})=5i-1;\ 1\le i\le n-1$$
$$f(v_iv_{i+1})=5i;1\le i\le n-2$$

Thus f provides a Super Heronian mean Labelling for $M(P_n)$.

Hence $M(P_n)$ is a Sper Heronian mean graph.

*2.6 Example*

Middle graph of path $P_5$ and its Super Heronian mean Labelling is displayed below.

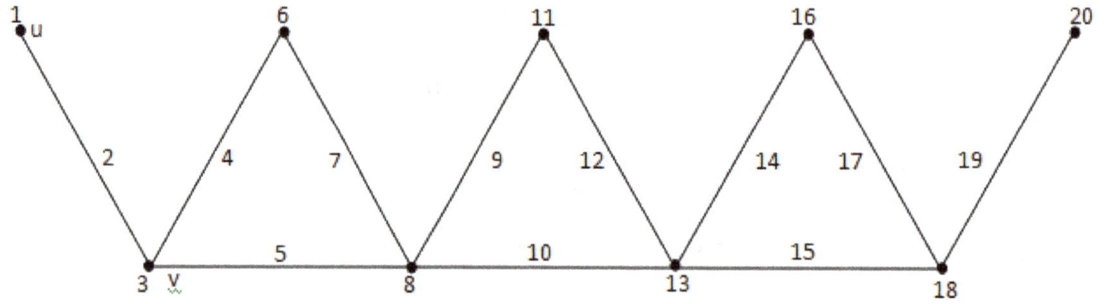


Figure 3

*2.7 Theorem*

Total graph of path $P_n$ is a Super Heronian mean graph.

**Proof:** Let $u_1$, $u_2$, --------, $u_n$ be the vertices and $v_1$, $v_2$, ----$v_{n-1}$ be the edges of path $P_n$ and $G=T(P_n)$ be the total graph of path $P_n$.Define a function $f:V(G)\rightarrow\{1, 2, ----, p+q\}$ by

$$f(u_1)=1;\ f(u_i)=6i-6;\ 2\le i\le n$$
$$f(v_i)=6i-2;\ 1\le i\le n-1$$

Edges are labeled with,

$$f(u_iv_i)=6i-4;\ 1\le i\le n-1;f(u_iv_{i-1})=6i-1;\ 1\le i\le n$$
$$f(v_iv_{i+1})=6i+1;\ 1\le i\le n-2;\ f(u_iu_{i+1})=6i-3;\ 1\le i\le n-1$$

The above defined function, f provides a Super Heronian mean Labelling for $T(P_n)$.

Hence $T(P_n)$ is a Super Heronian mean graph.

*2.8 Example*

Total graph of path $P_6$ and its Super Heronian mean labelling is displayed below.

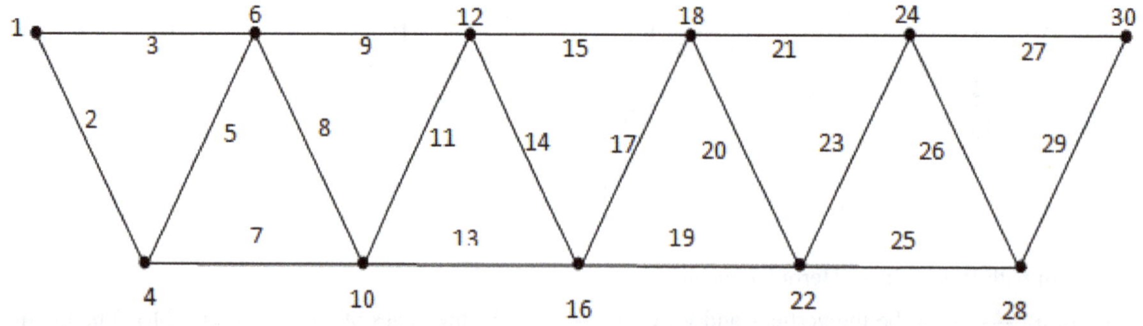


Figure 4

*2.9 Theorem*

$L_n \odot K_1$ is a Super Heronian mean graphs.

**Proof:** Let $L_n$ be a Ladder and also $w_i$ be the pendant vertex adjacent to $v_i$ and $x_i$ be the pendant vertex adjacent to $u_i$.

Define a function f: $V(L_n \odot K_1) \rightarrow \{1, 2, ..., p+q\}$ by,

$$f(u_i)=9i-4;\ 1\leq i\leq n,$$
$$f(v_i)=9i-6;\ 1\leq i\leq n,$$
$$f(w_i)=9i-8;\ 1\leq i\leq n$$
$$f(x_1)=8,$$
$$f(x_i)=9i-2;\ 2\leq i\leq n$$

Edges are labeled with,

$$f(u_iu_{i+1})=9i;\ 1\leq i\leq n-1,\ f(u_1v_1)=7,\ f(u_iv_i)=9i-1;\ 2\leq i\leq n$$
$$f(u_ix_i)=9i-3;\ 1\leq i\leq n,\ f(v_iw_i)=9i-7;\ 1\leq i\leq n$$
$$f(v_1v_2)=7,\ f(v_iv_{i+1})=9i-1;\ 2\leq i\leq n-1$$

Hence $L_n \odot K_1$ is a Super Heronian mean graph.

*2.10 Example*

A Super Heronian mean labelling of $L_5 \odot K_1$ is given below

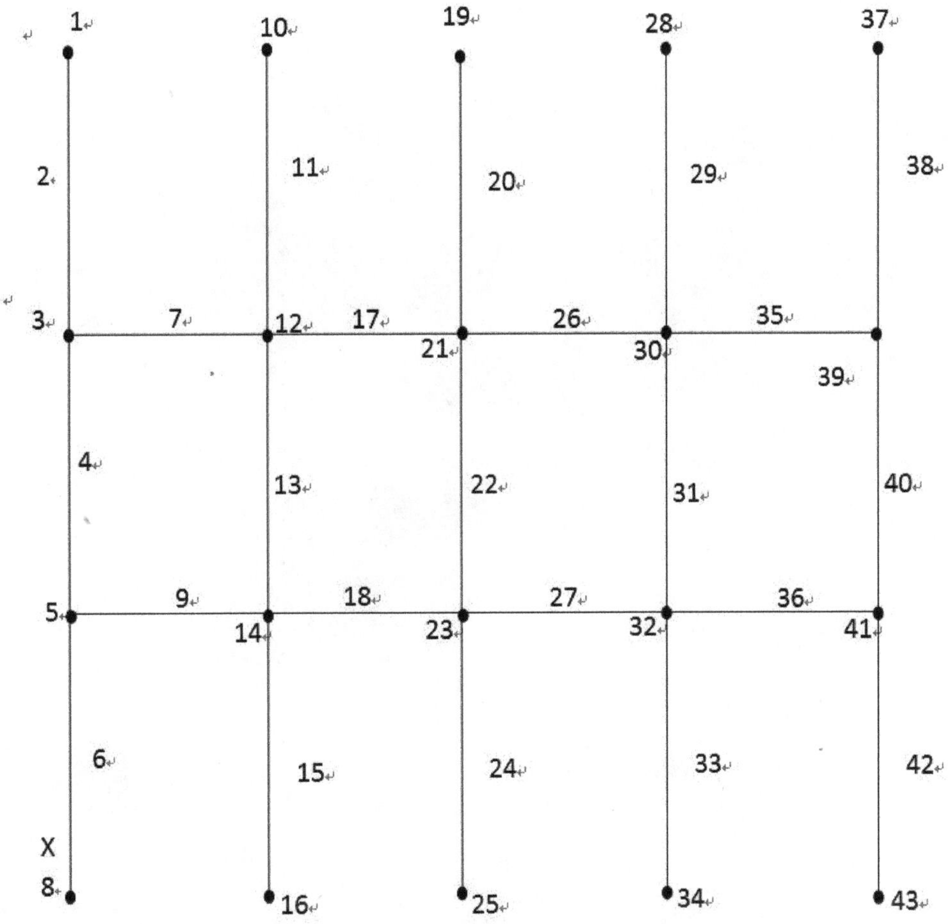


Figure 5

**3. Conclusion**

In this publication we discussed Some new results on Super Heronian Mean Labelling of graphs and we investigated on $Q_n \odot K_1$,$D(Q_n) \odot K_1$, Middle graph, total graph, $L_n \odot K_1$. Extending the study to other systematic formations of graph families is an open area of research.

**Acknowledgement**

The authors appreciate all kind suggestions and support of various anonymous referees.

**References**

Gallian, J. A. (2013). A Dynamic Survey of Graph labelling. *The Electronic journal of Combinatorics.*

Harary, F. (1988). *Graph Theory*. Narosa Publishing House Reading, New Delhi.

Jeyasekaran, C., Sandhya, S. S. & David, C. (2014). Some Results on Super Harmonic Mean Graphs. *International Journal of Mathematics trend and Tecnology, .6*(3), 215-224.

Sandhya, S. S., Ebin, E., Raja., Merly, & Jemi, G. D. Super Heronian Mean Labelling of Graphs. *International Journal of Mathematical Forum.*

Somasundaram, S., & Ponraj, R. Mean Labelling of Graphs. *National Ac*ademy of Science Letters, 26, 210-213.

Somasundaram, S., Ponraj, R., & Sandhya, S. S. Harmonic Mean Labelling of Graphs. *Journal of Combinatorial Mathematics and Combinatorial Computing.*

Somasundaram, S., Ponraj, R., & Vidhyarani, P, (2011). Geometric Mean Labelling of Graphs *Bulletin of Pure and Applied Sciences 30*, 153-160.

# 14

# Existence of Solutions to the Boundary Value Problems for a Class of P-Laplacian Equations at Resonance

Lina Zhou[1] & Weihua Jiang[2]

[1] College of Mathematics and Information Science, Hebei Normal University, Shijiazhuang 050024 Hebei, China

[2] College of Science,Hebei University of Science and Technology, Shijiazhuang 050018, Hebei, China

Correspondence: Lina Zhou. E-mail: lnazhou@163.com

**Abstract**

By the generalizing the extension of the continuous theorem of Ge and Ren and constructing suitable Banach spaces and operators, we investigate the existence- solutions to the boundary value problems for a class of p-Laplacian equations. Finally an example is given to illustrate our results.

**Keywords:** continuous theorem, resonance, p-Laplacian equations, boundary value problem

## 1. Introduction

In this paper,we will study the boundary value problem

$$\begin{cases} (\varphi_p(u''))'(t) = f(t, v, v', v'') \\ (\varphi_p(v''))'(t) = g(t, u, u', u'') \\ u(0) = u''(0) = 0, v(0) = v''(0) = 0 \\ u'(1) = \int_0^1 k_1(t)u'(t)dt, v'(1) = \int_0^1 k_2(t)v'(t)dt \end{cases} \tag{1.1}$$

where$\varphi_p(s) = |s|^{p-2}s, p > 1, \int_0^1 k_i(t)dt = 1, i = 1, 2$

In this paper,we will always suppose that

$(H_1)$ $k_i(t) \in L^1[0, 1]$ are nonnegative and$\| k_i \|_1= 1$, where$\| k_i \|_1= \int_0^1 | k_i(t) | dt, i = 1, 2$.

$(H_2) f(t, u, v, w), g(t, u, v, w)$is continuous in $[0, 1] \times R^3$

## 2. Preliminaries

**Definition 2.1** *Let $X$ and $Y$ be a two Banach spaces with norms $\| \cdot \|_X, \| \cdot \|_Y$,respectively.A continuous operator $M : X \cap domM \longrightarrow Y$is said to be quasi-linear if*

*(i)$ImM := M(X \cap domM)$ is a closed subset of $Y$,*

*(ii)$KerM := \{x \in X \cap domM : Mx = 0\}$ is linearly homeomorphic to $R^n, n < \infty$.*

*where $domM$ denote the domain of operator $M$.*

Let $X_1 = \mathrm{Ker}M$ and $X_2$ be complement space of $X_1$ in $X$,then $X = X_1 \bigoplus X_2$.Let $P : X \longrightarrow X_1$ be a projector and $\Omega \subset X$ an open and bounded set with the origin $\theta \in \Omega$

**Definition2.2** *Suppose $N_\lambda : \overline{\Omega} \to Y, \lambda \in [0, 1]$ is a continuous and bounded operator.Denote $N_1$ by $N$.Let $\Sigma_\lambda = \{x \in \overline{\Omega} \cap domM : Mx = N_\lambda x\}$.$N_\lambda$ is said M-quasi-compact in $\overline{\Omega}$ if there exists a vector subspace $Y_1$ of $Y$ satisfying $dimY_1 = dimX_1$ and two operators $Q, R$ with $Q : Y \to Y_1, QY = Y_1$,being continuous ,bounded,and satisfying $Q(I - Q) = 0$,$R : \overline{\Omega} \times [0, 1] \to X_2 \cap domM$ continuous and compact such that for $\lambda \in [0, 1]$,*
*(a) $(I - Q)N_\lambda(\overline{\Omega}) \subset ImM \subset (I - Q)Y$,*
*(b)$QN_\lambda x = \theta, \lambda \in (0, 1) \Leftrightarrow QNx = \theta$,*
*(c)$R(\cdot, 0)$ is zero operator and $R(\cdot, 0)|_\Sigma = (I - P)|_\Sigma$ ,*
*(d)$M[P + R(\cdot, 0)] = (I - Q)N_\lambda$*

**Theorem 2.1** *Let $X$ and $Y$ be two Banach spaces with norms $\| \cdot \|_X, \| \cdot \|_Y$,respectively,and $\Omega \subset X$ be an open and bounded nonempty set.Suppose*

$$M : X \cap domM \to Y$$

*is a quasi-linear operator and that $N_\lambda : \overline{\Omega} \to Y, \lambda \in [0,1]$ is M-quasi-compact.In addition,if the following conditions hold:*
*$(C_1) Mx \neq N_\lambda x, \forall x \in \partial\Omega \cap domM, \lambda \in (0,1)$,*
*$(C_2) deg\{JQN, \Omega \cap KerM, 0\} \neq 0$*
*then the abstract equation $Mx = Nx$ has at least one solution in $domM \cap \overline{\Omega}$,where $N = N_1, J : ImQ \to KerM$ is a homeomorphism with $J(\theta) = \theta$.*

*Proof.* The proof is similar to the one of lemma 2.1 and Theorem 2.1 in [Ge et al., 2004].

We can easily get the following inequalities.
**Lemma2.1** *For any $u, v \geq 0$,we have*
*(1)$\varphi_p(u,v) \leq \varphi_p(u) + \varphi_p(v), 1 < p \leq 2$.*
*(2)$\varphi_p(u,v) \leq 2^{p-2}(\varphi_p(u) + \varphi_p(v)), p \geq 2$.*

In the following,we will always suppose that $q$ satisfies $\frac{1}{p} + \frac{1}{q} = 1$.

**3. Main Results**

Let $X = C^2[0,1] \times C^2[0,1]$ with norm $\|(u,v)\| = \|u\| + \|v\|$ ,where$\|u\| = \max\{\|u\|_\infty, \|u'\|_\infty, \|u''\|_\infty\}$, $Y = C[0,1] \times C[0,1] \times C[0,1] \times C[0,1]$with norm $\|(y_1,y_2,y_3,y_4)\| = \max\{\|y_1\|_\infty, \|y_2\|_\infty, \|y_3\|_\infty, \|y_4\|_\infty\}$ with $\|y\|_\infty = \max_{t\in[0,1]}|y(t)|$.We know that $(X, \|\cdot\|)$ and $(Y, \|\cdot\|)$ are Banach spaces.

Define operators $M : X \cap domM \longrightarrow Y$,$N_\lambda : X \to Y$ as follows:

$$M(u,v) = \begin{pmatrix} (\varphi_p(u''))'(t) \\ (\varphi_p(v''))'(t) \\ T_1(\varphi_p(u''))'(t) \\ T_2(\varphi_p(v''))'(t) \end{pmatrix}, N_\lambda(u,v) = \begin{pmatrix} \lambda f(t, v(t) v'(t), v''(t)) \\ \lambda g(t, u(t), u'(t), u''(t)) \\ 0 \\ 0) \end{pmatrix}$$

where $T_i y = c_i, i = 1,2. y \in C[0,1], c_1, c_2$ satisfy

$$\int_0^1 k_i(t) \int_t^1 \varphi_q(\int_0^s (y(r) - c_i) dr) ds dt = 0 \qquad (0.1)$$

$$\text{dom}M = \{(u,v) \in X | \varphi_p(u''), \varphi_p(v'') \in C^1[0,1], u(0) = u''(0) = v(0) = v''(0) = 0\}$$

**Lemma3.1** *For $y \in C[0,1]$,there is only one constant $c_i \in R$ such that$T_i y = c_i$,with $|c_i| \leq \|y\|_\infty$,and $T_i : C[0,1] \to R$, $i = 1,2$. are continuous .*

The proof is similar to Lemma 3.1 in [Weihua, 2014].

It is clear that $(u,v) \in \text{dom}M$ is a solution if and only if it satisfies $M(u,v) = N(u,v)$ where $N = N_1$.For convenience,

let$(a,b,c,d)^L = \begin{pmatrix} a \\ b \\ c \\ d \end{pmatrix}$

**Lemma3.2** *M is a quasi-linear operator.*

*Proof.* It is easy to get Ker$M = \{(b_1 t, b_2 t) \mid b_1, b_2 \in R\} := X_1$.

For $(u,v) \in X \cap \text{dom}M$,if $M(u,v) = (y_1, y_2, c_1, c_2)^L$,then $c_1, c_2$ satisfy (3.1).On the other hand ,if $y_i \in C[0,1], T_i y_i = c_i$, $i = 1,2$,take

$$\begin{pmatrix} u(t) \\ v(t) \end{pmatrix} = \begin{pmatrix} \int_0^t (t-s)\varphi_q(\int_0^s y_1(r)dr)ds \\ \int_0^t (t-s)\varphi_q(\int_0^s y_2(r)dr)ds \end{pmatrix}.$$

We can get $(u,v) \in X \cap \text{dom}M$ and $M(u,v) = (y_1, y_2, c_1, c_2)^L$,then $c_1, c_2$.Thus

$$\text{Im}M = \{(y_1, y_2, c_1, c_2)^L \mid y \in C[0,1], c_1, c_2 \text{satisfy (3.1)}\}.$$

By the continuity of $T_i, i = 1,2$,we see that Im$M \subset Y$ is closed.So,$M$ is quasi-linear.The proof is completed.

**Lemma3.3** $T_i(c) = c, T_i(y + c) = T_i(y) + c, T_i(cy) = cT_i(y), i = 1,2, c \in R, y \in C[0,1]$

*Proof.* The proof is simple.Therfore we omit it.

Take a projector $P : X \to X_1$ and an operator $Q : Y \to Y_1$ as follows:

$$(P(u,v))(t) = (u'(0)t, v'(0)t), Q(y_u, y_v, y_1, y_2)^L = (0, 0, T_1y_1 - T_1y_u, T_2y_2 - T_2y_v)^L$$

where $Y_1 = \{(0, 0, c_1, c_2)^L \mid c_i \in R, i = 1.2\}$.Obviously $QY = Y_1$and $\dim Y_1 = \dim X_1$.

By the continuity and boundedness of $T_i, i = 1, 2$,we can easily see that $Q$ is continuous and bounded in$Y$.It follows Lemma3.3 that $Q(I - Q)(y_u, y_v, y_1, y_2)^L = (0, 0, 0, 0)^L, y_u, y_v, y_1, y_2 \in C[0, 1]$

Define a operator $R : X \times [0, 1] \to X_2$

$$R(u, v, \lambda)(t) = \begin{pmatrix} \int_0^t (t-s)\varphi_q(\int_0^s \lambda f(r, v(r), v'(r), v''(r))dr)ds \\ \int_0^t (t-s)\varphi_q(\int_0^s \lambda g(r, u(r), u'(r), u''(r))dr)ds \end{pmatrix},$$

where $\mathrm{Ker}M \bigoplus X_2 = X$.By $(H_2)$ and the Arzela-Asscoli theorem,we can get $R : \overline{\Omega} \times [0, 1] \to X_2$ is continuous and compact,where $\Omega \subset X$ is a bounded set.

**Lemma3.4** *Assume that $\Omega \in X$ is an open bounded set.Then $N_\lambda$ is $M$-quasi-compact in $\overline{\Omega}$.*

*Proof.* It is clear that $\mathrm{Im}P = \mathrm{Ker}M$, $QN_\lambda(u, v) = \theta \Leftrightarrow QN(u, v) = \theta$ and $R(\cdot, \cdot, 0) = 0$. for $(u, v) \in \overline{\Omega}$,

$$(I-Q)N_\lambda(u,v) = \begin{pmatrix} \lambda f(t, v(t), v'(t), v''(t)) \\ \lambda g(t, u(t), u'(t), u''(t)) \\ 0 \\ 0 \end{pmatrix} - \begin{pmatrix} 0 \\ 0 \\ T_1(\lambda f(t, v(t), v'(t), v''(t))) \\ T_2(\lambda g(t, u(t), u'(t), u''(t)) \end{pmatrix}$$

$$= \begin{pmatrix} \lambda f(t, v(t), v'(t), v''(t)) \\ \lambda g(t, u(t), u'(t), u''(t)) \\ T_1(\lambda f(t, v(t), v'(t), v''(t))) \\ T_2(\lambda g(t, u(t), u'(t), u''(t)) \end{pmatrix} \in \mathrm{Im}M.$$

Since $\mathrm{Im}M \subset \mathrm{Ker}Q$, and $y = Qy + (I - Q)y$,we obtain $\mathrm{Im}M \subset (I - Q)Y$.Thus $(I - Q)N_\lambda(\overline{\Omega}) \subset \mathrm{Im}M \subset (I - Q)Y$.
For $(u, v) \in \Sigma_\lambda = \{(u, v) \in \overline{\Omega} \cap \mathrm{dom}M \mid M(u, v) = N_\lambda(u, v)\}$

$$R(u, v, \lambda) = \begin{pmatrix} \int_0^t (t-s)\varphi_q(\int_0^s \lambda f(r, v(r), v'(r), v''(r))dr)ds \\ \int_0^t (t-s)\varphi_q(\int_0^s \lambda g(t, u(t), u'(t), u''(t))dr)ds \end{pmatrix}$$

$$= \begin{pmatrix} \int_0^t (t-s)\varphi_q(\int_0^s \varphi_p(u''))'(r)dr)ds \\ \int_0^t (t-s)\varphi_q(\int_0^s \varphi_p(v''))'(r)dr)ds \end{pmatrix}$$

$$= \begin{pmatrix} u(t) - u'(0)t \\ v(t) - v'(0)t \end{pmatrix} = (I - P)(u, v).$$

*i.e.* Definition 2.2(c)holds.For $u \in \overline{\Omega}$,we have

$$M[P(u,v) + R(u, v, \lambda)] = \begin{bmatrix} \lambda f(r, v(r), v'(r), v''(r)) \\ \lambda g(t, u(t), u'(t), u''(t)) \\ T_1(\lambda f(t, v(t), v'(t), v''(t))) \\ T_2(\lambda g(t, u(t), u'(t), u''(t)) \end{bmatrix} = (I - Q)N_\lambda(u, v)$$

Thus,Definition 2.2(d)holds.Therefore,$N_\lambda$ is $M$-quasi-compact in $\overline{\Omega}$.The proof is completed.

**Theorem 3.1** *Assume that the following conditions hold:*
*$(H_3)$There exist nonnegative constants $K_1, K_2$ such that one of (1)and (2) holds:*
*(1)*

$$B_1 f(t, A_1, B_1, C_1) > 0, \quad t \in [0, 1], |B_1| > K_1, A_1, C_1 \in R$$

*and*

$$B_2 g(t, A_2, B_2, C_2) > 0, \quad t \in [0, 1], |B_2| > K_2, A_2, C_2 \in R$$

*(2)*

$$B_1 f(t, A_1, B_1, C_1) < 0, \quad t \in [0,1], |B_1| > K_1, A_1, C_1 \in R$$

*and*

$$B_2 g(t, A_2, B_2, C_2) < 0, \quad t \in [0,1], |B_2| > K_2, A_2, C_2 \in R$$

$(H_4)$ *There exist nonnegative functions* $a_i(t), b_i(t), c_i(t), e_i(t) \in L^1[0,1], i = 1,2$ *such that*

$$| f(t,x,y,z) | \leq a_1(t)\varphi_p(| x |) + b_1(t)\varphi_p(| y |) + c_1(t)\varphi_p(| z |) + e_1(t)$$

*and*

$$| g(t,x,y,z) | \leq a_2(t)\varphi_p(| x |) + b_2(t)\varphi_p(| y |) + c_2(t)\varphi_p(| z |) + e_2(t)$$

*where* $\varphi_q(\|a_i(t)\|_1 + \|b_i(t)\|_1 + \|c_i(t)\|_1,) < 2^{2-q}$ *if* $1 < p \leq 2$ ;$\varphi_q(2^{p-2}\|a_i(t)\|_1 + 2^{p-2}\|b_i(t)\|_1 + \|c_i(t)\|_1,) < 1$ *if* $p \geq 2$
*The boundary value problem (1.1) has at least one solution.*

**Lemma3.5** *Suppose* $(H_3)$ *and* $(H_4)$ *hold.Then*

$$\Omega_1 = \{(u,v) \in domM \mid M(u,v) = N_\lambda(u,v)\}$$

*is bounded in X.*

*Proof.* For $(u,v) \in \Omega_1$,we have $QN_\lambda(u,v) = 0, i.e.$

$$T_1(\lambda f(t, v(t), v'(t), v''(t))) = 0, T_2(\lambda g(t, u(t), u'(t), u''(t)) = 0$$

By$H_3$,there exist constants $t_0, t_1 \in [0,1]$ such that $| u'(t_0) | \leq K_2$ and $| v'(t_1) | \leq K_1$.Since $u(t) = \int_0^t u'(s)ds$, $u'(t) = u'(t_0) + \int_{t_0}^t u''(s)ds, v(t) = \int_0^t v'(s)ds, v'(t) = v'(t_1) + \int_{t_1}^t v''(s)ds$,,then that

$$\begin{array}{l} | u(t) | \leq \| u' \|_\infty, | u'(t) | \leq K_2 + \| u'' \|_\infty, \quad t \in [0,1]. \\ | v(t) | \leq \| v' \|_\infty, | v'(t) | \leq K_1 + \| v'' \|_\infty, \quad t \in [0,1]. \end{array} \tag{0.2}$$

It follow from $M(u,v) = N_\lambda(u,v)$,$(H_4)$ and (3.2)that

$$\begin{array}{rcl} | u''(t) | & = & | \varphi_q(\int_0^t \lambda f(s, v(s), v'(s), v''(s))ds) | \\ & \leq & \varphi_q(\int_0^1 a_1(t)\varphi_p(| v |) + b_1(t)\varphi_p(| v' |) + c_1(t)\varphi_p(| v'' |) + e_1(t)dt) \\ & \leq & \varphi_q[(\|a_1\|_1 + \|b_1\|_1)\varphi_p(K_1 + \| v'' \|_\infty) + \|c_1\|_1\varphi_p(\| v'' \|_\infty) + \|e_1\|_1] \end{array}$$

If $1 < p \leq 2$,by Lemma 2.1,we get

$$\begin{array}{rcl} | u''(t) | & \leq & \varphi_q(D_1 + C_1\varphi_p \| v'' \|_\infty) \\ & \leq & 2^{q-2}[\varphi_q(D_1) + \varphi_q(C_1) \| v'' \|_\infty] \\ & \leq & 2^{q-2}[\varphi_q(D_1) + \varphi_q(C_1) \| v \|] \end{array}$$

thus

$$\| u'' \|_\infty \leq 2^{q-2}[\varphi_q(D_1) + \varphi_q(C_1) \| v \|]$$

where $C_1 = \|a_1\|_1 + \|b_1\|_1 + \|c_1\|_1, D_1 = (\|a_1\|_1 + \|b_1\|_1)\varphi_p(K_1) + \|e_1\|_1$. On the other hand, since $| u'(t) | \leq K_2 + \| u'' \|_\infty, t \in [0,1]$,we get $\| u' \|_\infty \leq K_2 + \| u'' \|_\infty$ since $| u(t) | \leq \| u' \|_\infty, t \in [0,1]$,we get$\| u \|_\infty \leq K_2 + \| u'' \|_\infty$.Thus

$$\| u \| \leq K_2 + 2^{q-2}[\varphi_q(D_1) + \varphi_q(C_1) \| v \|]. \tag{0.3}$$

Similarly,

$$\| v \| \leq K_1 + 2^{q-2}[\varphi_q(D_2) + \varphi_q(C_2) \| u \|] \tag{0.4}$$

where $C_2 = \|a_2\|_1 + \|b_2\|_1 + \|c_2\|_1, D_2 = (\|a_2\|_1 + \|b_2\|_1)\varphi_p(K_2) + \|e_2\|_1$ Take$M_1 = \max\{2^{q-2}\varphi_q(C_1), 2^{q-2}\varphi_q(C_2)\}$,we get

$$\| u \| \leq K_2 + 2^{q-2}\varphi_q(D_1) + M_1 \| v \|,$$

$$\| v \| \leq K_1 + 2^{q-2}\varphi_q(D_2) + M_1 \| u \| .$$

Since $\| (u, v) \| = \| u \| + \| v \|$, we get

$$\| (u, v) \| \leq \frac{K_1 + K_2 + 2^{q-2}\varphi_q(D_1) + 2^{q-2}\varphi_q(D_2)}{1 - M_1}$$

If $p > 2$,by Lemma 2.1,we get

$$| u''(t) | \leq \varphi_q(D_3 + C_3\varphi_p \| v'' \|_\infty) \leq \varphi_q(D_3) + \varphi_q(C_3) \| v'' \|_\infty$$
$$\leq \varphi_q(D_3) + \varphi_q(C_3) \| v \|,$$

thus

$$\| u'' \|_\infty \leq \varphi_q(D_3) + \varphi_q(C_3) \| v \| .$$

By (3.2),we get

$$\| u \| \leq K_2 + \varphi_q(D_3) + \varphi_q(C_3) \| v \| .$$

Similarly

$$\| v \| \leq K_1 + \varphi_q(D_4) + \varphi_q(C_4) \| v \| .$$

where $C_3 = 2^{p-2}(\|a_1\|_1 + \|b_1\|_1) + \|c_1\|_1, D_3 = 2^{p-2}(\|a_1\|_1 + \|b_1\|_1)\varphi_p(K_1) + \|e_1\|_1, C_4 = 2^{p-2}(\|a_2\|_1 + \|b_2\|_1) + \|c_2\|_1, D_4 = 2^{p-2}(\|a_2\|_1 + \|b_2\|_1)\varphi_p(K_2) + \|e_2\|_1$,Take $M_2 = \max\{\varphi_q(C_3), \varphi_q(C_4)\}$ ,we get

$$\| (u, v) \| \leq \frac{K_1 + K_2 + \varphi_q(D_3) + \varphi_q(D_4)}{1 - M_2}$$

So we obtain Lemma 3.5.

***Remark1*** If we take $\|(u, v)\| = \max\{\|u\|, \|v\|\}$,Lemmma 3.5 still holds.We only need (3.3) into (3.4),we can obtain the Lemma.

**Lemma3.6** *Assume* $(H_3)$ *holds,then*

$$\Omega_2 = \{(u, v) \in KerM \mid QN(u, v) = 0\}$$

*is bounded in X,where* $N = N_1$.

*Proof.* For $(u, v) \in \Omega_2$,we have $(u, v) = (b_1 t, b_2 t)$,then $T_1 f(t, b_2 t, b_2, 0) = 0, T_2 f(t, b_1 t, b_1, 0) = 0$.By $(H_3)$,we get$| b_1 | \leq K_2, | b_2 | \leq K_1$.So,$\Omega_2$ is bounded.

*Proof of theorem3.1* Let $\Omega = \{(u, v) \in X \mid \| (u, v) \| < r\}$,where $r$ is large enough such that $K_1 + K_2 < r < +\infty$ and $\overline{\Omega_1} \cup \overline{\Omega_2} \subset \Omega$.

By lemma3.5 and Lemma3.6,we can get if $(u, v) \in \text{dom}M \cap \partial\Omega$, then $M(u, v) \neq N_\lambda(u, v)$, if $(u, v) \in \text{Ker}M \cap \partial\Omega$, then $QN(u, v) \neq 0$.

Let

$$H(u, v, \delta) = \rho\delta(u, v)^L + (1 - \delta)JQN(u, v)^L . \delta \in [0, 1], (u, v) \in \text{Ker}M \cap \overline{\Omega}.$$

where $J : \text{ImQ} \to \text{KerM}$ is a homeomorphism with $J(0, 0, b_1, b_2)^L = (b_2 t, b_1 t)^L$,

$$\rho = \begin{cases} -1 , & \text{if}(H_3)(1)\text{holds} \\ 1 , & \text{if}(H_3)(2)\text{holds} \end{cases}$$

$$\text{sgn}(x) = \begin{cases} 1 , & x > 0 \\ -1 , & x < 0 \end{cases}$$

For $(u, v) \in \text{KerM} \cap \partial\Omega$,we have $(u, v) = (b_1 t, b_2 t) \neq (0, 0)$

$$H(u, v, \delta) = \rho\delta \begin{pmatrix} b_1 t \\ b_2 t \end{pmatrix} + (1 - \delta) \begin{pmatrix} -T_2 g(t, b_1 t, b_1, 0)t \\ -T_1 f(t, b_2 t, b_2, 0)t \end{pmatrix}$$

If $\delta = 1, h(u, v, 1) = \rho(b_1 t, b_2 t)^L \neq (0, 0)^L$.If $\delta = 0, h(u, v, 0) = JQN(b_1 t, b_2 t)^L \neq (0, 0)^L$. For $0 < \delta < 1$,we now prove that $H(u, v, \delta) \neq (0, 0)^L$.Otherwise,If $H(u, v, \delta) = (0, 0)^L$,then

$$\begin{pmatrix} T_2 g(t, b_1 t, b_1, 0) \\ T_1 f(t, b_2 t, b_2, 0) \end{pmatrix} = \begin{pmatrix} \frac{\rho\delta}{1-\delta} b_1 \\ \frac{\rho\delta}{1-\delta} b_2 \end{pmatrix}$$

Since $\|(u,v)\| = r > K_1 + K_2$,we have $|b_1| > K_2$ or $|b_2| > K_1$,If $|b_2| > K_1$we have

$$T_1(b_2 f(t, b_2 t, b_2, 0)) = b_2 T_1(f(t, b_2 t, b_2, 0)) = \frac{\rho\delta}{1-\delta} b_2^2$$

$$\mathrm{sgn}(T_1(b_2 f(t, b_2 t, b_2, 0))) = \mathrm{sgn}[b_2 f(t, b_2 t, b_2, 0)] = \mathrm{sgn}(\frac{\rho\delta}{1-\delta} b_2^2) = \mathrm{sgn}(\rho)$$

if $|b_1| > K_2$,we have

$$T_2(b_1 g(t, b_1 t, b_1, 0)) = b_1 T_2(g(t, b_1 t, b_1, 0)) = \frac{\rho\delta}{1-\delta} b_1^2$$

$$\mathrm{sgn}(T_2(b_1 g(t, b_1 t, b_1, 0))) = \mathrm{sgn}[b_1 g(t, b_1 t, b_1, 0)] = \mathrm{sgn}(\frac{\rho\delta}{1-\delta} b_1^2) = \mathrm{sgn}(\rho)$$

This is a contradiction with the definition of $\rho$. So,$H(u, v, \delta) \neq 0, (u, v) \in \mathrm{KerM} \cap \partial\Omega, \delta \in [0, 1]$.

By the homotopy of degree ,we get $\deg(JQN, \Omega \cap \mathrm{KerM}, 0) = \deg(H(\cdot,\cdot,0), \Omega \cap \mathrm{Ker}M, 0) = \deg(H(\cdot,\cdot,1), \Omega \cap \mathrm{Ker}M, 0) = \deg(\rho I, \Omega \cap \mathrm{Ker}M, 0) \neq 0$ By Theorem 2.1,we find that (1.1) has at least one solution in$\overline{\Omega}$ .The proof is completed.

***Remark2*** If $\|(u, v)\| = \max\{\|u\|, \|v\|\}$,Lemma 3.6 still holds.We can take $\max\{K_1, K_2\} < r < \infty$,we still get the Theorem 3.1.

## 4. Example

Let us consider the following boundary value problem

$$\begin{cases} (\varphi_p(u''))'(t) = \frac{t^3}{4}\sin x^5 + \frac{1}{8}y^5 + \frac{t^2}{4}\sin z^5 + \cos t \\ (\varphi_p(v''))'(t) = \frac{t^4}{8}\cos x^5 + \frac{1}{16}y^5 + \frac{t^3}{12}\cos z^5 + \sin t \\ u(0) = u''(0) = 0, v(0) = v''(0) = 0 \\ u'(1) = \int_0^1 2tu'(t)dt, v'(1) = \int_0^1 3t^2 v'(t)dt \end{cases} \tag{4.1}$$

where $p = 6$.

Corresponding to problem(1.1),we have $q = \frac{6}{5}, a_1(t) = \frac{t^3}{4}, b_1(t) = \frac{1}{8}, c_1(t) = \frac{t^2}{4}, e_1(t) = cost, k_1(t) = 2t,\ a_2(t) = \frac{t^4}{8}, b_2(t) = \frac{1}{16}, c_2(t) = \frac{t^3}{12}, e_1(t) = sint, k_2(t) = 3t^2$. Take $K_1 = 2, K_2 = 3$,we can get $(H_1) - (H_4)$ hold.By Theorem3.1,we have the problem(4.1)has at least one solution.

## References


DelPino, M, Elgueta, M, Mansevich, R. (1997). A homotopic deformation along p of a Leray-Schauder degree result and existence for $(|u'|^{p-2}u')' + f(t,u) = 0, u(0) = u(T) = 0, p > 1$. *J.Differ.Equ., 80*(1), 3227-3238.

Du, Z, Lin, X, & Ge, W. (2005). Some higher-order multi-point boundary value problem at resonance. *J.Comput.Appl.Math., 177*, 55-65. http://dx.doi.org/10.1016/j.cam.2004.08.003

Feng, W, & Webb, JRL. (1997). Solvability of m-point boundary value problems with nonlinear growth. *J.Math.Anal.Appl., 212*, 467-480. http://dx.doi.org/10.1006/jmaa.1997.5520

Garcia-Huidobro, M, Manasevich, R, & Zanolin, F. (1994). A Fredholm-like result for strongly nonlinear second order ODEs. *J.Differ.Equ., 114*, 132-167. http://dx.doi.org/10.1006/jdeq.1994.1144

Garcia-Huidobro, M, & Ubilla, P. (1997). Multiplicity of solutions for a class of non linear second order equations. *Nonlinear Anal, 28*(9), 1509-1520. http://dx.doi.org/10.1016/S0362-546X(96)00014-4

Ge, W, & Ren, J. (2004). An extension of Mawhin's continuation theorem and its application to boundary value problems with a p-Laplacian. *Nonlinear Anal,Theory Methods Appl, 58*, 477-488. http://dx.doi.org/10.1016/j.na.2004.01.007

Jiang, W. (2011). The existence of solutions to boundary value problems of fractional differential equations at resonance. *Nonlinear Anal., Theory Methods Appl. 74*, 1987-1994. http://dx.doi.org/10.1016/j.na.2010.11.005

Jiang, W. (2012). Solvability for a coupled system of fractional differential equations at resonance. *Nonlinear Anal.: Real World Appl., 13*, 2285-2292. http://dx.doi.org/10.1016/j.nonrwa.2012.01.023

Kosmatov, N. (2008). Multi-point boundary value problems on an unbounded domain at resonance. *Nonlinear Anal., 68*, 2158-2171. http://dx.doi.org/10.1016/j.na.2007.01.038

Kosmatov, N. (2010). A boundary value problem of fractional order at resonance. Electron. *J.Differ.Equ.*, 135.

Liu, B. (2003). Solvability of multi-point boundary value problem at resonance(II). *Appl.Math.Comput., 136,* 353-377. http://dx.doi.org/10.1016/s0096-3003(02)00050-4

Liu, Y, & Ge, W. (2004). Solvability of nonlocal boundary value problems for ordinary differential equations of higher order. *Nonlinear Anal., 57,* 435-458. http://dx.doi.org/10.1016/j.na.2004.02.023

Mawhin, J. (1979). Topological degree methods in nonlinear boundary value problems. In: NSFCBMS Regional Conference Series in Mathematics. *Am.Math.Soc.*, Providence. http://dx.doi.org/10.1090/cbms/040

Sun, W, & Ge, W. (2002). The existence of solutions to Sturm-Liouville BVPs with Laplacian-like operator. *Acta-Math.Appl.Sin., 18*(2), 341-348. http://dx.doi.org/10.1007/s102550200034

Weihua, Jiang. (2014). Solvability of boundary value problem with p-Laplacian at resonance. *Boundary value problems,* 36.

Zhang, X, Feng, M, & Ge, W. (2009). Existence result of second-order differential equations with integral boundary conditions at resonance. *J.Math.Anal.Appl., 353*, 311-319. http://dx.doi.org/10.1016/j.jmaa.2008.11.082

# 15

# A Double-indexed Functional Hill Process and Applications

Modou Ngom[1] & Gane Samb Lo[2]

[1] LERSTAD, Université Gaston Berger de Saint-Louis, SENEGAL

[2] LERSTAD, Université Gaston Berger de Saint-Louis, SENEGAL and LSTA, Université Pierre et Marie Curie, France (affiliated), gane-samb.

Correspondence: Gane Samb Lo, LERSTAD, Université Gaston Berger de Saint-Louis, SENEGAL and LSTA, Université Pierre et Marie Curie, France (affiliated). E-mail: lo@ugb.edu.sn

**Abstract**

Let $X_{1,n} \leq .... \leq X_{n,n}$ be the order statistics associated with a sample $X_1, ...., X_n$ whose pertaining distribution function (*df*) is $F$. We are concerned with the functional asymptotic behaviour of the sequence of stochastic processes

$$T_n(f,s) = \sum_{j=1}^{j=k} f(j)\left(\log X_{n-j+1,n} - \log X_{n-j,n}\right)^s, \tag{0.1}$$

indexed by some classes $\mathcal{F}$ of functions $f : \mathbb{N}^* \longmapsto \mathbb{R}_+$ and $s \in ]0, +\infty[$ and where $k = k(n)$ satisfies

$$1 \leq k \leq n, k/n \to 0 \text{ as } n \to \infty.$$

We show that this is a stochastic process whose margins generate estimators of the extreme value index when $F$ is in the extreme domain of attraction. We focus in this paper on its finite-dimension asymptotic law and provide a class of new estimators of the extreme value index whose performances are compared to analogous ones. The results are next particularized for one explicit class $\mathcal{F}$..

**Keywords:** Extreme values theory; asymptotic distribution; Functional Gaussian and nongaussian laws; Uniform entropy numbers; asymptotic tightness, stochastic process of estimators of extremal index; Slowly and regularly varying functions.

**2000 Mathematics Subject Classification**. Primary 62EG32, 60F05. Secondary 62F12, 62G20.

## 1. Introduction

### *1.1 General Introduction*

In this paper, we are concerned with the statistical estimation of the univariate extreme value index of a *df* $F$, when it is available. But rather than doing this by one statistic, we are going to use a stochastic process whose margins generate estimators of the extreme value index (SPMEEXI). To precise this notion, let $X_1, X_2, ...$ be a sequence of independent copies (s.i.c) of a real random variable (*rv*) $X > 1$ with *df* $F(x) = \mathbb{P}(X \leq x)$. $F$ is said to be in the extreme value domain of attraction of a nondegenerate *df* $M$ whenever there exist real and nonrandom sequences $(a_n > 0)_{n\geq 1}$ and $(b_n)_{n\geq 1}$ such that for any continuity point $x$ of $M$,

$$\lim_{n\to\infty} P(\frac{X_{n,n} - b_n}{a_n} \leq x) = \lim_{n\to\infty} F^n(a_n x + b_n) = M(x). \tag{1.1}$$

It is known that $M$ is necessarily of the family of the Generalized Extreme Value (GEV) *df* :

$$G_\gamma(x) = \exp(-(1+\gamma x)^{-1/\gamma}),\ 1 + \gamma x \geq 0,$$

parameterized by $\gamma \in \mathbb{R}$. The parameter $\gamma$ is called the extreme value index. There exists a great number of estimators of $\gamma$, going back to first of all of them, the Hill's one defined by

$$T_n(f,s) = k^{-1} \sum_{j=1}^{k} j(\log X_{n-j+1,n} - \log X_{n-j,n}),$$

where for each $n$, $k = k(n)$ is an integer such that

$$1 \leq k \leq n,\ k \to \infty,\ k/n \to 0 \text{ as } n \to \infty.$$

A modern and large account of univariate Extreme Value Theory can be found in Beirlant, Goegebeur and Teugels (2004), Galambos (1985), de Haan (1970) and de Haan and Feireira (2006), Embrechts *et al.* (1997) and Resnick (1987). One may estimate $\gamma$ by one statistic only. This is widely done in the literature. But one also may use a stochastic process of statistics $\{T_n(f), f \in \mathcal{F}\}$ indexed by $\mathcal{F}$, such that for any fixed $f \in \mathcal{F}$, there exists a sequence of nonrandom and positive real coefficients $(a_n(f))_{n\geq 1}$ such that $T_n^*(f) = T_n(f)/a_n(f)$ is an asymptotic estimator of $\gamma$. We name such families *Stochastic Procesess with Margins Estimating of the EXtreme value Index (SPMEEXI's).* Up to our knowledge, the first was introduced in Lo (1997) (see also Lo (2012)) as follows

$$T_n(p) = k^{-1} \sum_{h=1}^{p} \sum_{(s_1......s_h)\in\mathcal{P}(p,h)} \sum_{i_1=\ell+1}^{i_0} ... \sum_{i_h=\ell+1}^{i_{h-1}} i_h \prod_{i=i_1}^{i_h} \frac{(\log X_{n-i+1,n} - \log X_{n-i,n})^{s_i}}{s!},$$

for $1 \leq \ell < k < n$, $p \geq 1$, $i_0 = k$, where $\mathcal{P}(p,h)$ is the set of all ordered partitions of $p > 0$ into positive integers, $1 \leq h \leq p$ :

$$\mathcal{P}(p,h) = \{(s_1...s_h), \forall i,\ 1 \leq i \leq h, s_i > 0; s_1 + ... + s_h = p\}.$$

Further Lo *et al.* (2006, 2009) introduced continuous and functional forms described in (1.2) below. Meanwhile, without denoting it such that, Segers (2002) and others considered the Pickands process $\{P_n(s),\ \sqrt{k/n} \leq s \leq 1\}$, with

$$P_n(s) = \log \frac{X_{n-[k/s],n} - X_{n-[k/1],n}}{X_{n-[k/s^2],n} - X_{n-[k/s],n}},\ \sqrt{k/n} \leq s \leq 1.$$

Groeneboom (2003), proposed a family of kernel estimators, indexed by kernels. This family is surely a SPMEEXI although the authors did not consider a stochastic process view in the kernels $K$.

The main interest of *SPMEEXI*'s is first to have in hands an infinite class of estimators and especially, as shown in Segers (2002), to have the possibility to build discrete and continuous combinations of the margins as new and powerful estimators.

Number of the estimators of the extreme value index are either functions of consecutive log-spacings $\log X_{n-j+1,n} - \log X_{n-j,n}$ for $1 \leq j \leq k$, $1 \leq k \leq n$, or are functions of log-spacings from a threshold $\log t$ : $\log X_{n-j+1,n} - \log t$, $1 \leq j \leq k$, $1 \leq k \leq n$. In the last case, the threshold is usually taken as $t = X_{n-k,n}$. This simple remark teases the idea that taking functions of the log-spacings in place of the simple ones may lead to more general estimators. Dekkers *et al.* (1989) successfully experimented to the so-called moment estimator by using the power functions $h(x) = x^p$, $x \in \mathbb{R}_+$, $p = 1, 2$. Some available SPMEEX's are functions of these log-spacings as we will see soon. Here, In this paper, we aim at presenting a more general functional form in the following

$$T_n(f,s) = \sum_{j=1}^{k} f(j) \left(\log X_{n-j+1,n} - \log X_{n-j,n}\right)^s, \tag{1.2}$$

indexed by some classes $\mathcal{F}$ of functions $f : \mathbb{N}^* = \mathbb{N}\backslash\{0\} \longmapsto \mathbb{R}_+$, and by $s > 0$. We have two generalizations. First, for $s = 1$, we get

$$T_n(f,1)/k = \sum_{j=1}^{k} f(j) \left(\log X_{n-j+1,n} - \log X_{n-j,n}\right)/k,$$

which is the functional generalization of the Diop and Lo statistics (Diop and Lo (2006, 2009) for $f(j) = j^\tau$, for $0 < \tau$ and Deme *et al.* (2012). Secondly, if $f$ is the identity function and $s = 1$, we see that $T_n(Identity, 1)/k$ is Hill's statistic.

On the other hand, when utilizing the threshold method, we have, with the same properties of the parameters, the following statistic process :

$$S_n(f,s) = \sum_{j=1}^{k} f(j) \left(\log X_{n-j+1,n} - \log X_{n-k,n}\right)^s,. \tag{1.3}$$

This leads the couple of statistics

$$(M_{1,n}, M_{1,n}) = (S_n(\mathbf{1},1)/k, M_n(\mathbf{1},2)/k)$$

where **1** is the constant function $\mathbf{1}(x) = 1$ . From this couple of statistics Dekkers *et al.* (1989) deduced the following estimator of the extreme value index

$$D_n = M_{1,n}1 + (1 - (M_{2,n}/M_{1,n}^2))^{-1}/2.$$

Our objective is to show that these two stochastic processes (1.2) and (1.3) are SPMEEXI's. In this paper, we focus on the stochastic process $T_n(f, s)$ which uses sums of independant random variables. As to $S_n(f, s)'s$, to the contrary, it uses sums of dependent random variables. Its study will be done in coming up papers.

*1.2 Motivations and Scope of the Paper*

As announced, we focus on the stochastic process (1.2) here. We have been able to establish its finite-dimension asymptotic distribution. As already noticed in earlier works in Lo *et al.* (2006, 2009) and in Lo, Diop and Deme (2012), the limiting law may be Gaussian or non-Gaussian. In both cases, statistical tests may be implemented. In case of non-Gaussian asymptotic limits, the limiting distribution is represented through an infinite series of standard exponential random variables. Its law may be approximated through monte-Carlo methods, as showed in Fall *et al.* (2011)

Then we prove that it is a SPMEEXI in the sense of convergence in probability. Both for asymptotic distribution and convergence in probability, the used conditions are expressed with respect to an infinite series of standard exopnential randoms variables and through the auxiliary functions $a$ and $p$ in the representations of $df$'s in the extreme domain of attraction that will be recalled in the just next subsection. The conditions are next notably simplified by supposing that the $df$ $F$ is differentiable in the neighborhood of its upper endpoint.

To show how work the results for specific classes of functions $f$, we adapt them for $f_\tau(j) = j^\tau$, $\tau > 0$. It is interesting to see that although we have the existence of the asymptotic laws for any $\tau > 0$ and $s \geq 1$, we don't have an estimation of $\gamma$ in the region $\tau < s - 1$, when $s > 1$.

One advantage of using SPMEEXI's is that we may consider the best estimators, in some sense to be precised, among all margins. We show in Theorem 3 that $T_n(f_\tau, s)$ is asymptotically Gaussian for $\tau \geq s - 1/2$. When we restrict ourselves in that domain, we are able to establish that the minimum asymptotic variance is reached for $\tau = s$. Then we construc the best estimator $T_n^{(\tau)} = T_n(f_\tau, \tau)$, that is for $\tau = s$. This is very important since the Hill estimator is $T_n^{(1)}$ itself and, as a consequence, the Hill estimator is an element of a set of best estimators indexed by $\tau$. In fact, it is the best of all, that is $T_n^{(1)}$ has less asymptotic variance than $T_n^{(\tau)}, \tau > 1$.

It will be interesting to found out whether this minimim variance can be improved for other functional classes.

Even when we have a minimum asymptotic variance estimator, it is not sure that the performance is better for finite samples. This is why simulation studies mean reveal a best combination between bias and asymptotic variance. At finite sample size, the performance of an estimator is measured both by the bias and the variance and we don't know how the random value of the estimator is far from the exact value. We will see in the simulation Section 3 that the boundary case $\tau = s - 1/2$ gives performances similar to the optimal case.

Before we present the theoritical results and their consequences, we feel obliged to present a brief reminder of basic univariate extreme value theory and some related notation on which the statements of the results will rely on.

*1.3 Basics of Extreme Value Theory*

Let us make this reminder by continuing the lines of (1.1) above. If (1.1) holds, it is said that $F$ is attracted to $M$ or $F$ belongs to the domain of attraction of $M$, written $F \in D(M)$. It is well-kwown that the three possible nondegenerate limits in (1.1), called extreme value *df*, are the following :

The Gumbel *df* of parameter $\gamma = 0$,

$$\Lambda(x) = \exp(-\exp(-x)), \ x \in \mathbb{R}, \tag{1.4}$$

or the Fréchet *df* of parameter $\gamma > 0$,

$$\phi_\gamma(x) = \exp(-x^{-\gamma})\mathbb{I}_{[0,+\infty[}(x), \ x \in \mathbb{R} \tag{1.5}$$

or the Weibull *df* of parameter $\gamma < 0$ ,

$$\psi_\gamma(x) = \exp(-(x)^{-\gamma})\mathbb{I}_{]-\infty,0]}(x) + (1 - 1_{]-\infty,0]}(x)), \ x \in \mathbb{R}, \tag{1.6}$$

where $I_A$ denotes the indicator function of the set A. Now put $D(\phi) = \cup_{\gamma>0} D(\phi_\gamma)$, $D(\psi) = \cup_{\gamma>0} D(\psi_\gamma)$, and $\Gamma = D(\phi) \cup D(\psi) \cup D(\Lambda)$.

In fact the limiting distribution function $M$ is defined by an equivalence class of the binary relation $\mathcal{R}$ on the set of *df* $\mathcal{D}$ on $F$ defined as follows :

$$\forall (M_1, M_2) \in \mathcal{D}^2, (M_1 \, \mathcal{R} \, M_2) \Leftrightarrow \exists (a, b) \in \mathbb{R}_+ \backslash \{0\} \times \mathbb{R}, \forall (x \in \mathbb{R}),$$

$$M_2(x) = M_1(ax + b).$$

One easily checks that if $F^n\,(a_n x + b_n) \to M_1(x)$, then $F^n\,(c_n x + d_n) \to M_1(ax + b) = M_2(x)$ whenever

$$a_n/d_n \to a \text{ and } (b_n - d_n)/c_n \to b \text{ as } n \to \infty. \tag{1.7}$$

Theses facts allow to parameterize the class of extremal distribution functions. For this purpose, suppose that (1.1) holds for the three *df*'s given in (1.4), (1.5) and (1.6). We may take sequences $(a_n > 0)_{n\geq 1}$ and $(b_n)_{n\geq 1}$ such that the limits in (1.7) are $a = \gamma = 1/\alpha$ and $b = 1$ (in the case of Fréchet extremal domain), and $a = -\gamma = -1/\alpha$ and $b = -1$ (in the case of Weibull extremal domain). Finally, one may interprets $(1 + \gamma x)^{-1/\gamma} = exp(-x)$ for $\gamma = 0$ (in the case of Gumbel extremal domain). This leads to the following parameterized extremal distribution function

$$G_\gamma(x) = \exp(-(1 + \gamma x)^{-1/\gamma}), \; 1 + \gamma x \geq 0,$$

called the Generalized Extreme Value (GEV) distribution of parameter $\gamma \in R$.

Now we give the usual representations of $df's$ lying in the extremal domain in terms of the quantile function of $G(x) = F(e^x), x \geq 1$, that is $G^{-1}(1 - u) = \log F^{-1}(1 - u), 0 \leq u \leq 1$.

**Theorem 1.** *We have :*

1. *Karamata's representation (KARARE)*

   *(a) If $F \in D(\phi_{1/\gamma})$, $\gamma > 0$, then*

$$G^{-1}(1 - u) = \log c + \log(1 + p(u)) - \gamma \log u + (\int_u^1 b(t)t^{-1}dt), \; 0 < u < 1, \tag{1.8}$$

   *where* $\sup(|p(u)|, |b(u)|) \to 0$ *as* $u \to 0$ *and c is a positive constant and* $G^{-1}(1 - u) = \inf\{x, G(x) \geq u\}, 0 \leq u \leq 1$, *is the generalized inverse of G with* $G^{-1}(0) = G^{-1}(0+)$.

   *(b) If $F \in D(\psi_{1/\gamma})$, $\gamma > 0$, then $y_0(G) = \sup\{x, G(x) < 1\} < +\infty$ and*

$$y_0 - G^{-1}(1 - u) = c(1 + p(u))u^\gamma \exp\Big(\int_u^1 b(t)t^{-1}dt\Big), \; 0 < u < 1, \tag{1.9}$$

   *where c, $p(\cdot)$ and $b(\cdot)$ are as in (1.8)*

2. *Representation of de Haan (Theorem 2.4.1 in de Haan (1970)),*

   *If $G \in D(\Lambda)$, then*

$$G^{-1}(1 - u) = d - a(u) + \int_u^1 a(t)t^{-1}dt, \; 0 < u < 1, \tag{1.10}$$

   *where d is a constant and $a(\cdot)$ admits this KARARE :*

$$a(u) = c(1 + p(u)) \exp(\int_u^1 b(t)t^{-1}dt), \; 0 < u < 1, \tag{1.11}$$

   *c, $p(\cdot)$ anf $b(\cdot)$ being defined as in (1.8). We warn the reader to not confuse this function $a(.)$ with the function $a_n(.,.)$ which will be defined later.*

Finally, we shall also use the uniform representation of $Y_1 = \log X_1, Y_2 = \log X_2, ...$ by $G^{-1}(1 - U_1), G^{-1}(1 - U_2), ...$ where $U_1, U_2, ...$ are independent and uniform random variables on (0, 1) and where $G$ is the $df$ of $Y$, in the sense of equality in distribution (denoted by $=_d$)

$$\{Y_j, j \geq 1\} =_d \{G^{-1}(1 - U_j), j \geq 1\},$$

and hence

$$\{\{Y_{1,n}, Y_{2,n}, ...Y_{n,n}\}, n \geq 1\} \quad (1.12)$$

$$=_d \left\{\{G^{-1}(1 - U_{n,n}), G^{-1}(1 - U_{n-1,n}), ..., G^{-1}(1 - U_{1,n})\}, n \geq 1\right\}.$$

In connexion with this, we shall use the following Malmquist representation (see Shorack and Wellner (1986, p. 336) :

$$\{\log(\frac{U_{j+1,n}}{U_{j,n}})^j, j = 1, ..., n\} =_d \{E_{1,n}, ..., E_{n,n}\},$$

where $E_{1,n}, ..., E_{n,n}$ is an array of independent standard exponential random variables. We write $E_i$ instead of $E_{i,n}$ for simplicity sake. Some conditions will be expressed in terms of these exponential random variables. We are now in position to state our first results for finite distribution asymptotic normality.

## 2. Our Results

We need the following conditions. First define for $n \geq 1$, $f$ and $s$ fixed,

$$B_n(f, s) = \max\{f(j)j^{-s}/\sigma_n(f, s), 1 \leq j \leq k\},$$

$$a_n(f, s) = \Gamma(s + 1) \sum_{j=1}^{k} f(j)\, j^{-s}$$

and

$$\sigma_n^2(f, s) = \sum_{j=1}^{k} f^2(j)\, j^{-2s}.$$

We will use the two main conditions of $f$ and $s$ fixed :

$$\sum_{j=1}^{\infty} f(j)^2 j^{-2s} < \infty \quad (K1)$$

and

$$\sum_{j=1}^{\infty} f(j)^2 j^{-2s} = +\infty \text{ and } B_n(f, s) \to 0 \text{ as } n \to \infty. \quad (K2)$$

Further, any $df$ in $D(G_\gamma)$ is associated with a couple of functions $(p, b)$ as given in the representations (1.8), (1.9) and (1.11). Define then the following notation for $\lambda > 1$,

$$b_n(\lambda) = \sup\{|b(t)|, 0 \leq t \leq \lambda k/n\}$$

and

$$p_n(\lambda) = \sup\{|p(t)|, 0 \leq t \leq \lambda k/n\}$$

We will require below that, for some $\lambda > 1$,

$$b_n(\lambda) \log k \to 0, \text{ as } n \to +\infty. \quad (CR1)$$

From now, all the limits below are meant as $n \to \infty$ unless the contrary is specified.

Here are our fundamental results. First, we have marginal estimations of the extreme value index as expected. The conditions of the results are given in very general forms that allow further, specific hypotheses as particular cases. As well, although we focus here on finite-distribution limits, the conditions are stated in a way that will permit to handle uniform studies further.

**Theorem 2.** *Let $F \in D(G_\gamma)$, $0 \leq \gamma < +\infty$.*

*(A) Case $0 < \gamma < +\infty$.*

*1) Let $(K1)$ hold. If $a_n(f,s) \to \infty$ and for an arbitrary $\lambda > 1$*

$$a_n^{-1}(f,s)\left[\sum_{j=1}^{k} f(j)\, s \left\{\frac{\gamma}{j}E_j + \left(p_n(\lambda) + \frac{E_j}{j}b_n(\lambda)\right)\right\}^{s-1} \times \left(p_n(\lambda) + \frac{E_j}{j}b_n(\lambda)\right)\right] \to_{\mathbb{P}} 0, \tag{H0a}$$

*then*

$$(T_n(f,s)/a_n(f,s))^{1/s} \to_{\mathbb{P}} \gamma,$$

*where $\to_{\mathbb{P}}$ stands for convergence in probability.*

*2) Let $(K2)$ hold. If $a_n^{-1}(f,s)\sigma_n^{-1}(f,s) \to 0$ and for an arbitrary $\lambda > 1$*

$$\sigma_n^{-1}(f,s)\left[\sum_{j=1}^{k} f(j)\, s \left\{\frac{\gamma}{j}E_j + \left(p_n(\lambda) + \frac{E_j}{j}b_n(\lambda)\right)\right\}^{s-1} \times \left(p_n(\lambda) + \frac{E_j}{j}b_n(\lambda)\right)\right] \to_{\mathbb{P}} 0, \tag{H1a}$$

*then*

$$(T_n(f,s)/a_n(f,s))^{1/s} \to_{\mathbb{P}} \gamma.$$

*(B) case $\gamma = 0$.*

*1) Let $(K1)$ hold. If $a_n(f,s) \to +\infty$ and for an arbitrary $\lambda > 1$,*

$$a_n^{-1}(f,s)\left[\sum_{j=1}^{k} f(j)\, s \left\{j^{-1}E_j + \left(p_n(\lambda) + \frac{E_j}{j}p_n(\lambda) \vee b_n(\lambda)\log k\right)\right\}^{s-1} \times \left(p_n(\lambda) + \frac{E_j}{j}p_n(\lambda) \vee b_n(\lambda)\log k\right)\right] \to_{\mathbb{P}} 0, \tag{H0b}$$

*then*

$$\left(\frac{T_n(f,s)}{a_n(f,s)}\right)^{1/s} / a(k/n) \to_{\mathbb{P}} 1.$$

*2) Let $(K2)$ hold. If $a_n^{-1}(f,s)\sigma_n^{-1}(f,s) \to 0$ and for an arbitrary $\lambda > 1$, and*

$$\sigma_n^{-1}(f,s)\left[\sum_{j=1}^{k} f(j)\, s \left\{j^{-1}E_j + \left(p_n(\lambda) + \frac{E_j}{j}p_n(\lambda) \vee b_n(\lambda)\log k\right)\right\}^{s-1} \times \left(p_n(\lambda) + \frac{E_j}{j}p_n(\lambda) \vee b_n(\lambda)\log k\right)\right] \to_{\mathbb{P}} 0, \tag{H1b}$$

*then*

$$\left(\frac{T_n(f,s)}{a_n(f,s)}\right)^{1/s} / a(k/n) \to_{\mathbb{P}} 1.$$

**Theorem 3.** *Let $F \in D(G_\gamma)$, $0 \leq \gamma < +\infty$.*

*(A) Case $0 < \gamma < +\infty$.*

*1) If $(K1)$ and*

$$\sum_{j=1}^{k} f(j)\, s \left\{ \frac{\gamma}{j} E_j + \left( p_n(\lambda) + \frac{E_j}{j} b_n(\lambda) \right) \right\}^{s-1} \left( p_n(\lambda) + \frac{E_j}{j} b_n(\lambda) \right) \to_{\mathbb{P}} 0, \tag{H2a}$$

*then*

$$T_n(f,s) - \gamma^s a_n(f,s) \to \gamma^s \left\{ \Gamma(2s+1) - \Gamma(s+1)^2 \right\}^{1/2} \mathcal{L}(f,s),$$

*where*

$$\mathcal{L}(f,s) = \sum_{j=1}^{\infty} f(j)\, j^{-s} F_j^{(s)},$$

*and the $F_j^s$'s are independent and centred random variables with variance one.*

*2) If $(K2)$ and $(H2a)$ hold for an arbitrary $\lambda > 1$, then*

$$\sigma_n^{-1}(f,s)\left(T_n(f,s) - \gamma^s a_n(f,s)\right) \to \mathcal{N}(0, \gamma^{2s} \left\{ \Gamma(2s+1) - \Gamma(s+1)^2 \right\}).$$

*(B) case $\gamma = 0$.*

*1) If $(K1)$ and for an arbitrary $\lambda > 1$,*

$$\left[ \sum_{j=1}^{k} f(j)\, s \left\{ j^{-1} E_j + \left( p_n(\lambda) + \frac{E_j}{j} p_n(\lambda) \vee b_n(\lambda) \log k \right) \right\}^{s-1} \right. \tag{H2b}$$

$$\left. \times \left( p_n(\lambda) + \frac{E_j}{j} p_n(\lambda) \vee b_n(\lambda) \log k \right) \right] \to_{\mathbb{P}} 0,$$

*then*

$$\frac{T_n(f,s)}{a^s\left(\frac{k}{n}\right)} - a_n(f,s) \to \left\{ \Gamma(2s+1) - \Gamma(s+1)^2 \right\}^{1/2} \mathcal{L}(f,s)),$$

*where*

$$\mathcal{L}(f,s) = \sum_{j=1}^{\infty} f(j)\, j^{-s} F_j^{(s)}.$$

*2) If $(K2)$ and $(H2b)$ hold for an arbitrary $\lambda > 1$, then*

$$\sigma_n^{-1}(f,s)\left[ \frac{T_n(f,s)}{a^s(k/n)} - a_n(f,s) \right] \to \mathcal{N}\left(0, \left\{ \Gamma(2s+1) - \Gamma(s+1)^2 \right\}\right).$$

*2.1 Remarks and Applications.*

2.1.1 General Remarks on the Conditions

The conditions $(H0a)$, $(H0b)$, $(H1a)$, $(H1b)$, $(H2a)$ and $(H2b)$ hold if we show that the expectations of the $rv$'s of their right members tend to zero, for an arbitrary $\lambda > 1$, simply by the use of Markov's inequality. These expectations include intergrals $I(a,b,s) = \int_0^\infty (a+bx)^s e^{-x} dx$ and $J(a,b,c,s) = \int_0^\infty (a+cx)^s (a+bx) e^{-x} dx$ for real numbers $a$, $b$, $c$ and $s \geq 1$ computed in (4.1) and (4.2) of Section (4), ***for integer values of*** $s$, as

$$I(a,b,s) = \int_0^\infty (a+bx)^s e^{-x} dx = s! \sum_{h=0}^{s} b^h a^{s-h}/(s-h)!.$$

and

$$\begin{aligned} J(a,b,c,s) &= s!a \sum_{h=0}^{s-1} c^h a^{s-h}/(s-h)! + s!b \sum_{h=0}^{s-1} c^h I(a,c,s-h)/(s-h)! \\ &+ s!c^s I(a,b,1). \end{aligned}$$

Then the conditions $(H0a)$, $(H0b)$, $(H2a)$, $(H2b)$, $(H1a)$ and $(H1b)$ respectively hold when hold these ones

(HE0a) $a_n^{-1}(f,s)\sum_{j=1}^{k} f(j)I_n(1,j,s) \to 0;$
(HE0b) $a_n^{-1}(f,s)\sum_{j=1}^{k} f(j)I_n(2,j,s) \to 0;$
(HE1a) $\sigma_n^{-1}(f,s)\sum_{j=1}^{k} f(j)I_n(1,j,s) \to 0;$
(HE1b) $\sigma_n^{-1}(f,s)\sum_{j=1}^{k} f(j)I_n(2,j,) \to 0;$
(HE2a) $\sum_{j=1}^{k} f(j)I_n(1,j,s) \to 0;$
(HE2b) $\sum_{j=1}^{k} f(j)I_n(2,j,s) \to 0;$

with

$$I_n(1,j,s) = sJ(p_n(\lambda), b_n(\lambda)/j, (\gamma + b_n(\lambda))/j, s-1)$$

and

$$I_n(2,j,s) = sJ(p_n(\lambda), (p_n(\lambda) \vee b_n(\lambda)\log k)\,/\,j, (1 + (p_n(\lambda) \vee b_n(\lambda)\log k))/j, s-1)$$

*2.2 Weakening the Conditions for s Interger.*

When the distribution function $G$ admits an ultimate derivative at $x_0(G) = \sup\{x, G(x) < 1\}$, and this is the case for the usual *df*'s, one may take $p(u) = 0$, as pointed out in (2014). In that case, the conditions $(HE0x), (HE1x)$ and $(HE2x)$, for $x = a$ or $x = b$, are much simpler. We then have $I_n(1,j,s) = ss!b_n(\lambda)(\gamma + b_n(\lambda))^{s-1}j^{-s}$ and $I_n(2,j,s) = ss!(b_n(\lambda)\log k)(\gamma + b_n(\lambda)\log k)^{s-1}j^{-s}$. We get these simpler conditions :

(HE0a) $a_n^{-1}(f,s)\left(ss!b_n(\lambda)(\gamma + b_n(\lambda))^{s-1}\sum_{j=1}^{k} f(j)j^{-s}\right) \to 0;$
(HE0b) $a_n^{-1}(f,s)\left(ss!(b_n(\lambda)\log k)^{s-1}\sum_{j=1}^{k} f(j)j^{-s)}\right) \to 0;$
(HE1a) $\sigma_n^{-1}(f,s)\left(ss!b_n(\lambda)(\gamma + b_n(\lambda))^{s-1}\sum_{j=1}^{k} f(j)j^{-s}\right) \to 0;$
(HE1b) $\sigma_n^{-1}(f,s)\left(ss!(b_n(\lambda)\log k)^{s-1}\sum_{j=1}^{k} f(j)j^{-s)}\right) \to 0;$
(HE2a) $\left(ss!b_n(\lambda)(\gamma + b_n(\lambda))^{s-1}\sum_{j=1}^{k} f(j)j^{-s)}\right) \to 0;$
(HE2b) $\left(ss!(b_n(\lambda)\log k)(\gamma + b_n(\lambda)\log k)^{s-1}\sum_{j=1}^{k} f(j)j^{-s)}\right) \to 0.$

It is interesting to remark that all these conditions automaticaly hold whenever $b_n(\lambda) \to 0$, and/or $(CR1)$ holds. Indeed, we remark, by the Cauchy-Scharwz's inequality, that :

$$\sigma_n^{-1}(f,s)\sum_{j=1}^{k} f(j)j^{-s} \leq \sigma_n^{-1}(f,s)\left(\sum_{j=1}^{k} f^2(j)j^{-2s}\right)^{1/2} = 1.$$

Then for $\gamma > 0$, the corresponding conditions always hold since $b_n(\lambda) \to 0$ and for $\gamma = 0$, the corresponding conditions hold with $(CR1)$. This surely leads to powerfull results. It also happens that for the usual cases, we know the values of $b_n(\lambda)$, based on $b(u) = (G^{-1}(1-u))' + \gamma$ for $\gamma > 0$ and $b(u) = us'(u)/s(u)$ with $s(u) = u(G^{-1}(1-u))'$ (see for instance Fall and Lo (2011) or Segers (2002)).

*2.3 the Special Case of Diop-Lo*

Now it is time to see how the preceeding results work for the particular case the functions class $f_\tau(j) = j^\tau, \tau > 0$. This special study should be a model of how to apply the results for other specific classes. Here, we will replace $f$ by $\tau$ in all the notation meaning that $f = f_\tau$. We summarize the holding conditions depending on $\tau > 0$ and $s \geq 1$, in the following table

We may see the details as follows. First $\sum f(j)^2 j^{-2s} = \sum j^{-2(s-\tau)}$ is finite if and only if $2(s-\tau) > 1$. This gives the cases (I) and (II). For (III) in Table 1, we have

$$\sigma_n^2(\tau,s) = \sum_{j=1}^{k} j^{-2(s-\tau)} = \sum_{j=1}^{k} j^{-1} \sim (\log k),$$

by (4.3) in Section (4). Since for $1 \leq j \leq k$, $f(j)j^{-s}/\sigma_n(\tau,s) = j^{-1/2}/\sigma_n(\tau,s) \leq 1/\sigma_n(\tau,s)$, we have $B_n(\tau,s) \leq \sigma_n^{-1}(\tau,s) \to 0$ and then $(K2)$ holds. For (IV),

$$\sigma_n^2(\tau,s) = \sum_{j=1}^{k} j^{-2(s-\tau)} \sim k^{(2(\tau-s)+1))}/(2(\tau-s)+1)),$$

Table 1. Checking the conditions for the Diop-Lo class

| (I) | (II) | (III) | (IV) |
|---|---|---|---|
| $\tau < s-1$<br>$(K1)$ | $s-1 \le \tau < s-1/2$<br>$(K1)$ | $\tau = s-1/2$<br>$(K2)$ | $\tau > s-1/2$<br>$(K2)$ |
| $a_n$ bounded | $a_n \to \infty$ | $a_n \sim 2s!k^{1/2}$ | $a_n \sim s!\frac{k^{\tau-s+1}}{\tau-s+1}$ |
| $\sigma_n$ bounded | $\sigma_n$ bounded | $\sigma_n \sim (\log k)^{1/2}$ | $\sigma_n \sim \frac{k^{(\tau-s+1/2)}}{\sqrt{2(\tau-s)+1}}$ |

by (4.4). Next $1 \le j \le k$, $f(j)j^{-s}/\sigma_n(\tau, s) = j^{(\tau-s)}/\sigma_n(\tau, s) = C_0(j/k)^{(\tau-s)}k^{-1/2}$. Since $(\tau - s) > 0, B_n(\tau, s) \le C_0 k^{-1/2} \to 0$. Then $(K2)$ also holds. The lines above also explain the fourth row of the table. The third row is immediate since $a_n = \sum_{j=1}^{k} j^{-(s-\tau)} \to \infty$ for $(s-\tau) \le 1$ and remains bounded for $(s-\tau) > 1$.

It is worth mentioning that the case $\tau < s-1$ is not possible for $s = 1$. This unveils a new case comparatively with former studies of Deme *et al.* (2012) for $s = 1$.

Now, based on these facts, we are able to get the following estimations (results $(CR1)$ for $\gamma = 0$) :

1. For $s-1 \le \tau < s-1/2$, $a_n \to \infty$. Hence we have the estimation

$$\left(\frac{T_n(\tau, s)}{a_n(\tau, s)}\right)^{1/s} / a(k/n) \to_{\mathbb{P}} 1.$$

For $\gamma > 0$, $a(k/n) \to \gamma$. For $\gamma = 0$, $\left(\frac{T_n(\tau,s)}{a_n(\tau,s)}\right)^{1/s} \to \gamma = 0$ at the rate of $a_n(k/n)$.

2. $0 < \tau < s-1$. We do not have an estimation of $\gamma$.

For testing the hypothesis $F \in D(G_\gamma)$, $\gamma \ge 0$, we derive the following laws by the delta-method under $(CR1)$, especially for $\gamma = 0$.

Let $s-1 \le \tau < s-1/2$. For $\gamma > 0$,

$$a_n(\tau, s)\left\{\left(\frac{T_n(\tau, s)}{a_n(\tau, s)}\right) - \gamma^s\right\}$$

$$\to \gamma^s\left\{\Gamma(2s+1) - \Gamma(s+1)^2\right\}^{1/2} \mathcal{L}(\tau, s). \tag{2.1}$$

For $\gamma = 0$

$$a_n(\tau, s)\left\{a(k/n)^{-s}\left(\frac{T_n(\tau, s)}{a_n(\tau, s)}\right) - 1\right\} \to \left\{\Gamma(2s+1) - \Gamma(s+1)^2\right\}^{1/2} \mathcal{L}(\tau, s).$$

Let $\tau \ge s-1/2$. In this case $a_n(\tau, s)\sigma_n^{-1}(\tau, s) \to +\infty$. This enables the delta-method application to the limit in Theorem 3, case (A-2), that is

$$\frac{a_n(\tau, s)}{\sigma_n(\tau, s)}\left\{\left(\frac{T_n(\tau, s)}{a_n(\tau, s)}\right) - \gamma^s\right\} \to \mathcal{N}(0, \gamma^{2s}\left\{\Gamma(2s+1) - \Gamma(s+1)^2\right\}). \tag{2.2}$$

We derive

$$\frac{a_n(\tau, s)}{\sigma_n(\tau, s)}\left\{\left(\frac{T_n(\tau, s)}{a_n(\tau, s)}\right)^{1/s} - \gamma\right\} \to \mathcal{N}(0, (\gamma/s)\left\{\Gamma(2s+1) - \Gamma(s+1)^2\right\})$$

For $\gamma = 0$

$$\frac{a_n(\tau, s)}{\sigma_n(\tau, s)}\left\{a(k/n)^{-s}\left(\frac{T_n(\tau, s)}{a_n(\tau, s)}\right) - 1\right\} \to \mathcal{N}(0, \left\{\Gamma(2s+1) - \Gamma(s+1)^2\right\}).$$

For the new case $\tau < s-1$, we have for $\gamma > 0$,

$$T_n(\tau, s) - A(\tau, s)\gamma^s \to \gamma^s\left\{\Gamma(2s+1) - \Gamma(s+1)^2\right\}^{1/2} \mathcal{L}(\tau, s),$$

where $A(\tau, s) = \sum_{j=1}^{\infty} j^{-(s-\tau)} < \infty$, for $\gamma = 0$,

$$T_n(\tau, s)/a^s(k/n) - A(\tau, s) \to \left\{\Gamma(2s+1) - \Gamma(s+1)^2\right\}^{1/2} \mathcal{L}(\tau, s).$$

These two limiting laws also allow statistical tests based on Monte-Carlo methods as in Deme *et al.* (2012).

*2.4 Best Performance Estimators*

In pratical situations, we have to select a particular function $f$ from a particular class $\mathcal{F}$ of function $f$. A natural question is to select a couple $(f, s)$ for which the estimator is the best in some sense. Here we consider the class of Diop-Lo, $f(j) = j^{\tau}$ and we are interested in finding the best performance of the estimator $\left(\frac{T_n(\tau,s)}{a_n(\tau,s)}\right)^{1/s}$ of $\gamma > 0$. We place ourselves in the normality domain, that is $\tau > s - 1/2$. Straigthforward computations from (2.2) yield

$$V_n(\tau, s)\left[\left(\frac{T_n(\tau, s)}{a_n(\tau, s)}\right)^{1/s} - \gamma\right] \rightsquigarrow N(0, 1)$$

with

$$V_n(\tau, s) = \frac{a_n(\tau, s)}{\sigma_n(\tau, s)} \times \frac{s\gamma^{-1}}{\sqrt{\Gamma(2s+1) - \Gamma(s+1)^2}}.$$

So, finding the best performance is achieved for minimum value for the asymptotic variance $V_n(\tau, s)^{-2}$. We then have to find the greatest value of variance $V_n(\tau, s)$. But maximizing this function both in $s$ and $\tau$ might be tricky. However, for a fixed $s \geq 1$, we may find that the maximum value of $V_n(\tau, s)$ for $\tau \in [s - 1/2, +\infty[$. First, we have to isolate the boundary point $\tau = s - 1/2$. We prove in Subsection 4.4 below that the maximum value of $V_n(s, \tau)$ is reached when $\tau = s$. Using the formulae in Table 1, we see that, for $\tau \geq s - 1/2$, we have

$$V_n(\tau, \tau) = \tau!\sqrt{(k)} \times \frac{s\gamma^{-1}}{\sqrt{\Gamma(2\tau+1) - \Gamma(\tau+1)^2}} \tag{2.3}$$

and for $\tau = s - 1/2$, we get

$$V_n(\tau, \tau + 1/2) = 2s!\sqrt{(k/\log k)} \times \frac{s\gamma^{-1}}{\sqrt{\Gamma(2(\tau+1)) - \Gamma(\tau+3/2)^2}}. \tag{2.4}$$

We get as best estimator with least asymptotic variance

$$T_n(\tau)^{(\tau)} = \left(\frac{T_n(\tau, \tau)}{a_n(\tau, \tau)}\right)^{1/s}$$

for the normality zone $\tau \geq s - 1/2$. Its asymptotic variance (2.3) increases when $\tau$ decreases. This means that the Hill estimator is the best with respect to this sense. Now, let us move to the non-Gaussian zone, that is $0 \leq s-1 \leq \tau < s-1/2$, corresponding to the column II in the Table 1. We may easily derive from (2.1) that the asymptotic variance is of equivalent to

$$\gamma\frac{\Gamma(2s+1) - \Gamma(s+1)^2}{s(\tau - s + 1)},$$

which is still dominated by $V_n(\tau, \tau)^{-1}$. To sum up, we say that the Hill estimator has best asymptotic variance for all margins.

However for finite sample, we do not know how far the centered and normalized statistic is from the limiting Gaussian variable or the non-Gaussan limiting random. Here we are obliged to back on simulation studies. Let us consider

$$T_n^{(\tau)} = \left(\frac{T_n(\tau, \tau)}{a_n(\tau, \tau)}\right)^{1/\tau} \tag{2.5}$$

and

$$T_n^{(\tau+1/2)} = \left(\frac{T_n(\tau, \tau+1/2)}{a_n(\tau, \tau+1/2)}\right)^{1/(\tau+1/2)}$$

We get that these two estimators generally behave better than the Hill's and the Dekkers *et al.*'s ones. At least, they have equivalent performances. But absolutely, they seem to be more stable in a sense to precised later. This must result in lesser biases that constitute a compensation of their poorer performance regarding the asymptotic variance point of view. A full report of simulation studies are given in Section 3.

But we should keep in mind that the results presented here, go far beyond the Diop-Lo family for which the Hill's estimator demonstates to be the least asymptotic variance estimator.

Further, researches will be conducted on other functions families in order to possibly find out estimators with asymptotic variances better that $1/V_n(1,1)$.

*2.5 Proofs*

We will prove both theorems together. For both cases $\gamma > 0$ and $\gamma = 0$, we will arrive at the final statement based on the hypotheses $(K1)$ or $(K2)$, $(H1a)$ or $(H1a)$, $(H2a)$ in the case $\gamma > 0$ (A) , and on $(K1)$ or $(K2)$, $(H0b)$ or $(H1b)$, $(H2b)$ in the case $\gamma = 0$ $(B)$. In each case, an analysis will give the corresponding parts (1) and (2) for the two theorems. We begin with the A case.

Case (A) : Here, $F \in D(G_\gamma)$ with $0 < \gamma < +\infty$.

By using (1.12), we have

$$T_n(f,s) = \sum_{j=1}^{k} f(j)(G^{-1}(1-U_{j,n}) - G^{-1}(1-U_{j+1,n}))^s$$

By (1.8), we also have

$$G^{-1}(1-u) = \log c + \log(1+p(u)) - \gamma \log u + \int_u^1 b(t)t^{-1}dt, 0 < u < 1.$$

For $1 \leq j \leq k$,

$$\begin{aligned} G^{-1}(1-U_{j,n}) - G^{-1}(1-U_{j+1,n}) &= \log\left[\frac{1+p(U_{j,n})}{1+p(U_{j+1,n})}\right] \\ +\gamma \log\left(\frac{U_{j+1,n}}{U_{j,n}}\right) &+ \int_{U_{j,n}}^{U_{j+1,n}} b(t)t^{-1}dt. \end{aligned}$$

Put $p_n = \sup\{|p(t)|, 0 \leq t \leq U_{k+1,n}\}$ and $b_n = \sup\{|b(t)|, 0 \leq t \leq U_{k+1,n}\}$. Both $b_n$ and $p_n$ tend to zero in probability as $n \to +\infty$, since $U_{k+1,n} \to 0$ when $(n, k/n) \to (+\infty, 0)$. We then get

$$T_n(f,s) = \sum_{j=1}^{k} f(j)\left\{O_{\mathbb{P}}(p_n) + \frac{\gamma}{j}E_j + \frac{E_j}{j}O_{\mathbb{P}}(b_n)\right\}^s$$

Put

$$A_{n,j} = \left(O_{\mathbb{P}}(p_n) + \frac{\gamma}{j}E_j + \frac{E_j}{j}O_{\mathbb{P}}(b_n)\right)^s.$$

By the mean value Theorem, we have for $s \geq 1$

$$\begin{aligned} A_{n,j} - \left(\frac{\gamma}{j}E_j\right)^s &= s\left\{\frac{\gamma}{j}E_j + \theta_{n,j}\left(O_{\mathbb{P}}(p_n) + \frac{E_j}{j}O_{\mathbb{P}}(b_n)\right)\right\}^{s-1} \\ &\times \left(O_{\mathbb{P}}(p_n) + \frac{E_j}{j}O_{\mathbb{P}}(b_n)\right) \end{aligned}$$

where for $|\theta_{n,j}| \leq 1$. Put

$$\zeta_{n,j}(s) = s\left\{\frac{\gamma}{j}E_j + \theta_{n,j}\left(O_{\mathbb{P}}(p_n) + \frac{E_j}{j}O_{\mathbb{P}}(b_n)\right)\right\}^{s-1}\left(O_{\mathbb{P}}(p_n) + \frac{E_j}{j}O_{\mathbb{P}}(b_n)\right) \tag{2.6}$$

We have

$$T_n(f,s) = \gamma^s \sum_{j=1}^{k} f(j)j^{-s}E_j^s + \sum_{j=1}^{k} f(j)\zeta_{n,j}(s).$$

Recall that $\mathbb{E}\left(E_j^s\right) = \Gamma(s+1)$ and $\mathbb{V}(E_j^s) = \Gamma(2s+1) - \Gamma(s+1)^2$ and denote

$$\begin{aligned} V_n(f,s) &= \sum_{j=1}^{k} f(j)\, j^{-s}(E_j^s - s!) \\ &= \left\{\Gamma(2s+1) - \Gamma(s+1)^2\right\}^{\frac{1}{2}} \sum_{j=1}^{k} f(j)\, j^{-s} F_j^{(s)}(s), \end{aligned}$$

and for a fixed $s \geq 1$,

$$F_j^{(s)}(s) = (E_j^s - s!)/\left\{\Gamma(2s+1) - \Gamma(s+1)^2\right\}^{\frac{1}{2}}$$

is a sequence of independent mean zero random variables with variance one. We have

$$\begin{aligned} T_n(f,s) - \gamma^s a_n(f,s) &= \gamma^s\left\{\Gamma(2s+1) - \Gamma(s+1)^2\right\}\sum_{j=1}^{k} f(j)\, j^{-s}F_j^{(s)}(s) \\ &+ \sum_{j=1}^{k} f(j)\zeta_{n,j}(s). \end{aligned} \tag{2.7}$$

By (2.6),

$$\begin{aligned} \left|\zeta_{n,j}(s)\right| &= s\left\{\frac{\gamma}{j}E_j + \theta_{n,j}\left(|O_{\mathbb{P}}(p_n)| + \frac{E_j}{j}|O_{\mathbb{P}}(b_n)|\right)\right\}^{s-1} \\ &\times \left(|O_{\mathbb{P}}(p_n)| + \frac{E_j}{j}|O_{\mathbb{P}}(b_n)|\right) \end{aligned}$$

When $(K1)$ holds, Kolmogorov's Theorem on sums of centered random variables ensures that

$$\left\{\Gamma(2s+1) - \Gamma(s+1)^2\right\}^{1/2}\sum_{j=1}^{k} f(j)\, j^{-s}F_j^{(s)}(s)$$

converges to the *rv*

$$\left\{\Gamma(2s+1) - \Gamma(s+1)^2\right\}^{1/2}\sum_{j=1}^{+\infty} f(j)\, j^{-s}F_j^{(s)}(s) = \left\{\Gamma(2s+1) - \Gamma(s+1)^2\right\}^{1/2}\mathcal{L}(f,s)$$

which is centered and has variance one, as completely described in Lemma (1).

Now $(H0a)$ and Lemma (2) ensure that the second term of (2.7) tends to zero in probability. Then if $a_n(f,s) \to \infty$, we get that $T_n(f,s)/a_n(f,s) \to \gamma^s$. This proves Part (A)(1) of Theorem (2) for $\gamma > 0$. Further if $(H2a)$ holds, Lemma (2) and the first point yields that $T_n(f,s) - \gamma^s a_n(f,s)$ asymptotically behaves as $L(f,s) = \left\{\Gamma(2s+1) - \Gamma(s+1)^2\right\}^{1/2}\sum_{j=1}^{+\infty} f(j)\, j^{-s}F_j^{(s)}(s)$ since the second term of (2.7) is zero at infinity. This proves Part (A)(1) of Theorem (3) for $\gamma > 0$.

Now suppose that $(K1)$ does not hold and that $(K2)$ and $(H2a)$ both hold. Also $(H2a)$ implies via Lemma (1) that the second term, when divided by $\sigma_n(f,s)$, tends to zero in probability. Next by Lemma (2),

$$\sigma_n^{-1}(f,s)\left\{\Gamma(2s+1)-\Gamma(s+1)^2\right\}^{1/2}\sum_{j=1}^{k} f(j)\, j^{-s}F_j^{(s)}(s)$$

asymptotically behaves as a $\mathcal{N}(0,1)$ $rv$ under $(K2)$. It follows under these circumtances that

$$(a_n(f,s)/\sigma_n(f,s))(T_n(f,s)/a_n(f,s)-\gamma^s)\to N(0,1).$$

This ends the proof of Part (A)(2) of Theorem (2). Further, whenever $a_n(f,s)/\sigma_n(f,s)\to\infty$,

$$T_n(f,s)/a_n(f,s)\to\gamma^s,$$

which establishes Part (A) (2) of Theorem (2).

Case B : $F\in D(G_0)$, $\gamma=0$. Use representations (1.10) and (1.11) to get for $1\le j\le U_{k+1,n}$,

$$G^{-1}(1-U_{j,n})-G^{-1}(1-U_{j+1,n})=a\left(U_{j+1,n}\right)-a\left(U_{j,n}\right)+\int_{U_{j,n}}^{U_{j+1,n}} a(t)t^{-1}dt.$$

Remark that for $U_{1,2}<u,v<U_{k,n}$

$$\frac{v}{u}<\frac{U_{k,n}}{U_{1,n}}=\frac{U_{k,n}}{U_{k-1,n}}\times\frac{U_{k-1,n}}{U_{k-2,n}}\times\frac{U_{k-2,n}}{U_{k-3,n}}\times\cdots\times\frac{U_{2,n}}{U_{1,n}}$$

and

$$0\le\log\left(\frac{u}{v}\right)=\sum_{j=1}^{k-1}\log\left(\frac{U_{j+1,n}}{U_{j,n}}\right)=\sum_{j=1}^{k-1}j^{-1}E_j=\sum_{j=1}^{k-1}j^{-1}\left(E_j-1\right)+\sum_{j=1}^{k-1}j^{-1}.$$

By Kolmogorov's theorem for partial sums of independent and mean zero random variables, $\sum_{j=1}^{k-1}j^{-1}\left(E_j-1\right)$ converges in law to a finite $rv$ $E$. We have $\sum_{j=1}^{k-1}j^{-1}\sim\log k$ and this ensures that for $1\le u\le U_{k+1,n}$, $(1+p(u))/(1+p(U_{k+1,n}))-1=O_{\mathbb{P}}(p_n)$,

$$\exp\left(\int_u^1 b(t)t^{-1}dt\right)/\exp\left(\int_{U_{k+1,n}}^1 b(t)t^{-1}dt\right)-1=O_{\mathbb{P}}(b_n\log k)$$

both uniformly in $1\le u\le U_{k+1,n}$ and finally

$$a(U_{j+1,n})/a(U_{j,n})=(1+O_{\mathbb{P}}(p_n))\exp(O_{\mathbb{P}}(b_n)E_j/j),$$

But for $1\le j\le k$, $\left|j^{-1}E_j\right|\le\sum_{j=1}^{k-1}j^{-1}E_j=O_p(\log k)$. Then if $b_n\log k\to_{\mathbb{P}}0$,

$$a(U_{j+1,n})/a(U_{j,n})=(1+O_{\mathbb{P}}(p_n))(1+b_nE_j/j)$$
$$=1+O_{\mathbb{P}}(p_n)+O_{\mathbb{P}}(p_nb_nE_j/j)+O_{\mathbb{P}}(b_nE_j/j).$$

Finally, since $a(k/n)/a(U_{k+1,n})=1+O_{\mathbb{P}}(1)$, it follows that

$$a(u)/a(k/n)=1+O_{\mathbb{P}}((p_n\vee b_n)\log k)$$

uniformly in $1\le u\le U_{k+1,n}$. This finally leads to for $1\le j\le U_{k+1,n}$ ,

$$B_{j,n}(s)=\left(G^{-1}(1-U_{j,n})-G^{-1}(1-U_{j+1,n})\right)/a(k/n)$$
$$=(1+O_{\mathbb{P}}((p_n\vee b_n)\log k))\times\left\{O_{\mathbb{P}}(p_n)+O_{\mathbb{P}}(p_nb_nE_j/j)+O_{\mathbb{P}}(b_nE_j/j)\right\}$$

$$+\{1 + O_P((p_n \vee b_n)\log k)\}\, E_j/j$$

$$= E_j/j + R_{j,n}(s),$$

where

$$R_{j,n}(s) = (1 + O_P((p_n \vee b_n)\log k)) \times \left\{O_P(p_n) + O_P(p_n b_n E_j/j) + O_P(b_n E_j/j)\right\}$$

$$+O_P((p_n \vee b_n)\log k)E_j/j.$$

We can easily show that

$$R_{j,n}(s) = O_P(p_n) + O_P((p_n \vee b_n)\log k)E_j/j$$

and remark that $E_j/j = O_P(\log k)$ uniformly in $j \in \{1, ..., k\}$. This yields

$$\frac{T_n(f,s)}{a^s(k/n)} = \sum_{j=1}^{k} f(j)\left\{O_P(p_n) + O_P((p_n \vee b_n)\log k)E_j/j\right\}^s.$$

We have by the same methods used above

$$\frac{T_n(f,s)}{a^s(k/n)} = \sum_{j=1}^{k} f(j)\, j^{-s} E_j^s + \sum_{j=1}^{k} f(j)\, \xi_{n,j}(s)$$

$$\begin{aligned}\xi_{n,j}(s) &= s\left\{E_j j^{-1} + \theta_{n,j}\left(O_{\mathbb{P}}(p_n) + \frac{E_j}{j} O_{\mathbb{P}}(p_n \vee b_n \log k)\right)\right\}^{s-1} \\ &\times \left(O_{\mathbb{P}}(p_n) + \frac{E_j}{j} O_{\mathbb{P}}(p_n \vee b_n \log k)\right)\end{aligned}$$

And further

$$\frac{T_n(f,s)}{a^s(k/n)} - a_n(f,s) = \sum_{j=1}^{k} f(j)\, j^{-s}\left(E_j^s - s!\right) + \sum_{j=1}^{k} f(j)\, \xi_{n,j}(s).$$

$$\frac{T_n(f,s)}{a^s(k/n)} = \left\{\Gamma(2s+1) - \Gamma(s+1)^2\right\}^{1/2} \sum_{j=1}^{k} f(j)\, j^{-s} F_j^{(s)}(s) + \sum_{j=1}^{k} f(j)\, \xi_{n,j}(s). \quad (2.8)$$

When we compare Formulas (2.7) and (2.8), $(H0a) - (H2a) - (H1a)$ with $(H0b) - (H2b) - (H1b)$, we use Lemmas (1) and (2) and reconduct almost the same conclusion already done for the $\gamma > 0$ case to prove the parts (B) of Theorems (2) and (3).

## 3. Simulation Studies

Nowadays, simulation studies are very sophisticated and may be very difficult to follow. Here, we want give a serious comparison of our estimators with several analoges while keeping the study reasonably simple. Let us begin to explain the stakes before proceeding any further. The estimators of the extremal index generally use a number, say $k$ like in this text, of the greatest observations : $X_{n-j+1}$, $1 \leq j \leq k$. For almost all such estimators, we have a small bias and a great variance for large values of $k$, and the contrary happens for small values of $k$. This leads to the sake of an optimal value of $k$ keeping both the bias and the variance at a low level. A related method consists in considering a range of values $kv(j) = kmin + j(kmin - kmin)/ksize$, $1 \leq j \leq ksize$ over which the observed values of the statistic are stable and well approximate the index. This second method seems preferable when comparing two estimators with respect to the bias.

So we fix a sample size $n$ and consider the range of values as described above where $kmin$ and $kmax$ are suitably chosen. Thus checking the curves of two statistics over the interval $[mink, maxk]$ is a good tool for comparing their performances. Next, for each $j$, $1 \leq kmin \leq j \leq kmax$, we compute the mean square error of the estimated values of $\gamma$ for values of $k$ in a neighborhood of $kv(j)$, that is for $k \in [kv(j) - kstab, kv(j) + kstab]$, $1 \leq j \leq ksize$, where $kstab$ is also suitably fixed.

The mimimun of these MES's certainly corresponds to the most stable zone and may be taken as the best estimation when it is low enough.

Here, we compare our class of estimators, represented the optimal estimator (2.4) and the boundary form (2.5) for $s \in [1,5]$, with the estimators of Hill and Dekkers *et al.*. The estimators (2.4) and (2.5) for $s \in [1,5]$ fall in the asymptotic normality area $s - 1/2 \leq \tau$. In a larger study, we will include the Pickands' statistic and consider the nongaussian asymptotic area.

The study will only cover the heavy tail case, that $\gamma > 0$. The case $\gamma = 0$ will be part of a large simulation paper. And we consider a pure Pareto law (I), and two perturbed ones laws (II) and (III):

$$\text{(I) } F^{-1}(1-u) = u^{-\gamma}$$
$$\text{(II) } F^{-1}(1-u) = u^{-\gamma}(1-u^{\beta})$$
$$\text{(III) } F^{-1}(1-u) = u^{-\gamma}(1-(-1/\log u)^{\beta}).$$

*3.1 Simulations for $\gamma > 0$*

Let $f(j) = j^{\tau}$, $\tau > 0$. Our results say that $\left(\frac{T_n(\tau,s)}{a_n(\tau,s)}\right)^{1/s}$ is an estimator of $\gamma \geq 0$ if $s - 1 \leq \tau$.

Theoritically, we then have in hands an infinite class of estimators. We should be able to find values of $s$ and $\tau$ leading a lowest stable bias and hope that this bias will be lower than of the other analogues, or to be at their order at least. We know from Subsection 4.4 that our class of estimators for $f = f_{\tau}$ has miminal asymptotic variance for $s = \tau$. But it is not sure that this corresponds to the best performance for finite samples. And following the remark of Deme *et al.* (2012) who noticed that the boundary case, that discriminates the Gaussian and the non Gaussian asymptotic laws, behaves well, we include it also here, that is the case $s - 1/2 = \tau$ .

We fix the following values : $n = 1000$, $kmin = 105$, $kmax = 375$, $ksize = 100$, $kstab = 5$. And we fix the number of replications to $B = 1000$.

How to read our results? For each $j \in [1, ksize]$, we compue the mean square error $MSE(j)$ when $k$ spans $[kv(j) - kstab, kv(j) + kstab]$, that is

$$MSE(j)^2 = \frac{1}{2kstab+1} \sum_{k \in [kv(j)-kstab, kv(j)+kstab]} (T_n(k) - \gamma)^2.$$

Next we take the mimimum and the maximum values of these MSE(j)'s denoted as *Min* and *Max*. The difference *(df)* Diff=*Max-Min* is reported as well as the middle term $Mid = (Min + Max)/2$.

We classify the estimators with respect to both the values of *Mid* and *Diff*. If the values *Diff* are of the same order for two estimators, we will prefer the one with the minimum value of *Mid*. That mean that this latter estimator is more stable and then, is better.

Since the Hill estimator and the Dekkers *et al.* do not depend on parameters, we have conducted a series of simulations and the performances are of the order of values given in Table 2. Next, we conduct simulations on the performances of the boundary Double Hill and the optimal double Hill statistics for $s = 1$, ..., $s = 5$ in Tables 3, 4, 5, 6, 7, 8.

Table 2. Performances of Hill and Dekkers *et al.* estimators

| Values | Model I | | Model II | | Model III | |
|---|---|---|---|---|---|---|
| | Hill | Dekkers | Hill | Dekkers | Hill | Dekkers |
| Min | $1.529110^{-5}$ | $7.683510^{-3}$ | $3.575110^{-3}$ | $9.324510^{-5}$ | $1.042810^{-1}$ | $5.575110^{-2}$ |
| Max | $2.033810^{-3}$ | $1.515810^{-2}$ | $3.724210^{-2}$ | $1.648410^{-2}$ | $8.442510^{-1}$ | $5.456810^{-1}$ |
| Diff | $2.018510^{-3}$ | $7.474510^{-3}$ | $3.366710^{-2}$ | $1.639110^{-2}$ | $7.399610^{-1}$ | $4.901110^{-1}$ |
| Mid | $1.024510^{-3}$ | $1.142010^{-2}$ | $2.040810^{-2}$ | $8.288710^{-3}$ | $4.742710^{-1}$ | $3.008010^{-1}$ |

Table 3. Performances of the Boundary Double Hill statistics for s=1,...,5 with Model I

| Values | Model I Double Hill | | | | |
|---|---|---|---|---|---|
| | s=1 | s=2 | s=3 | s=4 | s=5 |
| Min | 6.203010$^{-4}$ | 1.111310$^{-3}$ | 2.598810$^{-2}$ | 5.237910$^{-3}$ | 6.655810$^{-3}$ |
| Max | 1.903710$^{-3}$ | 2.181410$^{-3}$ | 3.014210$^{-2}$ | 6.032410$^{-3}$ | 8.443310$^{-3}$ |
| Diff | 1.283410$^{-3}$ | 1.070110$^{-3}$ | 4.153710$^{-3}$ | 7.945410$^{-4}$ | 1.787410$^{-3}$ |
| Mid | 1.262010$^{-3}$ | 1.646310$^{-3}$ | 2.806510$^{-2}$ | 5.635210$^{-3}$ | 7.549610$^{-3}$ |

Table 4. Performances of the Optimal Double Hill statistics for s=1,...,5 with Model I

| Values | Model I Optimal Double Hill | | | | |
|---|---|---|---|---|---|
| | s=1 | s=2 | s=3 | s=4 | s=5 |
| Min | 4.712810$^{-3}$ | 1.404210$^{-4}$ | 9.436910$^{-7}$ | 5.456610$^{-3}$ | 3.849510$^{-3}$ |
| Max | 9.851410$^{-3}$ | 5.374710$^{-3}$ | 4.117010$^{-3}$ | 1.080210$^{-2}$ | 1.606810$^{-2}$ |
| Diff | 5.138610$^{-3}$ | 5.234210$^{-3}$ | 4.116010$^{-3}$ | 5.345410$^{-3}$ | 1.221810$^{-2}$ |
| Mid | 7.282110$^{-3}$ | 2.757510$^{-3}$ | 2.058910$^{-3}$ | 8.129310$^{-3}$ | 9.958910$^{-3}$ |

Table 5. Performances of the Boundary Double Hill statistics for s=1,...,5 with Model II

| Values | Model II Double Hill | | | | |
|---|---|---|---|---|---|
| | s=1 | s=2 | s=3 | s=4 | s=5 |
| Min | 1.825810$^{-3}$ | 1.322610$^{-3}$ | 2.981110$^{-2}$ | 1.562110$^{-5}$ | 8.124610$^{-6}$ |
| Max | 1.009710$^{-2}$ | 9.692610$^{-3}$ | 3.399510$^{-2}$ | 5.227810$^{-3}$ | 6.245110$^{-3}$ |
| Diff | 8.272010$^{-3}$ | 8.369910$^{-3}$ | 4.184010$^{-3}$ | 5.212210$^{-3}$ | 6.236910$^{-3}$ |
| Mid | 5.961810$^{-3}$ | 5.507610$^{-3}$ | 3.190310$^{-2}$ | 2.621710$^{-3}$ | 3.126610$^{-3}$ |

Table 6. Performances of the Optimal Double Hill statistics for s=1,...,5 with Model II

| Values | Model II Optimal Double Hill | | | | |
|---|---|---|---|---|---|
| | s=1 | s=2 | s=3 | s=4 | s=5 |
| Min | 1.843210$^{-3}$ | 1.345110$^{-3}$ | 1.495410$^{-3}$ | 1.562110$^{-5}$ | 8.124610$^{-6}$ |
| Max | 1.009710$^{-2}$ | 9.692610$^{-3}$ | 3.399510$^{-2}$ | 7.997210$^{-3}$ | 6.171210$^{-3}$ |
| Diff | 8.254610$^{-3}$ | 8.347410$^{-3}$ | 3.249910$^{-2}$ | 7.981610$^{-3}$ | 6.163010$^{-3}$ |
| Mid | 5.970510$^{-3}$ | 5.518810$^{-3}$ | 1.774510$^{-2}$ | 4.006410$^{-3}$ | 3.089610$^{-3}$ |

Table 7. Performances of the Boundary Double Hill statistics for s=1,...,5 with Model III

| Values | Model III Double Hill | | | | |
|---|---|---|---|---|---|
| | s=1 | s=2 | s=3 | s=4 | s=5 |
| Min | 3.351710$^{-2}$ | 3.144610$^{-2}$ | 3.144610$^{-2}$ | 2.493710$^{-2}$ | 2.550910$^{-2}$ |
| Max | 1.879410$^{-1}$ | 2.711010$^{-1}$ | 2.711010$^{-1}$ | 6.190110$^{-1}$ | 9.538710$^{-1}$ |
| Diff | 1.544210$^{-1}$ | 2.396510$^{-1}$ | 2.396510$^{-1}$ | 5.940710$^{-1}$ | 9.283610$^{-1}$ |
| Mid | 1.107210$^{-1}$ | 1.511210$^{-1}$ | 1.512710$^{-1}$ | 3.219710$^{-1}$ | 4.896910$^{-1}$ |

Table 8. Performances of the Optimal Double Hill statistics for s=1,...,5 with Model III

| Values | Model III Optimal Double Hill | | | | |
|---|---|---|---|---|---|
| | s=1 | s=2 | s=3 | s=4 | s=5 |
| Min | 3.197910$^{-2}$ | 3.182110$^{-2}$ | 3.182110$^{-2}$ | 2.536110$^{-2}$ | 2.606010$^{-2}$ |
| Max | 1.879210$^{-1}$ | 2.711010$^{-1}$ | 2.711010$^{-1}$ | 6.190110$^{-1}$ | 9.538710$^{-1}$ |
| Diff | 1.559410$^{-1}$ | 2.392810$^{-1}$ | 2.392810$^{-1}$ | 5.936410$^{-1}$ | 9.278110$^{-1}$ |
| Mid | 1.099510$^{-1}$ | 1.514610$^{-1}$ | 1.514610$^{-1}$ | 3.221810$^{-1}$ | 4.899610$^{-1}$ |

Now, we are able to draw a number of conclusions and remarks based on the Tables 2, 3, 4, 5, 6, 7, 8.

**(1.)** Model I : Compared to Hill's statistic and Dekkers *et al.*'s estimator, our Double Hill and Optimal Double Hill estimors are generally more stable and mostly better with regard to the middle value for $s = 1$, $s = 2$, $s = 4$.

**(2.)** Model II : we have simular results.

**(3.)** Model III : For $s \leq 3$. We also get similar results.

**(4.)** As a general conclusion, we say that our Double Hill estimator, for both cases of of boundary and minimum variance, behaves like these Hill's and Dekkers *et al.*'s estimator, are more stable and sloghtly better.

**(5.)** All these estimators present poorer performances for the very perturbated model III. But unlikely to the Hill's and Dekkers *et al.*'s estimators, we have in hand a SPMEEXI and hopefully we will be able to get better new estimators for model III through suitable combinations in future works. This is important since the Hill's and Dekkers *et al.*'s are the most used in the applications.

**(6.)** We may find theses patterns on Figure 1 and Figure 2.

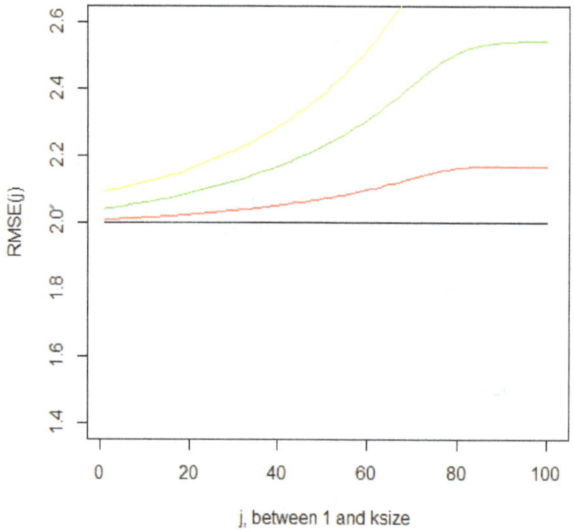


Figure 1. Curves of the mean values of the RMSE's at 100 values of k(j) (j=1,...,100) computed on eleven points around k(j) for the statistics : Hill [blue], Dekkers *et al.* [green], Boundary Double Hill [red] and Optimal Double Hill [yellow] for s=1

## 4. Technical Results

### *4.1 Technical Lemmas*

We begin by this simple lemma where we suppose that we are given a sequence of independent and identically mean zero $rv's$ $F_1, F_2, ..., F_k$ with variance unity. Denote

$$A(f, s) = \sum_{j=1}^{\infty} f(j)^2 j^{-2s}$$

and

$$\sigma_n^2(f, s) = \sum_{j=1}^{k} f(j)^2 j^{-2s}$$

**Lemma 1.** *Let*

$$V_n(f, s) = \sum_{j=1}^{k} f(j) j^{-s}(E_j^s - s!) =: \left\{\Gamma(2s+1) - \Gamma(s+1)^2\right\}^{1/2} \sum_{j=1}^{k} f(j) j^{-s} F_j^{(s)},$$

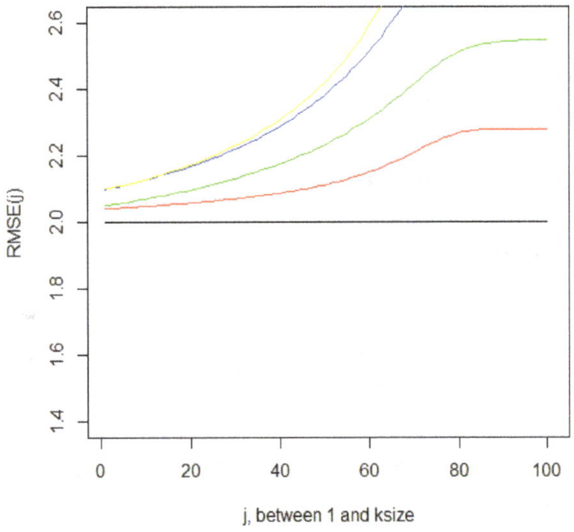


Figure 2. Curves of the mean values of the RMSE's at 100 values of k(j) (j=1,...,100) computed on eleven points around k(j) for the statistics : Hill [blue], Dekkers *et al.* [green], Boundary Double Hill [red] and Optimal Double Hill [yellow] for s=2

*wher the $F_j^{(s)}$ are independent centered random variables with variance one.*

*(1) If $(K1)\ :\ A(f,s)<\infty$, then*

$$V_n(f,s) \rightsquigarrow \left\{\Gamma(2s+1)-\Gamma(s+1)^2\right\}^{1/2}\mathcal{L}(f,s).$$

(2) If $(K2)$ : $B_n(f,s)=\max\{f(j)j^{-s}/\sigma_n(f,s), 1\leq j\leq k\}\rightarrow 0$, then

$$\sigma_n^{-1}(f,s)V_n(f,s)\rightsquigarrow \mathcal{N}(0,1)$$

*Proof.* Put

$$V_n^*(f,s)=\sigma_n(f,s)^{-1}V_n(f,s).$$

and suppose that $(K1)$ holds. Then Kolmogorov's Theorem for sums of zero mean indenpendent rv's applies. Since the series $\sum_{j\geq 1} Var(f(j)j^{-s}F_j^{(s)})$ is finite, we have :

$$V_n(f,s)\rightarrow\left\{\Gamma(2s+1)-\Gamma(s+1)^2\right\}^{-1/2}\sum_{j=1}^{\infty} f(j)j^{-s}F_j^{(s)}=\left\{\Gamma(2s+1)-\Gamma(s+1)^2\right\}^{1/2}\mathcal{L}(f,s),$$

Now suppose that $(K2)$ holds. Let us evaluate the moment generating function of $V_n^{**}(f,s)=\sigma_n(f,s)^{-1}\sum_{j=1}^{k} f(j)j^{-s}F_j^{(s)}$ :

$$\phi_{V_n^{**}(f,s)}(t)=\prod_{j=1}^{k}\phi_{F_j^{(s)}}(tf(j)j^{-s}\sigma_n(f,s)^{-1}). \tag{4.1}$$

We use the expansion common characteristic function $\phi(u)=\phi_{F_j^{(s)}}(u)=1-u^2/2+u^2\varepsilon(v)$, where $\varepsilon(v)\rightarrow 0$, uniformly in $|u|\leq v\rightarrow 0$. Remind that for $1\leq j\leq k$, $0\leq f(j)j^{-s}\sigma_n(f,s)^{-1}\leq B_n(f,s)\rightarrow 0$ and then

$$\phi_{V_n^{**}(f,s)}(t)=\prod_{j=1}^{k}(1-t^2f(j)^2j^{-2s}\sigma_n^{-2}(f,s)/2+t^2f(j)^2j^{-2s}\sigma_n^2(f,s)\varepsilon(B_n))$$

$$=\exp(\sum_{j=1}^{k}\log(1-t^2f(j)^2j^{-2s}\sigma_n^2(f,s)/2+t^2f(j)^2j^{-2s}\sigma_n^2(f,s)\varepsilon(B_n))).$$

By using a first order expansion the logarithmic function in the neighborhood of unity, we have

$$\phi_{V_n^*(f)}(t) = \exp(\sum_{j=1}^{k} -t^2 f(j)^2 j^{-2s}\sigma_n(f)^{-2}/2 + t^2 f(j)^2 j^{-2s}\sigma_n^{-2}(f)\varepsilon(B_n)),$$

where the function $\varepsilon(B_n)$ may change from one line to an other, but always tends to zero. Hence

$$\phi_{V_n^{**}(f,s)}(t) = \exp(\sum_{j=1}^{k} -t^2/2 + t^2\varepsilon(B_n)) \to \exp(-t^2/2).$$

and

$$V_n^{**}(f) \to \mathcal{N}(0,1).$$

and then

$$V_n^*(f) \to \mathcal{N}(0, \{\Gamma(2s+1) - \Gamma(s+1)^2\}).$$

□

**Lemma 2.** *If any of $(H1a)$, $(H1b)$, $(H2a)$ and $(H2b)$ holds with an abitrary $\lambda > 1$, then their analogues where $b_n(\lambda)$ is replaced with $b_n$ and $p_n(\lambda)$ is replaced with $p_n$ also hold.*

Proof. We have to prove this only for one case. The others are similarly done. We begin to recall the Balkema result, that is $\sqrt{n}((n/k)U_{k,n} - 1) \to_d N(0,1)$ which entails that $\sqrt{n}((n/k)U_{k+1,n} - 1) \to_d N(0,1)$ and next $(n/k)U_{k+1,n} \to_{\mathbb{P}} 1$. Then for any $\varepsilon > 0$, for any $\lambda > 1$, we have for large values of $n$, say $n \geq n_1$,

$$\mathbb{P}(U_{k+1,n} > \lambda n/k) \leq \varepsilon/3.$$

Next by the definition of $V_n = O_p(p_n)$ and $W_n = O_p(b_n)$, there exists $C_0$ such for large values of $n$, say $n \geq n_2$,

$$\mathbb{P}(|V_n| > C_0 p_n) \leq \varepsilon/3, P(|W_n| > C_0 b_n) \leq \varepsilon/3.$$

Then for $n \geq \max(n_1, n_2)$. Recall that $b_n(\lambda) = \{|b(t)|, t \leq \lambda k/n\}$, $p_n(\lambda) = \{|p(t)|, t \leq \lambda k/n\}$, $b_n = \{|b(t)|, t \leq U_{k+1,n}\}$ and $p_n = \{|p(t)|, t \leq U_{k+1,n}\}$. We have

$$\mathbb{P}(|V_n| \leq C_0 p_n, |W_n| \leq C_0 b_n, b_n \leq b_n(\lambda), p_n \leq p_n(\lambda)) \geq 1 - \varepsilon.$$

And next, since $s \geq 1$,

$$\mathbb{P}\left|s\left\{\frac{\gamma}{j}E_j + \theta_{n,j}\left(O_{\mathbb{P}}(p_n) + \frac{E_j}{j}O_{\mathbb{P}}(b_n)\right)\right\}^{s-1}\left(O_{\mathbb{P}}(p_n) + \frac{E_j}{j}O_{\mathbb{P}}(b_n)\right)\right|$$

$$\leq s\left|\frac{\gamma}{j}E_j + \left(C_0 p_n(\lambda) + \frac{E_j}{j}C_0 b_n(\lambda)\right)\right|^{s-1}\left(C_0 p_n(\lambda) + \frac{E_j}{j}C_0 b_n(\lambda)\right)\Bigg) \geq 1 - \varepsilon$$

Suppose that for an arbitrary $\lambda > 1$,

$$c_n(\lambda) = s\left|\frac{\gamma}{j}E_j + \left(C_0 b_n(\lambda) + \frac{E_j}{j}C_0 b_n(\lambda)\right)\right|^{s-1}\left(C_0 p_n(\lambda) + \frac{E_j}{j}C_0 b_n(\lambda)\right) \to_{\mathbb{P}} 0$$

and put

$$c_n = \left|s\left\{\frac{\gamma}{j}E_j + \theta_{n,j}\left(O_{\mathbb{P}}(p_n) + \frac{E_j}{j}O_{\mathbb{P}}(b_n)\right)\right\}^{s-1}\left(O_{\mathbb{P}}(p_n) + \frac{E_j}{j}O_{\mathbb{P}}(b_n)\right)\right|$$

, we have for any $\eta > 0$, and a fixed $\lambda > 1$ and for large values of $n$. This gives

$$\mathbb{P}(c_n > \eta) = P((c_n > \eta) \cap (c_n < c_n(\lambda))) + \mathbb{P}((c_n > \eta) \cap (c_n \geq c_n(\lambda)))$$

$$\leq (c_n(\lambda) > \eta) + \varepsilon.$$

Now letting $n \to +\infty$, we have

$$\limsup_{n\to\infty} \mathbb{P}(c_n > \eta) \leq \varepsilon.$$

By letting $\varepsilon \downarrow 0$, one achieves the proof, that is $c_n \to_{\mathbb{P}} 0$.

*4.2 Integral Computations*

Let $b \geq 1$, we get by comparing the area under the curve $x \longmapsto x^{-b}$ from $j$ to $k-1$ and those of the rectangles based on the intervals $[h, h+1]$, $h = j, .., k-2$, we get

$$\sum_{h=j+1}^{k-1} h^{-b} \leq \int_{j}^{k-1} x^{-b} dx \leq \sum_{h=j}^{k-2} h^{-b},$$

that is

$$\int_{j}^{k-1} x^{-b} dx + (k-1)^{-b} \leq \sum_{h=j}^{k-1} h^{-b} \leq \int_{j}^{k-1} x^{-b} dx + j^{-b}. \tag{4.2}$$

For $b = 1$, we get

$$\frac{1}{k-1} \leq (\sum_{h=j}^{k-1} \frac{1}{h}) - \log((k-1)/j) \leq \frac{1}{j}. \tag{4.3}$$

For $b = 2$,, we have

$$\frac{1}{j} - \frac{1}{k-1} + \frac{1}{(k-1)^2} \leq \sum_{h=j}^{k-1} h^{-2} \leq \frac{1}{j} - \frac{1}{k-1} + \frac{1}{j^2},$$

that is

$$\frac{1}{(k-1)^2} \leq \sum_{h=j}^{k-1} h^{-2} - \frac{1}{j}(1 - \frac{j}{k-1}) \leq \frac{1}{j^2}.$$

As well, we have for $b > 0$,

$$\sum_{h=j}^{k-2} h^{b} \leq \int_{j}^{k-1} x^{b} dx \leq \sum_{h=j+1}^{k-1} h^{b}$$

and then

$$\frac{1}{b+1}((k-1)^{b+1} - j^{b+1}) + j^{b} \leq \sum_{h=j}^{k-1} h^{b} \leq \frac{1}{b+1}((k-1)^{b+1} - j^{b+1}) + (k-1)^{b}. \tag{4.4}$$

Hence for $j$ fixed and $k \to \infty$, we get $\sum_{h=j}^{k-1} h^{b} \sim (k-1)^{b+1}/(b+1)$.

*4.3 Computation of J(a,b,c,s)*

Recall

$$J(a,b,c,s) = \int_{0}^{\infty} (a+cx)^{s}(a+bx)e^{-x} dx \text{ and } I(a,b,s) = \int_{0}^{\infty} (a+bx)^{s} e^{-x} dx.$$

4.3.1 Computation of I(a,b,s).

We have by integration by parts

$$I(a,b,s) = \int_{0}^{\infty} (a+bx)^{s} e^{-x} dx = [-e^{-x}(a+bx)^{s}]_{0}^{\infty} + bs \int_{0}^{\infty} (a+bx)^{s-1} e^{-x} dx,$$

that is, for any $s \geq 1$,

$$I(a,b,s) = a^{s} + sbI(a,b,s-1).$$

By induction for $s \geq 1$,this leads to

$$I(a,b,s) = s! \sum_{h=0}^{s} b^{h} a^{s-h}/(s-h)!.$$

4.3.2 Computation of J(a,b,c,s)

We have integration by parts

$$J(a,b,c,s) = a^{s+1} + bI(a,c,s) + csJ(a,b,c,s-1).$$

We get by induction for $\ell \geq 1$,

$$\begin{aligned} J(a,b,c,s) &= s!a\sum_{h=0}^{\ell} c^h a^{s-h}/(s-h)! + s!b\sum_{h=0}^{\ell} c^h I(a,c,s-h)/(s-h)! \\ &+ s!c^{\ell+1}J(a,b,c,s-\ell-1)/(s-\ell-1)! \end{aligned}$$

For $\ell + 1 = s$, we arrive at

$$\begin{aligned} J(a,b,c,s) &= s!a\sum_{h=0}^{s-1} c^h a^{s-h}/(s-h)! + s!b\sum_{h=0}^{s-1} c^h I(a,c,s-h)/(s-h)! \\ &+ s!c^s J(a,b,c,0). \end{aligned}$$

Since $J(a,b,c,0) = I(a,b,1)$, we finally get

$$\begin{aligned} J(a,b,c,s) &= s!a\sum_{h=0}^{s-1} c^h a^{s-h}/(s-h)! + s!b\sum_{h=0}^{s-1} c^h I(a,c,s-h)/(s-h)! \\ &+ s!c^s I(a,b,1) \end{aligned}$$

*4.4 Minimization of the Asymptotic Variance*

For $s \geq 1$ fixed, We have to maximize

$$V_n(\tau,s) = \frac{d_n(\tau,s)}{\sigma_n(\tau,s)} \times \frac{s\gamma^{-1}}{s!\sqrt{\Gamma(2s+1) - \Gamma(s+1)^2}}.$$

with respect of $\tau > s - 1/2$ for $s$, where $d_n(\tau,s) = \sum_{j=1}^{k} j^{\tau-s}$ . We denote $q(s) = \frac{s\gamma^{-1}}{s!\sqrt{\Gamma(2s+1)-\Gamma(s+1)^2}}$. Let us find critical points. It is easy to see that

$$\frac{\partial V_n(\tau,s)}{\partial\tau} = \frac{d_n'(\tau,s)\sigma_n(\tau,s) - d_n(\tau,s)\sigma_n'(\tau,s)}{(\sigma_n(\tau,s))^2} \times q(s).$$

A zero-point of $\frac{\partial V_n(\tau,s)}{\partial\tau}$ obviously, is a solution of the ordinary differential equation.

$$\frac{d_n'(\tau,s)}{d_n(\tau,s)} = \frac{\sigma_n'(\tau,s)}{\sigma_n(\tau,s)}.$$

Its general solution is given by

$$\log d_n(\tau,s) = \log\sigma_n(\tau,s) + C(s). \tag{4.5}$$

By taking the particular value of $\tau = s$, we find that $C(s) = (1/2)\log k$,and (4.5) becomes

$$\sum_{j=1}^{k} j^{\tau-s} = \left(k\sum_{j=1}^{k} j^{2(\tau-s)}\right)^{1/2}.$$

This is the equality form of Cauchy-Schwarz's inequality with respect to the usual scalar product in $\mathbb{R}^k$. Then there exists a constant $\lambda(s)$ such that $j^{\tau-s} = \lambda(s)$ for $1 \leq j \leq k$. The only solution is $\tau = s$. Now to show that $\tau$ is the global maximum point, it suffices to notice that

$$\frac{1}{s!q(s)}\frac{\partial^2 V_n(\tau,s)}{\partial^2\tau} = \frac{1}{k\sqrt{k}}\left(\left(\sum_{j=1}^{k}\log j\right)^2 - k\sum_{j=1}^{k}(\log j)^2\right) < 0.$$

$$\frac{1}{s!q\,(s)\ \sqrt{(}k)} \frac{\partial^2 V_n(\tau, s)}{\partial^2 \tau} = -\left(\left(k^{-1} \sum_{j=1}^{k} (\log j)^2_{(} k^{-1}) \sum_{j=1}^{k} \log j\right)^2\right) < 0.$$

since the left member is the opposite of a the empirical variance of $\log j$, $1 \leq j \leq k$. We conclude that the point $\tau = s$ is the unique local miximum point. Then the global maximum is reached at $\tau = s$.

**References**


Beirlant, J., Goegebeur, Y., & Teugels, J.(2004). *Statistics of Extremes Theory and Applications*. Wiley. (MR2108013)

de Haan, L. (1970). *On regular variation and its application to the weak convergence of sample extremes*. Mathematical Centre Tracts, 32, Amsterdam. (MR0286156)

de Haan, L., & Feireira A. (2006). *Extreme value theory: An introduction*. Springer. (MR2234156)

Dekkers, A. L. M., Einmahl, J. H. J., & Haan, L. D. (1989). A moment estimator for the index of an extreme value distribution. *Ann. Statist., 17*(4), 1833-1855.(MR1026315)

Dème, E., LO, G. S., & Diop, A. (2012). On the generalized Hill process for small parameters and applications. *Journal of Statistical Theory and Applications, 11*(4), 397-418. http://dx.doi.org/10.2991/jsta.2013.12.1.3.(MR3191797)

Diop, A., & Lo, G. S. (2006). Generalized Hill's Estimator. *Far East J. Theor. Statist., 20*(2), 129-149. (MR2294728)

Diop, A., & Lo, G. S. (2009). Ratio of Generalized Hill's Estimator and its asymptotic normality theory. Math. Method. *Statist., 18*(2), 117-133. (MR2537361)

Embrechts, P., Kűppelberg C., & Mikosh T. (1997). *Modelling extremal events for insurance and Finance*. Springer Verlag.

Fall, A. M., LO, G. S., Ndiaye, C. H., & Adekpedjou, A. (2014). Supermartingale argument for characterizing the Functional Hill process weak law for small parameters. Submitted. available at : http://arxiv.org/pdf/1306.5462

Galambos, J. (1985). *The Asymptotic theory of Extreme Order Statistics*. Wiley, Nex-York. (MR0489334)

Groeneboom, Lopuhaä, H. P., & Wolf, P. P.(2003). Kernel-Type Estimator for the extreme Values index. *Ann. Statist., 31*(6), 1956-1995.(MR2036396)

Lo, G. S.(1991). Caractérisation empirique des extrêmes et questions liées. Thèse d'Etat. Université de Dakar.

Lo, G. S, & Fall, A. M. (2011). Another look at Second order condition in Extreme Value Theory. *Afrik. Statist., 6*, 346-370.(zbl:1258.62058)

Lo, G. S. (2012). On a discrete Hill's statistical process based on sum-product statistics and its finite-dimensional asymptotic theory. available at : http://arxiv.org/abs/1203.0685

Resnick, S. I. (1987). *Extreme Values, Regular Variation and Point Processes*. Springer-Verbag, New-York. (M-R0900810)

Segers, J. (2002). Generalized Pickands Estimators for the Extreme Value Index J. *Statist. Plann. Inference, 128*(2), 381-396. (MR2102765)

Shorack, G. R., & Wellner J. A. (1986). *Empirical Processes with Applications to Statistics*. wiley-Interscience, New-York. (MR0838963)

# Flows on Discrete Traffic Flower

Alexander P. Buslaev[1] & Alexander G. Tatashev[2]

[1] Moscow Automobile and Road State Technical University, Moscow, Russia

[2] Moscow Technical University of Communications and Informatics, Moscow, Russia

Correspondence: Alexander P. Buslaev, Moscow Automobile and Road State Technical University, Moscow, Russia.E-mail: apal2006@yandex.ru

**Abstract** A discrete dynamical system is considered in this paper. There are $L$ contours, which have a common point. There are $N_i$ cells and $M_i$ particles located in cells. Each particle moves on its contour in accordance with a given rule. Velocities of particles and other characteristics of the system are investigated.

**Keywords:** discrete dynamical systems, traffic models, Markov process, particles flow velocity, self-organization

## 1. Introduction

A dynamical system is considered in the paper. The system contains a finite set of *cells.* Particles are located in cells. *Contours (cycles)* are elementary supporters. A contour is a closed nonself-intersecting sequence of cells, which determines the direction of particles movement. A *cell* is a place of location of a particle. A *particle* is a unit of mass. The supporter is an oriented graph. Cells are vertices of the graph. An arc (oriented edge) connects two vertices if particles can move from one of these vertices to the other vertex. There is just one common point of contours. Contours have either just one common cell (*node*) or just one common point located between neighboring cells (*alternating node*). A feature of the considered system is that there is only one common point of contours. This supporter is called a *flower.* The traffic model with "flower" supporter has been introduced in (Buslaev, & Yashina, 2009), (Buslaev, 2010), where an infinitesimal version of the system was considered.

A graph with alternating node is considered in the present paper, Figure 1. The common node is located between some pairs of cells on contours. In Figure 1, the node is shown as a square, and cells are shown as circles. The graph of the considered system is a bouquet graph. Therefore the considered system is called *BAN (Bouquet Alternance Node).* We discuss in what way the obtained results change in the case of *a flower with the common cell if this cell is joined with the node.*

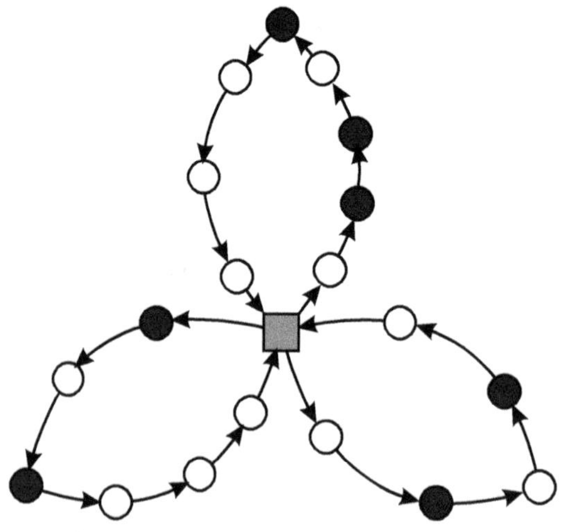

Figure 1. Bouquet with alternating node

Two types of traffic participants are considered. These types are *individual movement* and *total-connected movement.* In the case of *individual movement,* the particle shall be at the next time moment ($T + 1$) in the next cell if the next cell is vacant at the moment $T$. In the case of *total-connected movement,* Figure 2, the particle will be at the next time moment ($T + 1$) in the next cell if the next cell is vacant at the moment $T + 1$. Hence, in the case of the total-connected movement the particle comes to the next cell at the next moment if this cell is released. *In the case of total-connected movement* neighboring particles form *clusters.* A *cluster* is the maximum subset of particles without vacant cells between these particles. Cluster moves synchronously if the next cell in the direction of movement is vacant. In Figure 2, cells are shown as circles, and the node is shown as rectangle. Occupied cells are highlighted in black. Dotted outlines show clusters.

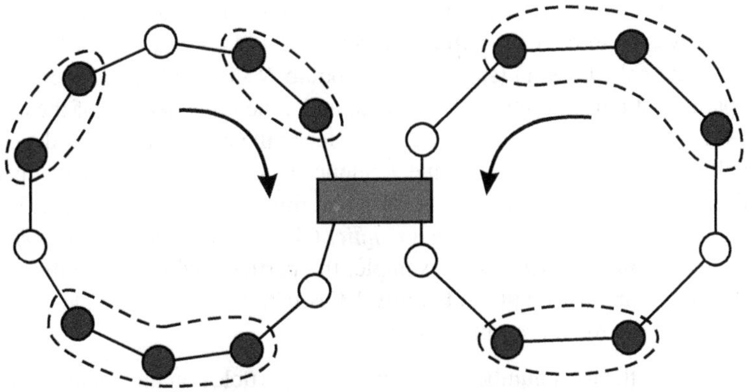

Figure 2. Total-connected movement

*More than one cluster cannot move through the node simultaneously.* The *FIFO (First Input First Output)* discipline acts at the node. Consider a cluster which has come to the node first. In accordance with the *FIFO discipline, this cluster moves through the node first when the node releases.* If *leaders of some clusters come to the node simultaneously, then one of these clusters* moves in accordance with the given *conflict resolution rule.* For example each of these clusters is chosen equiprobably (*the egalitarian rule*).

A traffic model was introduced in (Nagel, & Schreckenberg, 1992). In this model, *individual movement of a finite set of particles on a closed contour* takes place. A hypothesis has been formulated in (Schadschneider, & Schreckenberg, 1993). This hypothesis gives a formula that determines dependence of the particles velocity on the particles flow density. This formula has been proved in (Blank, 2000). The limit version of the model, considered in the present paper, is the continuous cluster model, considered in (Buslaev, & Yashina, 2009), (Buslaev, 2010) (incompressible clusters). Discrete traffic models with total-connected movement were considered in (Bugaev, Buslaev, Kozlov, Tatashev, & Yashina, 2013). In the case of the stochastic conflict resolution rule, considered in the present paper, the behavior of the system is s-tochastic. However, in present paper, we show that it is possible that the system behavior can *come, after time with a finite expectation, to the state* such that there will be no delays of particles movement in future. So, if the total number of particles is not greater than a critical number, then there are *no conflicts after a finite moment.* We say that *the synergy state (self-organization)* takes place. If the behavior of the system becomes deterministic, then the same sequence of states repeats periodically. These states form an *orbit* (in terms of dynamical systems theory (Halmos, 1956)), or a communicating class of states (in terms of Markov processes theory, (Feller, 1970), (Kemeny, & Snell, 1976), (Borovkov, 1986)). This self-organization is also characteristic for dynamical systems with regular periodic structures, which have been introduced and investigated in (Kozlov, Buslaev, & Tatashev, 2014), (Kozlov, Buslaev, Tatashev, & Yashina, 2014), (Kozlov, Buslaev, & Tatashev, 2015), (Kozlov, Buslaev, Tatashev, & Yashina, 2015).

Analysis of the system is related to the theory of *linear diophantine equations with two variables,* (Buchstab, 1972). If there are no solutions of the appropriate diophantine equation, then there will be no conflict of a considered pair of particle if no other conflicts take place. If there is a solution of the equation, then a conflict will be in future. Thus *analysis of diophantine equations gives a criterion of synergy.*

The system with individual movement is considered in Section 2. We formulate the necessary and sufficient condition for a state of the system to be a state of synergy. We formulate necessary conditions and sufficient conditions for a state of synergy to exist.

Individual movement of particles on BAN with contours of the same length is considered in Section 3. We find the necessary and sufficient condition for a state of synergy to exist.

The system with total-connected movement is investigated in Sections 4 and 5. We have considered the system with two contours of the same length and a single cluster on each contour. We have found a necessary and sufficient condition for this system to come to the state of synergy after a finite time. We describe an approach to calculation *of particles velocities* on contours of this system. In the case of BAN with contours of the same length and any number of clusters, finally statements have been obtained.

## 2. Individual Movement on BAN

*2.1 Formulation of Problem*

We consider a system (BAN) with $L$ *contours.* There are $N_i$ *cells* $(i,0),\dots,(i,N_i-1)$ and $M_i$ particles on the contour $i$, $M_i < N_i$, $i = 1,\dots,L$. Each particle is in one of the cells at any time moment $T = 0,1,2,\dots$ No more than one particle is in the cell. There is a common *alternating* node. The node is located such that a particle of the $i$th contour moves through the node when this particle moves from the cell $(i,0)$ to the cell $(i,N_i-1)$, $i = 1,\dots,L$. If a particle of the contour $i$ is in the cell $(i,j)$ at moment $T$, the cell $(i,j-1)$ (substraction modulo $N_i$) is vacant, and the node is not located between cells $(i,j)$, $(i,j-1)$ or no particle attempts to move through the node, then this particle of the contour $i$ will be in the cell $(i,j-1)$ at the moment $T+1$, $j = 1,2,\dots,N_i-1$, $i = 1,\dots,L$. Hence the node is ahead of the cell $(i,0)$, and the particle, occupying the cell $(i,j)$, will be in the cell $(i,0)$ after time $j$. Suppose the cell, located ahead a particle, is occupied, at the moment $T$. Then this particle will continue occupy the same cell at time $T+1$ $(T = 0,1,2,\dots)$. If more than one particle attempt to move through the node simultaneously, then a *conflict* takes place. One of these particles wins the conflict in accordance with a given *conflict resolution rule.* For example, the particles wins the conflict equiprobably (*egalitarian rule*) or the particle with the least index wins with probability 1 (*priority rule*). The particle, winning conflict, moves, and particles, losing the conflict, do not move at present moment.

Denote by $H_i(T)$ the expectation of the total number of transitions of particles on the contour $i$ in time interval $(0,T)$. The limit

$$v_i = \lim_{T\to\infty}\frac{H_i(T)}{TM_i},\ i = 1,\dots,L, \tag{1}$$

is called the *velocity of the ith contour particles* if this limit exists.

The value

$$v = \frac{M_1v_1+\cdots+M_Lv_L}{M_1+\cdots+M_L}.$$

is called *the average velocity of particles.*

The velocity of particles is a random value. The considered system is a finite Markov chain (Feller, 1970), (Kemeny, & Snell, 1976), (Borovkov, 1986). If the system comes from the state $i$ to the state $j$ after a finite time with positive probability, and the system cannot come from the state $j$ to the state $i$ after a finite time, then the state $i$ is called *inessential.* If a state is not inessential, then this state is called *essential.* If the system can come from the state $i$ to the state $j$ after a finite time, and the system can come from the state $j$ to the state $i$ after a finite time, then the states $i$ and $j$ are called *communicated with each other.* The set of essential states is divided into *communicating classes.* If there is only one communicating class, then there exist steady state probabilities, and the value of particles velocity is the same with probability 1. Suppose there are more than one communicating class. If the initial state belongs to one of communicating classes, then the value of particles velocity depends on the class. If the initial state is inessential, then the chain comes to one of communicating class after time with a finite expectation. The chain comes, with positive probabilities, to different communicating classes, and values of velocities depend on the communicating class. Thus the limit (1) exists with probability 1, this limit is a random value, and the distribution of this value depends on the initial state of the system.

*2.2 Conditions of Synergy (Self-organization) on BAN*

The system is in the state of *synergy* after a moment $T$ if, after this moment, all particles move at every moment.

Suppose

$$(k_{11},\dots,k_{1M_1};\dots;k_{L1},\dots,k_{LM_L})$$

is a vector such that $k_{i1},\dots,k_{iM_i}$, $0 \le k_{i1} < \cdots < k_{iM_i} \le N_i-1$, are indexes of cells, where particles of the $i$th contour are located, $i = 1,\dots,L$. Then this vector is called a *state of the system.*

Denote by $GCD(N_1,\dots,N_L)$ the greatest common divisor of the numbers $N_1,\dots,N_L$; $GCD(N_{i_1},N_{i_2})$ is the $GCD-$ divisor of the numbers $N_{i_1}$ and $N_{i_2}$, $i_1 \ne i_2$, $1 \le i_1,i_2 \le L$.

Consider the equation

$$ax + by + c = 0, \tag{2}$$

where $a$ and $b$ are any integer numbers not equal to 0, and $c$ is any integer number. Suppose the numbers $a$ and $b$ are coprime. Indeed, if $a$ and $b$ are not coprime, then we divide both parts of (2) by the $GCD$ of the numbers $a$ and $b$. Then we obtain a similar equation, and $a$, $b$ are coprime. There exist *integer solutions of (2) if and only if $GCD(a,b)$ is the divisor of the number $c$ (Buchstab, 1972).* Suppose that this condition is fulfilled.

Suppose $x = x_0$, $y = y_0$ is an integer solution of Equation (2). Then the formulas

$$x = x_0 - bt,\ y = y_0 + at$$

give all solutions of Equation (2), $t = 0, \pm 1, \pm 2, \ldots$

**Theorem 1 (Criterion of Synergy).** *A necessary and sufficient condition for the state*

$$(k_{11}, \ldots, k_{1,M_1}; \ldots; k_{L1}, \ldots, k_{L,M_L})$$

*of BAN with individual movement to be a state of synergy is that the following conditions be fulfilled*

$$(1)\ \forall l,\ 1 \le l \le L,\ \forall i, 1 \le i \le M_l,$$

$$k_{l,i+1} - k_{li} > 1;$$

$$(2)\ \forall i_1, i_2,\ 1 \le i_1, i_2 \le L,\ i_1 \ne i_2,\ \forall j_1, j_2,\ 1 \le j_1 \le M_{i_1},\ 1 \le j_2 \le M_{i_2},$$

*$GCD(N_{i_1}, N_{i_2})$ is not a divisor of the number $(k_{i_1 j_1} - k_{i_2 j_2})$.*

*Proof.* If there are particles in neighboring cells of a contour, then one of these particles does not move at the present moment. Therefore the system is not in the state of synergy. Suppose there are no particles in neighboring cells on any contour. Let us show that the considered state is a state of synergy if and only if, for any $i_1, i_2, j_1, j_2$ ($i_1 \ne i_2$, $1 \le j_1 \le M_{i_1}$, $1 \le j_2 \le M_{i_2}$), there exist no nonnegative integer solutions $x, y$,

$$N_{i_1} x - N_{i_2} y = k_{i_1 j_1} - k_{i_2 j_2}. \tag{3}$$

Indeed, suppose the system is in the state

$$(k_{11}, \ldots, k_{1M_1}, \ldots, k_{11}, \ldots, k_{LM_L}).$$

If, for some $i_1 \ne i_2$, $j_1$, $j_2$ ($1 \le j_1 \le M_{i_1}$, $1 \le j_2 \le M_{i_2}$), there exists a solution $(x_0, y_0)$ of Equation (3), then there will be a conflict after a finite time. If for any $i_1$, $i_2$, $j_1$, $j_2$ ($1 \le j_1 \le M_{i_1}$, $1 \le j_2 \le M_{i_2}$), there exists no integer solution of Equation (3), then there will be no delays, i,e. the system is in the state of synergy. There exists an integer solution of Equation (3) if and only if $GCD(N_{i_1}, N_{i_2})$ is a divisor of $k_{i_1 j_1} - k_{i_2 j_2}$. Thus we obtain the statement of Theorem 1.

We shall formulate a necessary condition for at least one state of synergy to exist. The ratio of the particles of the contour $i$ to the number of cells of this contour is called *the density of particles of the contour i* and is denoted by $r_i$,

$$r_i = \frac{M_i}{N_i},\ i = 1, \ldots, L.$$

Denote by

$$A = LCM(N_1, \ldots, N_L)$$

the least common multiple of $N_1, \ldots, N_L$.

**Theorem 2.** *Suppose movement on BAN is individual. Then a necessary condition for at least one state of synergy to exist is that (1) $\forall i, j,\ 1 \le i \ne j \le L$*

$$GCD(N_i, N_j) > 1; \tag{4}$$

*(2) $\forall i,\ 1 \le i \le L$*

$$r_i \le \frac{1}{2}; \tag{5}$$

*(3)*

$$r_1 + r_2 + \cdots + r_L \le 1. \tag{6}$$

*Proof*

(1) Consider the system state

$$(k_{11}, \ldots, k_{1M_1}; \ldots; k_{L1}, \ldots, k_{L,M_L}).$$

If (4) is not true, i.e., numbers $N_{i_1}$, $N_{i_2}$ are coprime, then there exists a solution of Equation (3) for any $k_{i_1 j_1}$, $k_{i_2 j_2}$. From this, taking into account Theorem 1, we obtain the first statement of Theorem 2.

(2) If (5) is not true, then, at any time, the number of vacant cells of the contour is less than the number of particles. Therefore there exist particles such that these particles cannot move at present time. From this we obtain the second statement of Theorem 2.

(3) Suppose that the system is in the state of synergy. Then, in any time interval of length $A$, there are $r_iA$ moments such that a particles of $i$th contour moves through the node. Since more one particle cannot move through the node simultaneously, then $r_1A + \cdots + r_LA \leq A$, i.e., (6) is true. Theorem 2 has been proved.

**Theorem 3.** *Let $L$ be the number of contours, $N_i$ be the number of cells of the contour $i$, $M_i$ be the number of particles of the contour $i$, $i = 1, \cdots, L$. Suppose the movement on BAN is individual. Then a sufficient condition for at least one state of synergy to exist is that at least one of the following conditions be fulfilled.*

*(1) There is only one particle on each contour, $M_1 = \cdots = M_L = 1$, and*

$$GCD(N_1, \ldots, N_L) \geq L. \tag{7}$$

*(2) $L$ is divider of numbers $N_1, N_2, \ldots, N_L$ and $r_i \leq \frac{1}{L}$ for all $i = 1, \ldots, L$.*

*Proof.* (1) Suppose (7) is true and, at time $T_0$, the system is in the state such that the particle of the contour $i$ is in the cell $(i, k_i)$, and the remainder of the division of $k_i$ by $GCD(N_1, \ldots, N_L)$ equals $i$, $i = 1, \ldots, L$. Such state exists in accordance with (7). Suppose a conflict of particles of contours $i$ and $j$ takes place at moment $T \geq T_0$. Then there exist integer non-negative solutions $x = x_0$, $y = x_0$ of the equation

$$N_ix - N_jy = k_j - k_i.$$

If such solution $x_0$, $y_0$ exists, then $GCD(N_i, N_j)$ is a divisor of $k_j - k_i$, and therefore $GCD(N_1, \ldots, N_k)$ is also a divisor of $k_j - k_i$. However the remainders of the division of numbers $k_i$ and $k_j$ by $GCD(N_1, \ldots, N_L)$ are not equal to each other. Therefore $GCD(N_1, \ldots, N_k)$ is not a divisor of $k_i - k_j$. This contradiction proves the first statement of the theorem.

(2) Suppose $L$ is a divisor of the numbers $N_1, N_2, \ldots, N_L$, and $r_i \leq \frac{1}{L}$ for all $i = 1, \ldots, L$. Assume that particles of the contour $i$ are in cells such that remainders of the division of cells indexes by $L$ are equal to $i$ ($i = 1, \ldots, L$). We shall prove that this state is a state of synergy. Suppose a conflict of the contours $i$ and $j$ takes place at the moment $T \geq T_0$. Then there exist integer non-negative solutions $x = x_0$, $y = x_0$ of the equation

$$N_ix - N_jy = k_j - k_i.$$

If such solution $x_0$, $y_0$ exists, then $GCD(N_i, N_j)$ is a divisor of $k_i - k_j$. Since $L$ is a divisor of $N_i$ and $N_j$, we see that $L$ is also a divisor of $k_j - k_i$. However remainders of the division of $k_i$, $k_j$ by $L$ are not equal to each other. This contradiction proves the second statement of the theorem. Theorem 3 has been proved.

**Remark 1.** *Parameters of BAN with individual movement can be chosen such that both necessary conditions of Theorem 2 are fulfilled, and there is no state of synergy.*

We give an example. Suppose $L = 2$, $N_1 = 6$, $N_2 = 3$, $M_1 = 3$, $M_2 = 1$. Then we have $r_1 + r_2 = \frac{M_1}{N_1} + \frac{M_2}{N_2} = \frac{5}{6} < 1$, i.e. both conditions of Theorem 2 are fulfilled. However we shall prove that there exists no state of synergy. Without loss of generality, we assume that the cell $(1, 1)$ is occupied. If $(k_{11}, k_{12}, k_{13}; k_2)$ is the state of synergy, then $k_{11} = 1$, $k_{12} = 3$, $k_{13} = 5$. In accordance of Theorem 1, in the case of state of synergy, $GCD(3, 6) = 3$ is not a divisor of $k_{11} - k_2$, $k_{12} - k_2$, $k_{13} - k_2$. However this condition is not fulfilled for none of states $(1, 3, 5; 0)$, $(1, 3, 5; 1)$, $(1, 3, 5; 2)$. Thus there is no state of synergy.

## 3. BAN with Contours of the Same Length Necessary and Sufficient Condition for Self-organization of Individual Movement

*Let $L$ is the number of contours, $N_i$ is the number of cells of the contour $i$, $M_i$ is the number of particles of the contour $i$, $i = 1, \cdots, L$. Assume that the number of cells on each contour is the same, $N_1 = \cdots = N_L = N$.*

**Theorem 4.** *Let movement be individual. A necessary and sufficient condition for at least one state of synergy to exist is that*

*(1) $\forall i$, $1 \leq i \leq L$*

$$r_i \leq \frac{1}{2};$$

*(2)*

$$M_1 + M_2 + \cdots + M_L \leq N.$$

*Proof.* In accordance with statements 2, 3 of Theorem 2, we have that the condition is necessary.

Let us show that the condition is sufficient. Assume that $M_1 \geq M_i$, $i = 2, \ldots, L$. Put particles of the contour 1 into cells with indexes $0, 2, 4, \ldots, 2(M_1 - 1)$, put particles of the contour 2 into cells with indexes $2M_1, 2M_1 + 2, \ldots, 2(M_1 + M_2) - 1$, etc. If all cells with even indexes are already occupied, then we put remaining particles into cells with odd indexes. In accordance with the second condition of the theorem all particles will be located in cells with different indexes, and there will be no particles in neighboring cells of any contour. Thus there exists a state of synergy. Theorem 4 has been proved.

## 4. Total-connected Movement on BAN

*Let $L$ be the number of contours, $N_i$ be the number of cells of the contour $i$, $M_i$ be the number of particles of the contour $i$, $i = 1, \cdots, L$.*

### *4.1 Synergy of BAN with total-connected movement*

**Theorem 5.** *Let the movement be total-connected. Then a state $(k_{11}, \ldots, k_{1M_1}, \ldots, k_{L1}, \ldots, k_{LM_L})$ is a state of synergy if and only if, $\forall\, i_1, i_2, j_1, j_2$ $(i_1 \neq i_2, 1 \leq j_1 \leq M_{i_1}, 1 \leq j_2 \leq M_{i_2})$, $GCT(N_{i_1}, N_{i_2})$ is not a divisor of $(k_{i_1 j_1} - k_{i_2 j_2})$.*

The proof of Theorem 5 is similar to the proof of Theorem 1.

Suppose there is an ordered set of clusters on each contour. There are $l_i$ clusters on the contour $i$, and the $s$th cluster contains $M_{is}$ particles, $s = 1, \ldots, l_i$, $M_{i1} + \cdots + M_{i,l_i} = M_i$, $M_i < N_i$, $i = 1, \ldots, L$.

**Theorem 6.** *Let the movement be total-connected. Then a necessary condition for at least one state of synergy to exist is that*

*(1)*

$$r_1 + \cdots + r_L \leq 1;$$

*(2)* $\forall\, i, j \neq \{1, \ldots, L\}$

$$\max(M_{i1}, \ldots, M_{iN_i}) + \max(M_{j1}, \ldots, M_{jN_j}) \leq GCD(N_i, N_j).$$

The proof of the first statement of Theorem 6 is similar to the proof of the third statement of Theorem 2.

In accordance with Theorem 5, for the state of synergy, the sum of two clusters of different contours cannot be greater than $GCD(N_i, N_j)$. From this, the second statement follows. Theorem 6 has been proved.

**Theorem 7.** *Let the movement be total-connected. Then a sufficient condition for at least one state of synergy to exist is that*

$$M_1 + \cdots + M_L \leq GCD(N_1, \ldots, N_L).$$

The proof of Theorem 7 is similar to proof of the first statement of Theorem 3.

The behavior of BAN depends on the initial state and on the realization of the process if the initial state is fixed. Let us give examples.

In the following example, *the initial state is fixed, but, with positive probabilities, the velocities and limit numbers of clusters on contours are different, and either the system behavior will be deterministic after a time interval with finite expectation or the system will come to a set of states such that deterministic behavior is not possible.*

Suppose $L = 2$, $N_1 = 7$, $N_2 = 3$, $M_1 = 2$, $M_2 = 1$. Each particle of the contour 1 wins every conflict with probability $\frac{2}{3}$. Assume that the initial state is $(0, 5; 0)$. There is a conflict in this state. If the particle of the contour 1 wins the conflict, then particles of this contour form a cluster, the behavior of system will be deterministic, and the following sequence of states repeats periodically

$$(5,6;0) \to (4,5;1) \to (3,4;0) \to (2,3;1) \to$$
$$\to (1,2;0) \to (0,1;1) \to (6,1;0) \to (5,6;0)\ldots$$

The values of velocities are

$$v_1 = 1,\ v_2 = \frac{6}{7},\ v = \frac{2v_1 + v_2}{3} = \frac{20}{21}.$$

If the particle of the contour 2 wins the conflict in state $(0, 5; 0)$, then particles of this contour do not form a cluster, the behavior of system will never be deterministic, and, with probability 1, the values of velocities are

$$v_1 = \frac{35}{36},\ v_2 = \frac{8}{9},\ v = \frac{17}{18}.$$

The following example shows that, *in the case of total-connected movement on contours of the same length, it is possible that the system will not come to the state of synergy, and the limit number of clusters on a contour is not equal to 1.*

Suppose $N_1 = N_2 = 25$, $M_1 = M_2 = 13$ and the initial state is

$$(1,2,3,4,10,11,12,13,14,18,19,20,21;4,5,6,7,8,9,15,16,17,22,23,24,25).$$

Then we have deterministic movement with period 26. Velocities of particles are

$$v_1 = v_2 = \frac{25}{26}.$$

The limit number of clusters on each contour equals 3.

*4.2 Synergy and Velocity of Total-connected Movement on BAN with Contours of the Same Length*

Suppose that the length of each contour is the same, $N_1 = \cdots = N_L = N$. We give the definition of a *group of clusters.* A set of clusters (these clusters can be located on different contours) is called a *group of clusters* if the set of indexes of cells, where these clusters are located, is a connected set on the integer numbers segment $[0, \dots, N-1]$. Group of the maximum length is called a *maximum group (maximum cluster).* If clusters of a maximum group do not contain any cells with the same numbers, then this group is called a *maximum simple group (maximum simple cluster).* It is evident that the length of the supporter of a maximum simple group is equal to the sum of lengths of all clusters such that these clusters are contained in the maximum group.

Suppose values $L, M_1, \dots, M_L, N$ are given. We formulate the necessary and sufficient condition of synergy.

**Theorem 8.** *A necessary and sufficient condition for BAN with total-connected movement to come to the state of synergy after a time with a finite expectation is that*

$$M_1 + \cdots + M_L \leq N. \tag{8}$$

**Lemma 1.** *Suppose the movement is total-connected and $N_1 = \cdots = N_L$. Then any maximum group cannot become thinner, i.e., this group cannot lose any clusters or their particles.*

*Proof.* Suppose clusters of a maximum group $G$ do not hinder the movement of clusters, not contained in the maximum group, and clusters, not contained in the maximum group, do not hinder the movement of clusters of $G$. Then the set of clusters of the group does not change. Suppose a cluster $C$ does not belong to the group $G$, and this cluster delays the movement of clusters of $G$, or clusters of $G$ delays the movement of the cluster $C$. Then a group is formed such that this group contains both the clusters of the group $G$ and the cluster $C$. In all these cases, the maximum group does not lose both clusters and particles. Lemma 1 has been proved.

**Lemma 2.** *Let the movement be total-connected. Suppose $N_1 = \cdots = N_L$. At any moment, the set of clusters on BAN is a union of nonself-intersecting maximum clusters.*

The statement of Lemma 2 is obvious.

**Lemma 3.** *Assume that the movement is total-connected and $N_1 = \cdots = N_L = N$. Suppose the velocity and configuration of a maximum cluster do not change during $N$ time units (indexes of cells, which are occupied by the cluster, shift onto 1 per time unit). Then this maximum cluster is simple.*

*Proof.* Lemma 3 follows from the definition.

**Lemma 4.** *Let the movement be total-connected. Suppose $N_1 = \cdots = N_L = N$, the system is not in the state of synergy and there is no maximum cluster, containing more than $N-1$ particles. Then, for time interval of duration $N$, either some maximum clusters merge or supporters of some clusters, which are not simple, increase.*

*Proof.* Suppose $N_1 = \cdots = N_L = N$, the system is not in the state of synergy and there is no maximum cluster, containing more than $N-1$ particles. If no cluster is delayed during $N$ time units, then a sequence of $N$ states will repeat periodically. If a delay takes place, then some groups, which are not simple, increase, and it is possible that some maximum clusters merge. Lemma 4 has been proved.

**Lemma 5.** *Suppose the movement is total-connected and $N_1 = \cdots = N_L = N$. If inequality (8) is fulfilled, then, for all $T_1$, there exists a moment $T_2$ such that no particle moves through the node at this moment, and $T_1 \leq T_2 < T_1 + N$.*

*Proof.* Any particle cannot move through the node more than one time during a time interval of duration $N$. Since the number of particles is less than $N$, we obtain Lemma 5.

**Lemma 6.** *Suppose $N_1 = \cdots = N_L = N$, the movement is total-connected and (8) is true. Then all clusters will be maximum simple clusters after a finite time. We have the process of movement without conflicts, with minimum configuration, and maximum velocity.*

*Proof* Suppose (8) is true. Then, in accordance with Lemma 4, no delays of clusters will be in future or a maximum cluster increases. However maximum cluster cannot increase more than a finite times. From this, Lemma 6 follows.

*Proof of Theorem 8.* The condition is necessary in accordance with Theorem 6. Let us show that the condition is sufficient. In accordance with (8) and Lemma 4, after a finite moment, any cluster will be in a maximum group such that clusters of other groups (if there are other groups) do not hinder clusters of the group. Theorem 8 has been proved.

Suppose

$$M_1 + \cdots + M_L > N. \tag{9}$$

**Lemma 7.** *Suppose $N_1 = \cdots = N_L = N$, the movement is total-connected and (9) is true. Then there exists a moment $T_0$ such that after this moment just one particle moves through the node.*

*Proof.* Assume that $N_1 = \cdots = N_L = N$. Suppose there is a delay of a maximum group $G$ cluster at the moment $T_2 < T_1 + N$. Then either a group of greater length forms, and this group contains all clusters of the group $G$, or, there exists a moment $T_0$ such that, for all $j \in \{0, 1, \ldots, N-1\}$, the cell $j$ is occupied at least on one contour. Hence, $\forall T_1 > T_0$ the cell $j$ is occupied at least one contour, $j \in \{0, 1, \ldots, N-1\}$. Indeed, assume that it is not true. Then the cell $j+1$ (addition modulo $N$) of any contour was not occupied at the preceding moment. At least one particle moves through the node at each moment after the moment $T_0$. However more than one particle cannot move through the node simultaneously. Lemma 6 has been proved.

Suppose (9) is true. Let us consider the system behavior after moment $T_0$ such that this moment satisfies condition of Lemma 7.

**Lemma 8.** *Let the movement be total-connected. Assume that $N_1 = \cdots = N_L = N$. Suppose a particle $P$ is the leader of a cluster, and, at moment $T_1 > T_0$, this particle moves through the node to the cell $N-1$ of the contour. Then this particle moves at the moments $T_1 + 1 \ldots, T_1 + N - 1$ and the particle will be in the cell before the node (cell 0) at the moment $T_1 + N - 1$.*

*Proof.* Since a sequence of system states repeats periodically after the moment $T_0$, we see that clusters cannot merge and form clusters of greater length. Therefore a cluster cannot be delayed unless the leader of the cluster is before the node, and the cluster moves $N$ time units after crossing the node. From this, Lemma 8 follows.

**Lemma 9.** *Assume that $N_1 = \cdots = N_L = N$. Suppose the movement is total-connected, and a particle $P$ moves through the node at time moments $T_1$ and $T_2$, $T_0 < T_1 < T_2$. Then $T_2 - T_1 \geq M_1 + \cdots + M_L$.*

*Proof.* The particle $P$ is located in the cell $N-1$ at moment $T_1$. No other particle can be in the cell $N-1$ of the contour. Indeed, if a particle occupies the cell $N-1$ at the moment $T_1$, then, in accordance with Lemma 8, this particle moves through the node at this moment. However two particles cannot move through the node simultaneously. Therefore it is true the following. Suppose a particle is not the particle $P$. Then, at the moment $T_1$, either this particle is in the queue before the node or, in accordance with Lemma 8, this particle comes to the node earlier than the particle $P$. Therefore any particle (not the particle $P$) moves through the node between moments $T_1$ and $T_2$. From this, taking into account that more one particle cannot move through the node simultaneously, we obtain Lemma 9.

**Lemma 10.** *Assume that $N_1 = \cdots = N_L = N$. Let the movement be total-connected. Suppose a particle $P$ and moments $T_1, T_2$ satisfy Lemma 9. Then the equality $T_2 - T_1 = M_1 + \cdots + M_L$ is true.*

*Proof.* From Lemma 9, it follows that the inequality $T_2 - T_1 \geq M_1 + \cdots + M_L$ is true.

Since just one particle moves through the node at each moment, we have see that the inequality $T_2 - T_1 > M_1 + \cdots + M_L$ can be true only if at least one particle crosses the node more than one time during the time segment $[T_1, T_2]$. However this is impossible. Suppose a particle moves through the node in this time segment. Then, in accordance with Lemma 8, this particle returns to the node later than the particle $P$, and therefore the priority of the considered particle is lower than priority of the particle $P$. This contradiction proves Lemma 10.

**Lemma 11.** *Assume that $N_1 = \cdots = N_L = N$. Let the movement be total-connected. Suppose (9) is true. Then time intervals of particle movement and waiting alternate. Any particle moves for $N$ time units and waits for $M_1 + \cdots + M_L - N$ times units.*

*Proof.* If a particle is a leader of a cluster, then Lemma 11 is true for this particle in accordance with Lemmas 8, 10. Since all particles of any cluster move simultaneously, we obtain Lemma 11.

**Theorem 9.** *Assume that $N_1 = \cdots = N_L = N$. Suppose the movement is total-connected. Then velocities of particles on all contours are the same, and*

$$v_1 = \cdots = v_L = \begin{cases} 1, & M_1 + \cdots + M_L \le N, \\ \frac{N}{M_1 + \cdots + M_L}, & M_1 + \cdots + M_L > N. \end{cases}$$

**Proof.** If (8) is true, then synergy takes place. Therefore,

$$v_1 = \cdots = v_L = 1.$$

If (9) is true, then, in accordance of Lemma 10,

$$v_1 = \cdots = v_L = \frac{N}{M_1 + \cdots + M_L}.$$

Theorem 9 has been proved.

**Corollary 1.** *Assume that $N_1 = \cdots = N_L$ and the movement is total-connected. Let values of $L, N, M_1, \ldots, M_L$ be given. Then velocities of particles on any contour do not depend on the initial state of the system.*

## 5. Total-connected Movement on BAN with One Cluster on Every Contour

*5.1 Synergy of Two Clusters*

Assume that the number of contours equals 2, and there is only one cluster on each contour. We formulate a necessary and sufficient condition of synergy.

**Theorem 10.** *Suppose $L = 2$, and there is a cluster on each contour. Then a necessary and sufficient condition for the system to come to the state of synergy after a finite time is that*

$$M_1 + M_2 \le GCD(N_1, N_2). \tag{10}$$

**Proof.** Suppose the system is in the state such that $(k_1, k_2)$ are coordinates of leaders clusters, $(0 \le k_1 \le N_1 - 1$, $0 \le k_2 \le N_2 - 1)$. If (10) is not true, then $GCD(N_1, N_2)$ is a divisor of numbers $M_1 + 2, \ldots, k_2 - k_1, \ldots, k_2 - k_1 + M_2 - 2$, $k_2 - k_1 + M_2 - 1$. Taking into account Theorem 7, we see that the condition is sufficient. If (10) is true, then either the system is in the state of synergy, or a delay of a cluster takes place after a finite time. This cluster begins to move again when the system is in the state $(0, N_2 - M_2)$ or $(N_1 - M_1, 0)$. From (10), it follows that these states satisfy the condition of synergy (Theorem 5). From this, Theorem 10 follows.

*5.2 Velocity of Particles on BAN with Two Contours and a Cluster on Each Contour*

Assume that there are two contours and a cluster on each contour. We shall describe an approach to velocities calculation. Suppose the condition of synergy (10) is not fulfilled. Denote by $A$ a set of system states such that a state belongs to this set if and only if one of clusters do not moves in this state. After being in the set $A$, the system comes either into the state $(0, N_2 - M_2)$, if there was a delay of the cluster 1, or into the state $(N_1 - M_1, 0)$, if there was a delay of the cluster 2. For each of these two states, analyzing the appropriate linear equation with two variables, we can calculate after what time interval the system comes to a state of the set $A$ and to what state of this set $A$ the system comes. Taking into account results of this analysis, we can calculate velocities $v_1, v_2$. Depending on values of system parameters, there is a pair of values $(v_1, v_2)$ such that these values are taken with probability 1, or there are two pairs of values $(v_1, v_2)$ such that these values are taken with positive probabilities. In the latter case, the velocities depend on the initial state, and there exist initial states such that each pair of values is taken with positive probability.

## 6. Comments, and Further Research

*(6.1)* We have found a necessary and sufficient condition of synergy and formulas for velocities in the case of contours of the same length and in the case of two contours of any length. We have found some necessary conditions and some sufficient conditions of synergy on a flower with contours of different lengths. *In general case, problems of formulation of necessary and sufficient condition of synergy and calculation of velocities, in cases of total-connected or individual condition have not been solved.*

*(6.2)* For analysis of general characteristics of the behavior of dynamical systems on networks, it is interesting to investigate systems *with regular structures such that it is possible to find analytical characteristics of their behavior.*

*(6.3)* An alternating flower (BAN) is considered in the present paper. Some results can be reformulated for the flower with a common cell, which is also the common node. The condition of synergy (Theorem 1) changes. This condition takes

into account that a delay of one particles takes place not only if particles come to the node simultaneously, but also if one of the particles comes at next moment. In accordance with this, the inequality $GCD(N_i, N_j) > 3$ substitutes (4), and the inequality

$$r_1 + r_2 + \cdots + r_L \leq \frac{1}{2}$$

substitutes (6).

*(6.4)* We intend to investigate generalizations of the system, considered in the present paper, e.g., the system with several neighboring (Figure 3) or equidistant from each other *joined nodes and cells.*

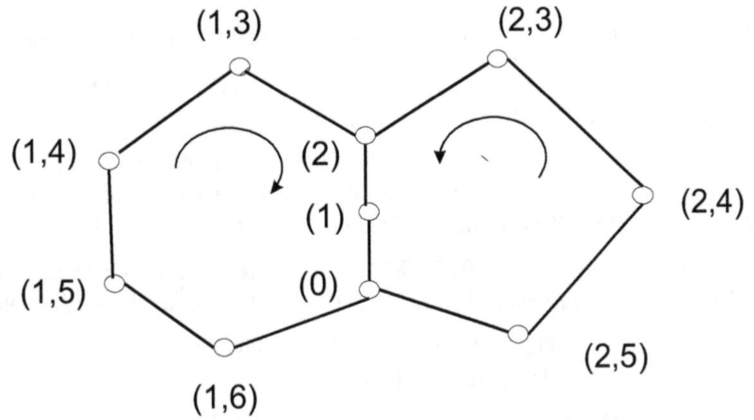


Figure 3. Contours with three common cells joined with nodes

*(6.5) There exist initial states such that particles velocities (in general case, probability distribution of velocities) are different.*

*(6.6)* As we mentioned in Section 1, in [1] – [2], a dynamical system has been introduced. In this system, clusters move on a "flower" ("bouquet"). The state space and time scale of this system are continuous. We can reformulate results of the present paper for this continuous system. For example, consider a bouquet with two closed contours. They are the contour 1 of length $L_1$, and the contour 2 with length $L_2$, where $L_1$ and $L_2$ are positive numbers. The cluster of length $l_i$ ($l_i < L_i$) moves on the cluster $i$, $i = 1, 2$. The point $(0, 0)$ is the node, i.e., common point of these contours. Each state of the system is $(x_1, x_2)$, where $x_i$ is the rear point of the cluster, $0 \leq x_i < L_i$, $i = 1, 2$. Two clusters cannot move through the node simultaneously. In the state $(0, 0)$, a conflict of clusters takes place. Denote by $T_i$ the duration of time interval such that, during this time interval, the cluster $i$ passes the distance equal to the contour length if there are no delays: $T_i = l_i/v_i$. Consider a time interval such that, during this time interval, the cluster $i$ passes through the node. The duration of this interval is equal to $l_i/v_i$, $i = 1, 2$. *If the ratio $T_1/T_2$ is an irrational number, then the system cannot come to the state of synergy.* Suppose that the ratio $T_1/T_2$ is a rational number. Denote by $GCD(T_1, T_2)$ the greatest divisor of $T_1$ and $T_2$. *If $T_1/T_2$ is rational, then a necessary and sufficient condition for the system to come to the system of synergy after a finite time is that*

$$\frac{l_1}{v_1} + \frac{l_2}{v_2} \leq GCD(T_1, T_2). \tag{11}$$

*(11) is similar to (10).*

## References

Blank, M. (2000). Exact analysis of dynamic systems appearing in traffic flows models. *Uspekhi matematicheskikh nauk, 55*(3), 167 –168.

Borovkov, A. A. (1986). *Probability theory* (2th ed.) Moscow: Nauka. (In Russian.)

Buchstab, A. A. (1972). *Number theory.* Moscow, Nauka. (In Russian.)

Bugaev, A. S., Buslaev, A. P., Kozlov, V. V., Tatashev, A. G., & Yashina, M. V. (2013). Traffic modeling: Monotonic random walks on network. *Matematicheskoye modelirovaniye, 25*(8), 3 –21.

Buslaev, A. P. (2010). Traffic flower with *n* petals *Journal of Applied Functional Analysis (JAFA) 5*(1), 85 – 99.

Buslaev, A.P., & Tatashev, A. G. (2016). Bernoulli algebra on common fractions and generalized oscillations *Journal of Mathematics Research, 8*(3), 82 – 93. http://dx.doi.org/105539/jmr.v8n3p82

Buslaev, A. P., & Tatashev, A. G. (2015). Generalized real numbers pendulums and transport logistic applications. In *New Developments in Pure and Applied Mathematics*, 388 – 392. Vienna. www.inase.org/library/2015/vienna/bypaper/MAPUR/MAPUR-63.pdf

Buslaev, A. P., & Yashina, M. V. (2009). About flows on a traffic flower with control. MCV. *The World Congress in Computer Science, Computer Engineering, and Applied Computing (WORLDCOMP'09), LAS Vegas, Nevada USA (July 13 – 16, 2009) in Proc. of the 2009 International Conference on Modelling Simulation and Visualization, CSREAS Press*, 254 – 257.

Feller, W. (1970). *An introduction to probability theory and its applications, 1.* New York, John Willey.

Halmos, P. R. *Lectures on Ergodic Theory*. Publ. Math. Soc. Japan, Tokyo, 1956; reprinted Chelsea, New York, 1960.

Kemeny, J. G., & Snell, J. L. (1976). Finite Markov chains. New York, Heidelberg, Tokyo: Springer Verlag.

Kozlov, V. V., Buslaev, A. P., & Tatashev, A. G. (2014). Behavior of pendulums on a regular polygon. *Journal of Communication and Computer, 11*, 30 – 38.

Kozlov, V. V., Buslaev, A. P., Tatashev, A.G., & Yashina, M. V. (2014). Monotonic walks of particles on a chainmail and coloured matrices. *Proceedings of the 14th International Conference on Computational and Mathematical Methods in Science and Engineering, CMSSE 2014, Cadiz Spain, June 3 – 7 2014, 3*, 801 – 805.

Kozlov, V. V., Buslaev, A. P., & Tatashev, A. G. (2015). Monotonic walks on a necklace and coloured dynamic vector. *International Journal of Computer Mathematics, 92*(9), 1910 – 1920. http://dx.doi.org/1080/00207160.2014/915964

Kozlov, V. V., Buslaev, A. P., Tatashev, A. G., Yashina, M. V. (2015). Dynamical systems on honeycombs. *Traffic and Granular Flow '13.* Springer - Verlag Heidelberg, 441 – 452.

Nagel, K., & Schreckenberg, M. (1992). A cellular automation models for freeway traffic. *J. Phys. I. France, 2*, 2221 – 2229.

Schadschneider, A., Schreckenberg, M. (1993). Cellular automation models and traffic flow. *J. Phys. A. Math. Gen., 51*, L679 – L683.

# Degree Splitting of Heronian Mean Graphs

S. S. Sandhya[1], E. Ebin Raja Merly[2] & S. D. Deepa[3]

[1] Department of Mathematics, Sree Ayyappa College for Women, Chunkankadai, India

[2] Department of Mathematics, Nesamony Memorial Christian College, Marthandam, India

[3] Research Scholar, Nesamony Memorial Christian College, Marthandam, India

Correspondence: S. D. Deepa, Research Scholar, Nesamony Memorial Christian College Marthandam, India. E-mail: christodeepadit@gmail.com

**Abstract**

In this paper, we prove Heronian Mean labeling of some degree splitting graphs. Already we have proved Heronian Mean labeling for some standard graphs. Here we prove that degree splitting of Path $P_3$, Path $P_4$, $P_3 \odot K_1$, $P_2 \odot K_{1,2}$, $P_2 \odot K_{1,3}$, $P_2 \odot K_3$ are Heronian Mean graphs.

**Keywords:** Heronian Mean graph, degree Splitting graphs, union of graphs, Path.

**AMS Subject Classification: 05C78**

## 1. Introduction

By a graph we mean a finite undirected graph without loops or parallel edges. For all detailed survey of graph labeling, we refer to J. A. Gallian (Gallian, 2013). For all other standard terminology and notations we follow Harary (Harary, 1988). The concept of Mean labeling was introduced in (Somasundaram & Ponraj, 2003). The concept of Harmonic Mean labeling was introduced in (Somasundaram, Ponraj, & Sandhya). The concept of Harmonic Mean labeling on Degree Splitting graph was introduced in (Sandhya, Jeyasekharan, & David). Motivated by the above results and by the motivation of the authors we study the Heronian Mean labeling on Degree Splitting graphs. Heronian Mean labeling was introduced in (Sandhya, Merly, & Deepa) and the Heronian Mean labeling of some standard graphs was proved in (Sandhya, Merly, & Deepa).

We shall make frequent references to the definitions and theorems that are useful for our present study. **A Path $P_n$** is a walk in which all the vertices are distinct.

***Definition 1.1:***

A graph **G=(V,E)** with p vertices and q edges is said to be a **Heronian Mean graph** if it is possible to label the vertices $x \in V$ with distinct labels $f(x)$ from **1,2,…,q+1** in such a way that when each edge $e = uv$ is labeled with,

$$f(e = uv) = \left\lceil \frac{f(u) + \sqrt{f(u)f(v)} + f(v)}{3} \right\rceil \text{(OR)} \left\lfloor \frac{f(u) + \sqrt{f(u)f(v)} + f(v)}{3} \right\rfloor$$

then the edge labels are distinct. In this case **f** is called a **Heronian Mean labeling** of G.

***Definition 1.2:***

Let G=(V,E) be a graph with $V = S_1 \cup S_2 \cup \dots . \cup S_t \cup T$, Where each $S_i$ is a set of vertices having atleast two vertices and $T = V - \cup S_i$. The degree splitting graph of G is denoted by DS(G) and is obtained from G by adding vertices $w_1, w_2, \dots . w_t$ and joining $w_i$ to each vertex of $S_i$ $(1 \leq i \leq t)$. The graph G and its degree splitting graph DS(G) are given in figure:1.

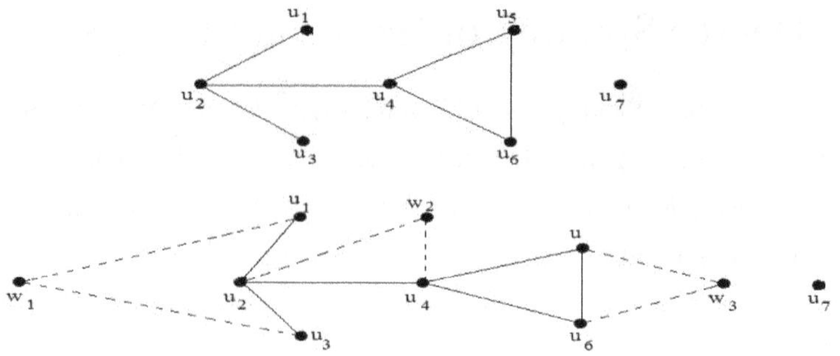


Figure 1.

***Definition 1.3:***

The union of two graphs $G_1 = (V_1, E_1)$ and $G_2 = (V_2, E_2)$ is a graph $G = G_1 \cup G_2$ with vertex set $V = V_1 \cup V_2$ and the edge set $E = E_1 \cup E_2$.

***Theorem 1.4:*** Any Path $P_n$ is a Heronian mean graph.

***Remark 1.5:***

Any graph G is a subgraph of DS(G). If G has atleast two vertices, then G contains atleast two vertices of the same degree. Hence $G = K_1$ is the only graph such that G=DS(G).

***Remark 1.6:***

If G is regular, then DS(G)= $G + K_1$.

**2. Main Results**

***Theorem 2.1:***

$nDS(P_3)$ is a Heronian mean graph.

***Proof:***

The graph $DS(P_3)$ is shown in figure:2

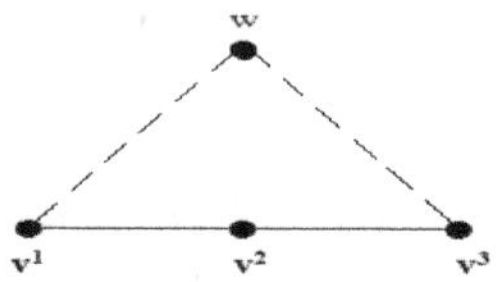


Figure 2.

Let $G = nDS(P_3)$. Let the vertex set of G be $V = V_1 \cup V_2 \cup \ldots . \cup V_n$,

Where $V = \{V_i^1, V_i^2, V_i^3, w_i / 1 \leq i \leq n\}$ } is the vertex set of $i^{th}$ copy of $DS(P_3)$

Define a function $f: V(G) \rightarrow \{1,2, \ldots \ldots q+1\}$ by

$$f(V_i^1) = 4i - 3, 1 \leq i \leq n$$

$$f(V_i^2) = 4i - 2, 1 \leq i \leq n$$

$$f(V_i^3) = 4i - 1, 1 \leq i \leq n$$

$$f(w_i) = 4i, 1 \leq i \leq n$$

Then the edges are labeled with

$$f(V_i^1 V_i^2) = 4i - 3, 1 \leq i \leq n$$

$$f(V_i^2 V_i^3) = 4i - 1, 1 \leq i \leq n$$

$$f(V_i^{1}w_i) = 4i - 2, 1 \leq i \leq n$$

$$f(V_i^{3}w_i) = 4i, 1 \leq i \leq n$$

Hence by definition 1.1, G is a Heronian mean graph.

***Example 2.2:*** Heronian mean labeling of $4DS(P_3)$ is shown in figure 3.

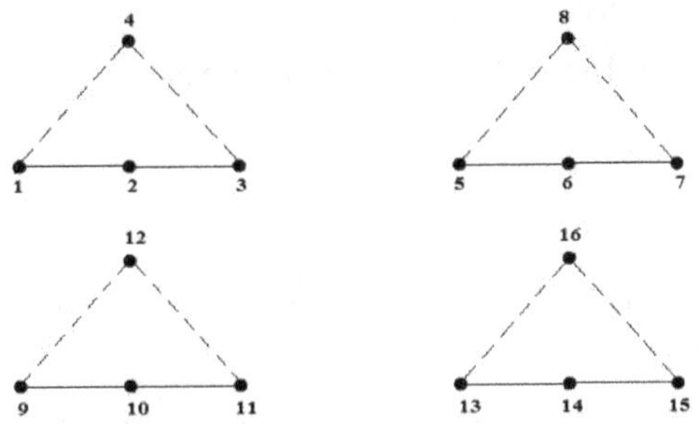


Figure 3.

***Theorem 2.3:***

$nDS(P_4)$ is a Heronian mean graph.

***Proof:***

The graph $DS(P_4)$ is shown in figure 4.

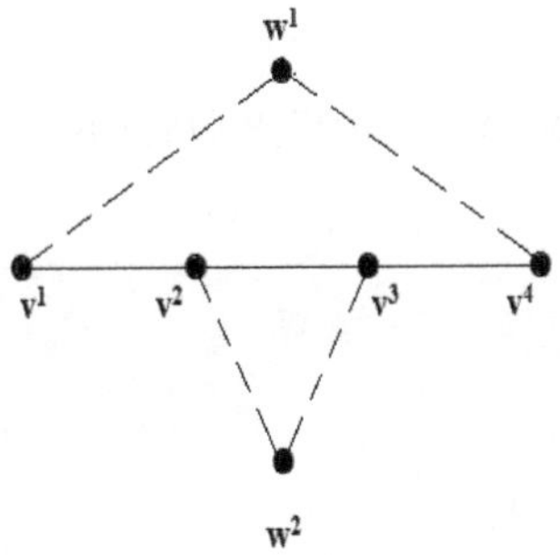


Figure 4.

Let $G = nDS(P_4)$. Let the vertex set of G be $V = V_1 \cup V_2 \cup \ldots . \cup V_n$,

Where $V = \{V_i^{1}, V_i^{2}, V_i^{3}, V_i^{4}, w_i^{1}, w_i^{2} / 1 \leq i \leq n\}$ is the vertex set of $i^{th}$ copy of $DS(P_4)$

Define a function $f: V(G) \rightarrow \{1,2, \ldots\ldots q+1\}$ by

$$f(V_i^{1}) = 7i - 5, 1 \leq i \leq n$$

$$f(V_i^{2}) = 7i - 3, 1 \leq i \leq n$$

$$f(V_i^{3}) = 7i - 1, 1 \leq i \leq n$$

$$f(V_i^{4}) = 7i - 4, 1 \leq i \leq n$$

$$f(w_i^{1}) = 7i - 6, 1 \leq i \leq n$$

$$f(w_i^{2}) = 7i, 1 \leq i \leq n$$

Then the edges are labeled with

$$f(V_i^{1}V_i^{2}) = 7i - 4, 1 \leq i \leq n$$

$$f(V_i^{2}V_i^{3}) = 7i - 2, 1 \leq i \leq n$$

$$f(V_i^{3}V_i^{4}) = 7i - 3, 1 \leq i \leq n$$

$$f(V_i^1 w_i^1) = 7i - 6, 1 \le i \le n$$

$$f(V_i^2 w_i^2) = 7i - 1, 1 \le i \le n$$

$$f(V_i^3 w_i^2) = 7i, 1 \le i \le n$$

$$f(V_i^4 w_i^1) = 7i - 5, 1 \le i \le n$$

Hence by definition 1.1, G is a Heronian mean graph.

***Example 2.4:*** Heronian mean labeling of $4DS(P_4)$ is shown in figure 5.

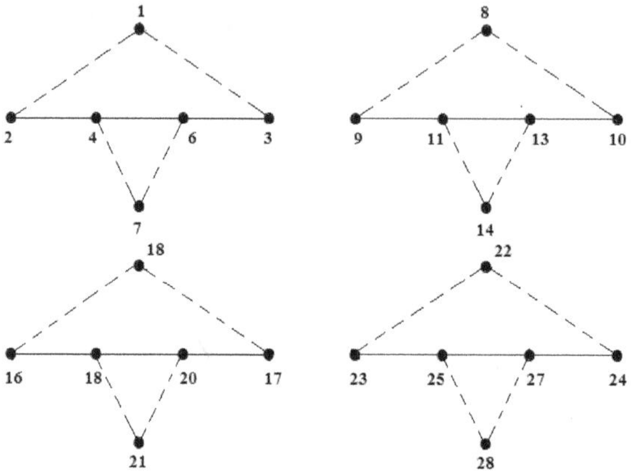


Figure 5.

***Remark 2.5:***

We know that $P_2 \odot K_1 = P_4$. Hence using theorem:2.3, $nDS(P_2 \odot K_1)$ is a Heronian mean graph.

***Theorem 2.6:***

$nDS(P_3 \odot K_1)$ is a Heronian mean graph.

***Proof:***

The graph $DS(P_3 \odot K_1)$ is shown in figure:6

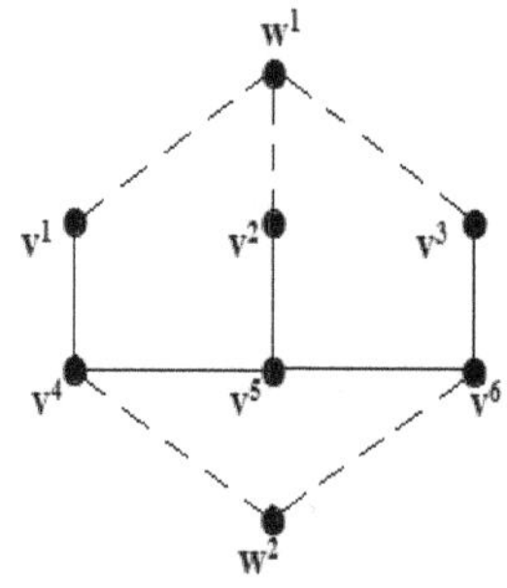


Figure 6.

Let $G = nDS(P_3 \odot K_1)$ . Let the vertex set of G be $V = V_1 \cup V_2 \cup \ldots . \cup V_n$,

Where $V = \{V_i^1, V_i^2, V_i^3, V_i^4, V_i^5, V_i^6, w_i^1, w_i^2 / 1 \le i \le n\}$ is the vertex set of $i^{th}$ copy of $DS(P_3 \odot K_1)$

Define a function $f: V(G) \to \{1, 2, \ldots \ldots q + 1\}$ by

$$f(V_i^1) = 10i - 3, 1 \le i \le n$$

$$f(V_i^2) = 10i - 2, 1 \le i \le n$$

$$f(V_i^3) = 10i - 1, 1 \le i \le n$$

$$f(V_i^4) = 10i - 8, 1 \leq i \leq n$$

$$f(V_i^5) = 10i - 5, 1 \leq i \leq n$$

$$f(V_i^6) = 10i - 6, 1 \leq i \leq n$$

$$f(w_i^1) = 10i - 6, 1 \leq i \leq n$$

$$f(w_i^2) = 10i - 9, 1 \leq i \leq n$$

Then we get distinct edge labels from **{1,2,…,q}** Hence G is a Heronian mean graph.

***Example 2.7:*** Heronian mean labeling of $4DS(P_3 \odot K_1)$ is shown in figure 7.

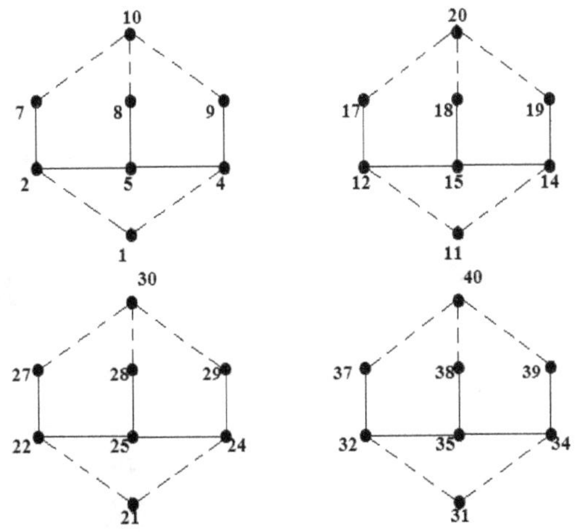


Figure 7.

***Theorem 2.8:***

$nDS(P_2 \odot K_{1,2})$ is a Heronian mean graph.

***Proof:***

The graph $DS(P_2 \odot K_{1,2})$ is shown in figure:8

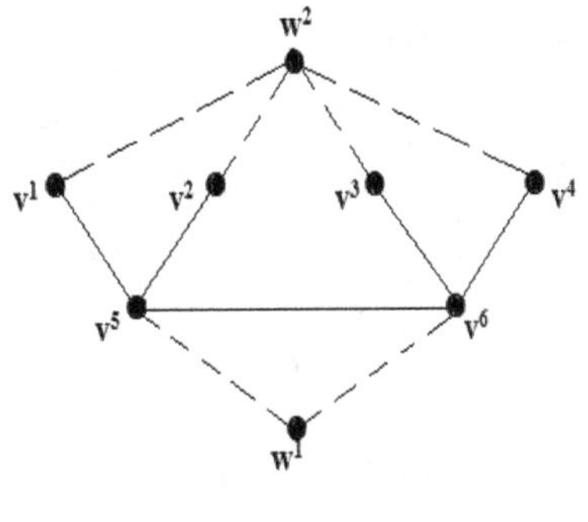


Figure 8.

Let $G = nDS(P_2 \odot K_{1,2})$. Let the vertex set of G be $V = V_1 \cup V_2 \cup \ldots . \cup V_n$, Where $V = \{V_i^1, V_i^2, V_i^3, V_i^4, V_i^5, V_i^6, w_i^1, w_i^2 / 1 \leq i \leq n\}$ is the vertex set of $i^{th}$ copy of $DS(P_2 \odot K_{1,2})$

Define a function $f: V(G) \to \{1,2, \ldots \ldots q+1\}$ by $f(V_i^1) = 11i - 5, 1 \leq i \leq n$

$$f(V_i^2) = 11i - 3, 1 \leq i \leq n$$

$$f(V_i^3) = 11i - 2, 1 \leq i \leq n$$

$$f(V_i^{4})= 11i-1, 1 \leq i \leq n$$

$$f(V_i^{5})= 11i-8, 1 \leq i \leq n$$

$$f(V_i^{6})= 11i-6, 1 \leq i \leq n$$

$$f(w_i^{1})= 11i, 1 \leq i \leq n$$

$$f(w_i^{2})= 11i-10, 1 \leq i \leq n$$

Then we get distinct edge labels from $\{1,2,\ldots,q\}$ Hence G is a Heronian mean graph.

***Example 2.9:*** Heronian mean labeling of $4DS(P_2 \odot K_{1,2})$ is shown in figure 9.

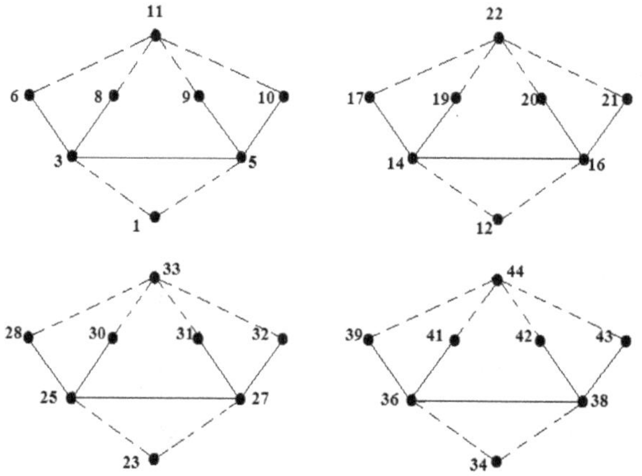


Figure 9.

***Theorem 2.10:***

$nDS(P_2 \odot K_{1,3})$ is a Heronian mean graph.

***Proof:***

The graph $DS(P_2 \odot K_{1,3})$ is shown in figure:10

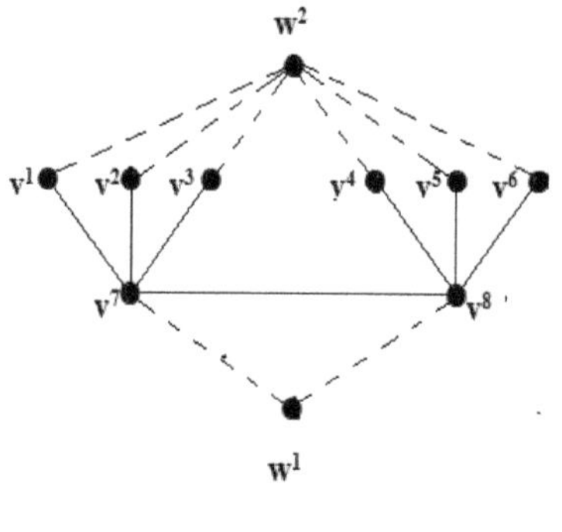


Figure 10.

Let $G = nDS(P_2 \odot K_{1,3})$ . Let the vertex set of G be $V = V_1 \cup V_2 \cup \ldots \cup V_n$,

Where $V = \{V_i^{1}, V_i^{2}, V_i^{3}, V_i^{4}, V_i^{5}, V_i^{6}, V_i^{7}, V_i^{8}, w_i^{1}, w_i^{2} / 1 \leq i \leq n\}$ is the vertex set of $i^{th}$ copy of $DS(P_2 \odot K_{1,3})$

Define a function$f: V(G) \rightarrow \{1,2,\ldots\ldots q+1\}$ by

$$f(V_i^{1})= 15i-9, 1 \leq i \leq n$$

$$f(V_i^{2})= 15i-7, 1 \leq i \leq n$$

$$f(V_i^{3})= 15i-6, 1 \leq i \leq n$$

$$f(V_i^4) = 15i - 5, 1 \le i \le n$$

$$f(V_i^5) = 15i - 3, 1 \le i \le n$$

$$f(V_i^6) = 15i - 1, 1 \le i \le n$$

$$f(V_i^7) = 15i - 2, 1 \le i \le n$$

$$f(V_i^8) = 15i - 10, 1 \le i \le n$$

$$f(w_i^1) = 15i - 14, 1 \le i \le n$$

$$f(w_i^2) = 15i, 1 \le i \le n$$

Then we get distinct edge labels from **{1,2,…,q}** Hence G is a Heronian mean graph.

***Example 2.11:*** Heronian mean labeling of $4DS(P_2 \odot K_{1,3})$ is shown in figure 11.

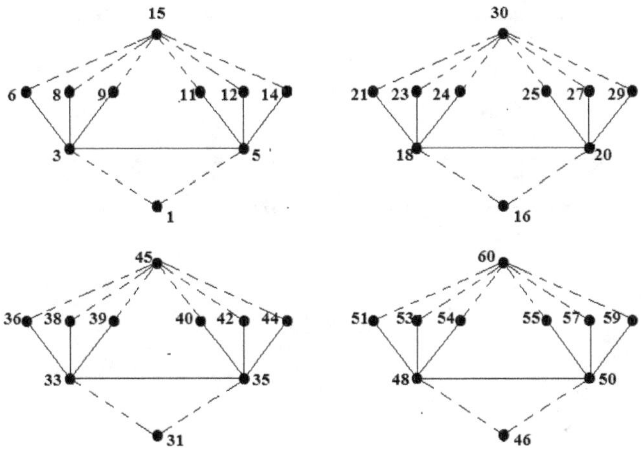


Figure 11.

***Theorem 2.12:***

$nDS(P_2 \odot K_3)$ is a Heronian mean graph.

***Proof:***

The graph $DS(P_2 \odot K_3)$ is shown in figure:12

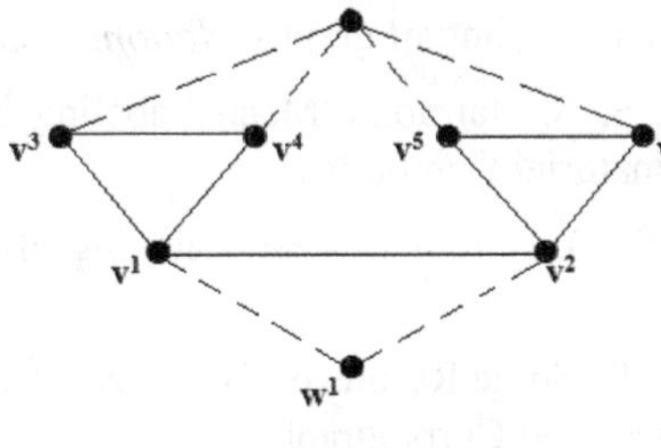


Figure 12.

Let $G = nDS(P_2 \odot K_3)$ . Let the vertex set of G be $V = V_1 \cup V_2 \cup \ldots . \cup V_n$,

Where $V = \{V_i^1, V_i^2, V_i^3, V_i^4, V_i^5, V_i^6, w_i^1, w_i^2 / 1 \le i \le n\}$ is the vertex set of $i^{th}$ copy of $DS(P_2 \odot K_3)$

Define a function $f: V(G) \to \{1,2, \ldots\ldots q+1\}$ by $f(V_i^1) = 13i - 10, 1 \le i \le n$

$$f(V_i^2) = 13i - 8, 1 \le i \le n$$

$$f(V_i^3) = 13i - 7, 1 \le i \le n$$

$$f(V_i^4) = 13i - 5, 1 \leq i \leq n$$

$$f(V_i^5) = 13i - 3, 1 \leq i \leq n$$

$$f(V_i^6) = 13i - 1, 1 \leq i \leq n$$

$$f(w_i^1) = 13i - 12, 1 \leq i \leq n$$

$$f(w_i^2) = 13i, 1 \leq i \leq n$$

Then we get distinct edge labels from **{1,2,...,q}** Hence G is a Heronian mean graph.

***Example 2.13:*** Heronian mean labeling of $4DS(P_2 \odot K_3)$ is shown in figure 13.

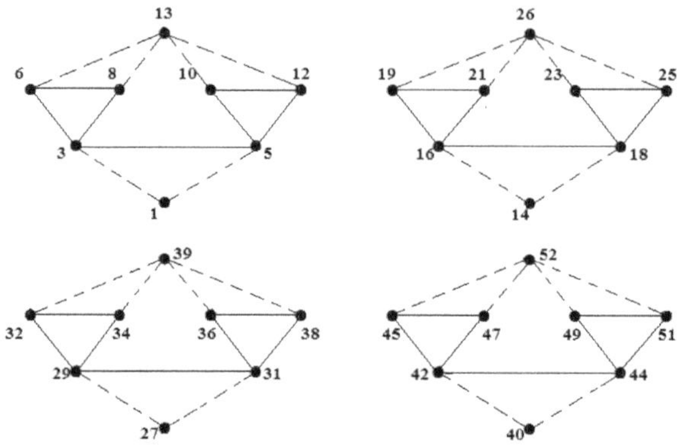


Figure 13.

## 3. Conclusion

In this paper, we studied the degree splitting behavior of some standard Heronian mean graphs. The authors are of the opinion that the study of Heronian mean labeling of degree splitting graphs will lead to newer and different results.

**Acknowledgement**

The authors are thankful to the referee for their valuable comments and suggestions.

**References**

Gallian, J .A. (2013). A Dynamic Survey of Graph Labeling. *The Electronic Journal of combinatorics.*

Harary. F (1988). *Graph Theory.* Narosa publishing House, New Delhi.

Somasundaram, S., & Ponraj, R. (2003). Mean Labeling of graphs. *National Academy of Science Letters, 26*, 210-213.

Somasundaram, S., Ponraj, R. & Sandhya, S. S. Harmonic Mean Labeling of Graphs. Communicated to Journal of Combinatorial Mathematics and Combinatorial Computing.

Sandhya, S. S., Merly E. E. R., & Deepa, S. D. Heronian Mean Labeling of Graphs. Communicated to International Journal of Mathematical Form.

Sandhya, S. S., Merly E. E. R., & Deepa, S. D. Some Results on Heronian Mean Labeling of Graphs. communicated to Journal of Discrete Mathematical Sciences and Cryptography.

Sandhya, S. S., Jeyasekharan, C., & David, R. C. Harmonic Mean Labeling of Degree Splitting Graph. Communicated to Bulletin of Pure and Applied Sciences-Mathematics.

# 18

# Computational Algorithms for Solving Spectral/*hp* Stabilized Incompressible Flow Problems

Rakesh Ranjan[1,2], Anthony Theodore Chronopoulos[1,3], Yusheng Feng[1,2]

[1] Center for Simulation, Visualization and Real-Time Prediction (SiViRT)

[2] Department of Mechanical Engineering, University of Texas, San Antonio, USA

[3] Department of Computer Science, University of Texas, San Antonio, USA

Correspondence: Yusheng Feng. E-mail: yusheng.feng@utsa.edu

**Abstract**

In this paper we implement the element-by-element preconditioner and inexact Newton-Krylov methods (developed in the past) for solving stabilized computational fluid dynamics (CFD) problems with spectral methods. Two different approaches are implemented for speeding up the process of solving both steady and unsteady incompressible Navier-Stokes equations. The first approach concerns the application of a scalable preconditioner namely the element by element LU preconditioner, while the second concerns the application of Newton-Krylov (NK) methods for solving non-linear problems. We obtain good agreement with benchmark results on standard CFD problems for various Reynolds numbers. We solve the Kovasznay flow and flow past a cylinder at Re-100 with this approach. We also utilize the Newton-Krylov algorithm to solve (in parallel) important model problems such as flow past a circular obstacle in a Newtonian flow field, three dimensional driven cavity, flow past a three dimensional cylinder with different immersion lengths. We explore the scalability and robustness of the formulations for both approaches and obtain very good speedup. Effective implementations of these procedures demonstrate for relatively coarse macro-meshes the power of higher order methods in obtaining highly accurate results in CFD. While the procedures adopted in the paper have been explored in the past the novelty lies with applications with higher order methods which have been known to be computationally intensive.

**Keywords:** spectral/*hp* methods, element by element preconditioning, Newton Krylov methods, scalability, 3D flow past a cylinder

## 1. Introduction

In both computational research and practical applications there is continuing interest in solving incompressible fluid flow problems efficiently. Different approaches have been used for solving the Navier-Stokes equations. The finite element community has been beset with various difficulties such as satisfaction of the Ladyshenskaya Babuska Brezzi (LBB) condition between the velocity and pressure spaces (Reddy, 2010), presence of the zero pressure block in the weak form Galerkin formulation, spurious pressure and velocity fluctuations, along with non-convergence for moderate to high Reynolds number flows. Large scale applications with usage of LBB compatible finite element pairs however can be found in Elman (2005) and Benzi (2005) among others who also introduce techniques to solve the saddle point problem.

To alleviate some of the above mentioned difficulties, various approaches have been proposed. One formulation is the penalty finite element method, which treats the continuity equation as a constraint and adds this restriction on to the momentum equations. Penalty finite element methods provide acceptable results on the velocity fields. The pressures are post computed from the velocities via the incompressibility constraint. Using such techniques however to solve complex problems are beset with difficulties primarily resulting from the high condition numbers associated with the discrete form of the penalty terms which can engender convergence issues with high values of the penalty parameter. More specifically, for large scale problems, a high value of the Penalty parameter is required to satisfy the incompressibility condition and this leads to non-convergence for complex problems. Attractive alternatives that exist for solving incompressible flow utilize splitting techniques of Chorin type (Chorin, 1968) which split the equations pre-discretization. Post discretization splitting techniques requiring the usage of *Yosida method* (Dembo, 1982) have also been found to perform very well for large scale problems.

A separate formulation namely the Least squares finite element procedures (LSFEM) has recently been proposed as an alternative formulation for solving Navier-Stokes equations. LSFEM formulation provides a variational setting for the

Navier-Stokes equations, and a symmetric positive definite matrix, which can be solved with conjugate gradient solvers combining Jacobi or multi-grid preconditioning (Quarteroni, 1999). Another advantage of the LSFEM is that it provides a parameter free formulation for solving incompressible flow. One of the main drawbacks of LSFEM is it often results in extensive loss of mass. LSFEM procedures also fail for problems in contraction regions resulting in spurious results. An attractive alternative to the techniques mentioned above (other than those highlighted), is a stabilized formulation in the context of streamline upwind Petrov Galerkin (SUPG), and pressure stabilized Petrov Galerkin formulations (PSPG) for incompressible flow (often mentioned as SUPS formulation in literature). The problems associated with the ill-conditioning of the discrete form of the Galerkin finite element formulation are alleviated with the addition of consistent terms into the variational formulation for solving the Navier-Stokes equations (Brooks, 1982; Brooks, 1980; Hughes, 1979; Hughes, 1982). In general the SUPS formulation for solving incompressible flow can be interpreted as a special case of the generic residual based variational multiscale (RBVMS) formulation of Hughes (1995) with the omission of two terms. Since the SUPS formulation has been studied extensively and provides excellent results for solutions of CFD problems we adhere to this specialized formulation in this article. We utilize the $p$-version FEM or the spectral/$hp$ method for solving CFD problems. Stabilized finite element formulations have been used for solving incompressible flow with higher order spectral approximations in (Gervasio, 1998; Ranjan, 2016). However effective preconditioning techniques for these methods and efficient methods to solve large scale problems were not discussed.

Lower order finite element formulations have been used extensively to solve incompressible flow problems. There has been an interest lately to usage of higher order spectral approximations to solve CFD problems since they provide spectral (exponential) accuracy to known analytical solutions available. This allows the usage of coarse macro meshes for solving problems rather than extensively refined low order finite element meshes to capture evolving fields of interest. Due to the increase in the number of degrees of freedom of spectral element dicretizations, higher order spectral methods take a long time to converge and consequently have high computation times even on multiprocessors. On the other hand lower order finite element implementations converge faster, and it is known $p$ version of the finite element method is slow to converge. Further higher order spectral methods have problems in maintaining monotonicity in the presence of singularities.

For two dimensional problems spectral implementations the quadrature step requires an $O(p^6)$ operation where as for complicated three dimensional solutions, there is an increase in the quadrature requirements (due to increase in the degrees of freedom per element). To control the relatively high computation times for solving complicated problems, one has to resort to non-linear solvers and effective preconditioning techniques. In this context using effective preconditioners and Newton-Krylov algorithms are of interest for solving large linear systems generated from spectral discretizations of the Navier-Stokes equations.

## 2. Literature Review

A review of sparse preconditioners has been provided in Meister et al. (2001) for hyperbolic conservation laws. Details on the construction of the incomplete block LU preconditioner are provided by Sturler et al. (2005). Issues of the singularity of approximate inverse preconditioners have been outlined in Wang et al. (2005). Performances of the incomplete LU and Cholesky preconditioned iterative methods on GPU's have been explored by Naumov et al. (2011). Parallel implementations with Domain decomposition algorithms are presented by Barth et al. (1998). Theory and procedural development of non-linear solution techniques with the Newtons method is presented in Dembo et al. (1982) and Eisenstat et al. (1996). Applications of the Inexact Newton-Krylov methods for solving large computational fluid dynamics problems can be found in Shadid et al. (1997) where they solved the thermal convection in a square cavity, lid driven cavity flow at Reynolds number (Re) 10,000 and backward facing step problems at Re-800 with stabilized methods. Elias et al. (2006) solved the two dimensional problems of flow inside a borewall well and flow through an abrupt contraction region with Inexact Newton-Krylov methods. Further developments in the area include solutions of the lid driven cavity at a range of Re-100 through Re-1000, backward facing step for Reynolds numbers of Re-100 and Re-500. Elias et al. (2006) also obtained the solutions of viscoplastic flows with the inexact Newton-Krylov methods for three-dimensional flows through a channel with a sudden expansion and the three-dimensional driven cavity problem (2006). Edge based data structures were found to perform better in solving complex CFD problems with stabilized methods. However scalability of edge based data structures on large processor counts was found to deteriorate and for the case of large processor counts an improvement in performance as compared to an element by element procedure was found to be marginal at best. Parallel implementations for solving Inexact Newton-Krylov procedures have been demonstrated in Hwang et al. (2005). They implemented the backtracking procedure for solving the driven cavity problem at Re-10,000 and the backward facing step problem at Re-100, Re-500, Re-700 and Re-800.

The approaches described above have been mostly confined to either a finite difference or a lower order finite element method implementation for solving Navier-Stokes equations other than multigrid preconditioning that was implemented for solving LSFEM incompressible flow by Ranjan (2012). In this paper we apply established algorithms for solving high order spectral discretized equations which are computationally intense problems otherwise. While the procedures that are employed are standard the effort is directed towards the development of efficient computational frameworks for obtaining accurate results for Navier-Stokes equations on relatively coarse levels of discretization with higher order methods which have been known to be computationally intense problems.

### 3. Incompressible Flow Equations

Let us consider the flow of a Newtonian fluid with density $\rho$, and viscosity $\mu$. Let $\Omega \subseteq R^N$ and $t \in [0, T]$ be the spatial and temporal domains respectively, where $N$ is the number of space dimensions. Let $\Gamma$ denote the boundary of $\Omega$. The equations of fluid motion that describe the unsteady Navier-Stokes governing incompressible flows are provided by

$$\rho\left(\frac{\partial \mathbf{u}}{\partial t} + \mathbf{u} \cdot \nabla\mathbf{u} - \mathbf{f}\right) - \nabla \cdot \sigma = 0 \; on \; \Omega \; \forall \; t \in [0, T] \tag{1}$$

$$\nabla \cdot \mathbf{u} = 0, \; on \; \Omega \; \forall \; t \in [0, T] \tag{2}$$

where, $\rho$ and $\mathbf{u}$, $\mathbf{f}$, and $\sigma$ are the density, velocity, body force, and stress tensor respectively. The stress tensor is written as a sum of the pressure and viscous components as

$$\sigma = -p\mathbf{I} + \mathbf{T} = -p\mathbf{I} + 2\mu \, \epsilon(\mathbf{u}) \tag{3}$$

$$\epsilon(\mathbf{u}) = \frac{1}{2}\left(\nabla\mathbf{u} + \nabla\mathbf{u}^T\right) \tag{4}$$

Here, $I$ is the identity tensor, $\mu = \rho v$, $p$ is the pressure and $\mathbf{u}$ is the fluid velocity. The part of the boundary at which the velocity is assumed to be specified is denoted by $\Gamma_g$:

$$\mathbf{u} = \mathbf{g} \; on \; \Gamma_g \; \forall t \in \; [0, T] \tag{5}$$

The above set of equations require a divergence free initial velocity field to complete the problem specification.

### 4. Finite Element Formulation

Consider a spectral/$hp$ element discretization of $\Omega$ into subdomains, $\Omega_e$, $k = 1, 2, ...., n_{el}$ where $n_{el}$ is the number of spectral/$hp$ elements into which the domain is divided. Based on this discretization, for velocity and pressure, we define the trail discrete function spaces $\Im_u^{hp}$ and $\Im_p^{hp}$, and the weighting function spaces; $W_u^{hp}$ and $P_p^{hp}$ defined above. These spaces are selected, by taking the Dirichlet boundary conditions into account, as subspaces of $\left[H^{1h}(\Omega)\right]^{n_{el}}$ and $\left[H^{1h}(\Omega)\right]$, where $\left[H^{1h}(\Omega)\right]$ is the finite-dimensional function space over $\Omega$. We write the stabilized Galerkin formulation of eq 1-2 as follows, find $u^{hp} \in \Im_u^{hp}$ and $p^{hp} \in \Im_p^{hp}$ such that $\forall \; w^{hp} \in W_u^{hp}$, $q^{hp} \in P_p^{hp}$;

$$\begin{aligned} &\int_\Omega \mathbf{w^{hp}}\rho \cdot \left(\frac{\partial \mathbf{u^{hp}}}{\partial t} + \mathbf{u^{hp}} \cdot \nabla\mathbf{u^{hp}} - \mathbf{f}\right) d\Omega - \int_\Omega \mathbf{w^{hp}} \cdot \nabla \cdot \sigma(p^{hp}, \mathbf{u^{hp}}) d\Omega + \\ &\int_\Omega q^{hp} \nabla \cdot \mathbf{u^{hp}} d\Omega + \sum_{e=1}^{n_{el}} \int_{\Omega^e} \frac{1}{\rho} \tau \left(\rho\mathbf{u^{hp}} \cdot \nabla\mathbf{w^{hp}} + \nabla q^{hp}\right) \cdot \\ &\left[\rho \frac{\partial \mathbf{u^{hp}}}{\partial t} + \rho\left(\mathbf{u^{hp}} \cdot \nabla\mathbf{u^{hp}} - \mathbf{f}\right) - \nabla \cdot \sigma(p^{hp}, \mathbf{u^{hp}})\right] d\Omega^e + \\ &\sum_{e=1}^{n_{el}} \int_{\Omega^e} \delta\nabla \cdot \mathbf{w^{hp}} \rho \nabla \cdot \mathbf{u^{hp}} d\Omega^e = \int_{\Gamma_h} \mathbf{w^{hp}} \cdot \mathbf{h^{hp}} d\Gamma \end{aligned} \tag{6}$$

As can be seen from Eqn. 6, three stabilizing terms are added to the standard Galerkin formulation. In Eqn. 6 the first three terms and the right hand side constitute the classical Galerkin formulation of the problem. The surface integrals on the right are obtained from the weak form development as the natural boundary conditions. The first series of element-level integrals comprise the SUPS stabilization terms, which are added to the variational formulation (Mittal, 1997; Sampath, 2001). The second series of integrals comprise the least squares terms on the incompressibility. The stabilization parameter $\tau$ is defined as

$$\tau = \left[\left(\frac{2}{\Delta t}\right)^2 + \left(\frac{2\|u^h\|}{h}\right)^2 + \left(\frac{4v}{h^2}\right)^2\right]^{-\frac{1}{2}} \tag{7}$$

Here, $h$ denotes the element size, which is computed based on the equivalent diameter of the element. We have tested using the stabilization parameters for lower order finite element implementations for the solutions of incompressible Navier-Stokes equations in the interest of lower computation times. Stabilization parameters typically reduce in magnitude for higher order discretizations and increase the time the linear solver takes to converge. The second stabilization parameter $\delta$ on the continuity equation was set to zero. The spatial discretization of Eq. 6 leads to the following set of non-linear algebraic equations,

$$[\mathbf{M}(\mathbf{v}) + \mathbf{M}_{\kappa_1}(\mathbf{v})]\,\dot{\mathbf{v}} + [\mathbf{N}(\mathbf{v}) + \mathbf{N}_{\kappa_1}(\mathbf{v})]\,\mathbf{v} + [\mathbf{K} + \mathbf{K}_{\kappa_1}]\,\mathbf{v} - [\mathbf{G} + \mathbf{G}_{\kappa_1}]\,\mathbf{p} = [\mathbf{F} + \mathbf{F}_{\kappa_1}] \tag{8}$$

$$[\mathbf{M}_{\kappa_2}(\mathbf{v})]\,\dot{\mathbf{v}} + \mathbf{G}^{\mathrm{T}}\mathbf{v} + \mathbf{N}_{\kappa_2}(\mathbf{v})\mathbf{v} + \mathbf{K}_{\kappa_2}\mathbf{v} + \mathbf{G}_{\kappa_2}\mathbf{p} = \mathbf{E} + \mathbf{E}_{\kappa_2} \tag{9}$$

Here, $\mathbf{v}$ is the vector of unknown nodal values of $\mathbf{v}^{\mathbf{hp}}$, and $p$ is the vector of nodal values of $p^{hp}$, and the mass matrices have been specified with derivatives of the velocity components. The matrices $\mathbf{N}(\mathbf{v})$, $\mathbf{K}$, and $\mathbf{G}$ are derived, respectively, from the advective, viscous, and pressure terms. The vectors $\mathbf{F}$ and $\mathbf{E}$ are due to boundary conditions. The subscripts $\kappa_1$ and $\kappa_2$ identify the SUPG and PSPG contributions, respectively. The various matrices forming the discrete finite element equations are provided. The definition of the terms in the stiffness matrices are outlined below;

$$(\mathbf{M}(\mathbf{v}) + \mathbf{M}_{\kappa_1}(\mathbf{v})) = \int_{\Omega}\left[\rho\mathbf{w}^{\mathbf{hp}}\frac{\partial\mathbf{u}^{\mathbf{hp}}}{\partial\mathbf{t}} + \sum_{e=1}^{n_{el}}\kappa_1\rho\frac{\partial\mathbf{u}^{\mathbf{hp}}}{\partial t}\right]d\Omega \tag{10}$$

$$(\mathbf{N}(\mathbf{v}) + \mathbf{N}_{\kappa_1}(\mathbf{v})) = \int_{\Omega}\left[\rho\mathbf{w}^{\mathbf{hp}}\cdot(\mathbf{u}^{\mathbf{hp}}\cdot\nabla\mathbf{u}^{\mathbf{hp}}) + \sum_{e=1}^{n_{el}}\kappa_1\cdot(\mathbf{u}^{\mathbf{hp}}\cdot\nabla\mathbf{u}^{\mathbf{hp}})\right]d\Omega \tag{11}$$

$$(\mathbf{K} + \mathbf{K}_{\kappa_1}) = \int_{\Omega}\left[\nabla\cdot\mathbf{w}^{\mathbf{hp}}\cdot 2\mu\epsilon(\mathbf{u}^{\mathbf{hp}}) + \sum_{e=1}^{n_{el}}\kappa_1\cdot 2\mu\epsilon(\mathbf{u}^{\mathbf{hp}}) + \delta\nabla\cdot\mathbf{w}^{hp}\nabla\cdot\mathbf{u}^{hp}\right]d\Omega \tag{12}$$

$$(\mathbf{G} + \mathbf{G}_{\kappa_1}) = \int_{\Omega^e}\nabla\cdot\mathbf{w}^{\mathbf{hp}}p^{hp}d\Omega + \sum_{e=1}^{n_{el}}\int_{\Omega^e}\kappa_1\nabla\cdot p^{hp}d\Omega \tag{13}$$

$$\mathbf{M}_{\kappa_2}(\mathbf{v}) = \sum_{e=1}^{n_{el}}\kappa_2\rho\frac{\partial\mathbf{u}^{\mathbf{hp}}}{\partial t}d\Omega \tag{14}$$

$$\mathbf{G}^T = \int_{\Omega^e}q^{hp}\nabla\cdot\mathbf{u}^{\mathbf{hp}}d\Omega \qquad \mathbf{G}_{\kappa_2} = \sum_{e=1}^{n_{el}}\kappa_2\cdot\nabla\cdot p d\Omega \tag{15}$$

$$\mathbf{N}_{\kappa_2}(\mathbf{v}) = \sum_{e=1}^{n_{el}}\kappa_2\cdot(\mathbf{u}^{\mathbf{hp}}\cdot\nabla\mathbf{u}^{\mathbf{hp}})\rho d\Omega \quad \mathbf{K}_{\kappa_2} = \int_{\Omega^e}\sum_{e=1}^{n_{el}}\kappa_2\cdot\nabla\cdot 2\mu\epsilon(\mathbf{u})\Omega \tag{16}$$

And the forcing functions are defined as

$$\mathbf{F} + \mathbf{F}_{\kappa_1} = \int_{\Omega^e}\left\{\mathbf{w}^{\mathbf{hp}}\rho\cdot\mathbf{f} + \sum_{e=1}^{n_{el}}\kappa_1\cdot\rho\mathbf{f}\right\}d\Omega \tag{17}$$

$$\mathbf{E} + \mathbf{E}_{\kappa_2} = \int_{\Omega_e}\sum_{e=1}^{n_{el}}\kappa_2\cdot\rho\mathbf{f}d\Omega \tag{18}$$

This stabilization of the Galerkin formulation presented in this paper is a generalization of the Galerkin Least squares (GLS) formulation, and the SUPS procedure employed for incompressible flows. With such stabilization procedures described above, it is possible to use elements that have equal order interpolation functions for the velocity and pressure, and are otherwise unstable (Brooks, 1982). There are various ways to linearise the non-linear terms in stabilized finite element methodology. We follow the approach adopted in (Tezduyar, 1992) where the non-linear terms are allowed to lag behind one non-linear iteration. This is mentioned as the iteration lagged stabilization terms update.

**5. Preconditioning**

In this section, we present the use of preconditioners in solving a large scale element structured linear system:

$$A(x)\{x\} = \{b\} \tag{19}$$

using Krylov subspace methods. In the above equation $\{x\}$ denotes the solution of the linear system. We say that the element $A$ admits an element structure $\varepsilon$ if

$$A := \sum_{e\in\varepsilon} P_e A_e P_e^T \tag{20}$$

Here $\varepsilon$ is a set of elements, $A_e$ is a dense square coefficient matrix (element matrix), $P_e$ is a the boolean connectivity matrix that maps the coordinates of the element stiffness matrix to the global matrix $A$. Note that $A(x)$ is a non-linear operator in general which has been linearized prior to the construction of the linear system. It has been mentioned in Benzi et al. (1999) that effective preconditioners are required to solve sparse systems efficiently using Krylov-subspace methods. Some preconditioners exploit knowledge of the element structure. These so-called element by element (EBE) preconditioners operate exclusively on the element matrices (Gustafsson, 1986; Nour-Omid, 1985). The global matrix $A$ is never assembled (computed). Typically, one computes an approximation to $A^{-1}$ by computing the $LU$-factorizations of agglomerates of elements, and then by combining the results. However, these pre-conditioners are block element by element because they entail the factorizations of blocks of elements. They can be applied to a vector using high-throughput dense vector multiplications. Unfortunately, their quality is unsatisfactory to tackle more difficult linear systems.

The total time spent on preconditioning, excluding the factorization, is the product of the number of iterations and the execution time per iteration. That implies that it is *equally* important to reduce the number of iterations as increasing the computational performance. A compromise should be reached between reducing the iteration count, by forming a more accurate factorization for Incomplete LU (ILU) based factorizations and increasing the computational performance, to allow for dense operations. Current research appears to be focused on reducing the iteration count (Vannieuwenhoven, 2010). In this work, we seek to develop a true element-by-element preconditioner which completes the LU factorization of the element stiffness matrix in its dense form and applies it to the huge linear system obtained. Another problem with the development of preconditioners is the scalability of the factorizations while computing the agglomerates of elements.

Let us denote the non-linear problem comprised of the element stiffness matrix contributions by equation 20. Then the preconditioning operation is defined as

$$M^{-1}[A(x)]\{x\} = M^{-1}\{b\} \tag{21}$$

Here, $M$ is the preconditioning matrix. Let us denote the global stiffness matrix with

$$[A] = ([L][U])\{x\} = [L]([U]\{x\}) = \{b\} \tag{22}$$

The preconditioning operation is obtained as a series of backsolves after the factorizations of the matrix has been achieved. It is to be noted that the factorizations have been achieved for the non-linear operator that has been linearized prior to the factorization step.

Let us denote by the subscript $pe$ the element stiffness matrix for a preconditioning element matrix in the domain to generate the non-linear system as

$$[A_{pe}(x_{pe})]\{x_{pe}\} = \{b_{pe}\} \tag{23}$$

where $\{x_{pe}\}$ is the restriction of the non-linear unknown vector, $\{x\}$ on the element nodes and $\{b_{pe}\}$ is the restriction of the residual on the element. In the construction of the non-linear terms for the element stiffness matrix, we utilize a damping of the non-linear system. Thus, the elements of the stiffness matrices are obtained as

$$\{x_{pe}\} = \alpha \cdot \{x\} \tag{24}$$

In the above equation $\alpha$ denotes a constant which is used to damp the non-linear solution vector. We varied the damping coefficient $\alpha$ from 0.40-0.50. We consider a spectral element with a relatively high $p_{level}$ as a domain for which the preconditioning matrix was constructed. The storage of the matrix elements in sparse matrix format allows for affordable memory allocations, as well as a reduction in the operation count in construction of the LU factors that can be expensive for high $p_{levels}$ ($p^6$) operation in 2D or 3D). Linearization of the linear system of equations was achieved with the help of the method of successive substitution (or Picard method). Picard method has been known to be a robust technique which has a large radius of convergence and is based on the fixed point (contraction mapping) theorem. The specification of Dirichlet boundary conditions on the interface nodes between different spectral elements allowed for a communication free implementation with the preconditioned value contributions at the interface replaced by the corresponding residuals.

**Kovasznay Flow**

We consider a two-dimensional, steady flow in $\Omega = [-0.5, 1.5] \times [-0.5, 1.5]$. We use the Kovasznay et al. **?** solution to the problem. Kovasznay provided the analytical solution to the Navier- Stokes equations in two dimensions. The equations were linearised by introducing the stream function. Following the introduction of the stream function and its exponential functional representation, the quadratic terms in the Navier-Stokes equations vanish. Further in the derivation, it was proposed that the two constants in the general solutions, $\lambda_1$ and $\lambda_2$, tend to $2\pi$ when the Reynolds number is very small. We utilize the solution provided with the above assumptions as the analytical solutions for the Navier-Stokes equations. The analytical solutions are provided by

$$\begin{aligned} u(x,y) &= 1 - e^{\lambda x}cos(2\pi y) \\ v(x,y) &= \frac{\lambda}{2\pi}e^{\lambda x}sin(2\pi y) \\ p(x,y) &= p_0 - \frac{1}{2}e^{2\lambda x} \end{aligned} \tag{25}$$

where $\lambda = Re/2 - (Re^2/4 + 4\pi^2)^{\frac{1}{2}}$ and $p_0$ is the reference pressure (an arbitrary constant). We assigned the reference pressure to zero for the purpose of this simulation. The SUPS spectral element formulation was used to formulate the problem and the discretization was performed with the spectral/$hp$ element method. Nodal expansions (spectral elements) were used to obtain the discrete model. The exact solution was prescribed on the boundaries for the $u$ and $v$ components of the velocities. Pressure was assigned based on the reference pressure, set on the whole left face. Nonlinear convergence was declared when the nonlinear residual was reduced to a value of $10^{-4}$ for both the velocities and the pressure. An under-relaxation factor of 0.50 was found to be necessary to affect convergence. For this problem the resulting system is ill-conditioned. The number of non-linear iterations for convergence was typically 14. The mesh used for the analysis is presented in Figure 1($a$). A uniform polynomial expansion of order 5 was used in each element. The discrete mesh resulted in 25,921 nodes, with three degrees of freedom per node for the finest spectral/$hp$ mesh. Figure 1($a$) and 1($b$) present the spectral/$hp$ mesh, $u$ velocity contour plot over the domain of interest, and 1($c$) and 1($d$) present the $v$-velocity and pressure contour plots over the domain for a Reynolds number of 40. Spectral or exponential convergence of the velocity components ($u$ & $v$) and pressure ($p$) are presented in Figure 2. Higher order convergence is evident for the three variables. For faster convergence and in the interest of low solve times we adhere to the stabilization parameter presented in eq. 7.

The iterations required to convergence for the un-preconditioned element-by-element solution (EBE) to the problem vs. the LU preconditioned (EBE-LU) operator is provided in Table 1. Notations for Tables 1- 3 are as follows: Index denotes the non-linear iteration, Re the Reynolds number, $Np$ for the number of processors, EBE the number of iterations required by the EBE procedure, EBE-LU for number of iterations required by EBE-LU preconditioned problem, and Time for the execution time (in secs). Ndof denotes the number of degrees of freedom. '-' is used to denote cases where one algorithm converged to the non-linear tolerance specified earlier than the other and thus those non-linear iterations are redundant for the first algorithm (or in cases where there is no comparison possible). The degrees of freedom for the problem

Table 1. Kovaszany Flow Iterations to Convergence

| Index | Re | Np | EBE | Time(s) | EBE-LU | Time(s) |
|---|---|---|---|---|---|---|
| 1 | 40 | 16 | 422 | 386 | 337 | 334 |
| 2 | 40 | 16 | 1012 | 937 | 471 | 687 |
| 3 | 40 | 16 | 1079 | 1061 | 678 | 696 |
| 4 | 40 | 16 | 1143 | 1236 | 853 | 888 |
| 5 | 40 | 16 | 1015 | 1003 | 653 | 721 |
| 6 | 40 | 16 | 1059 | 1269 | 723 | 715 |
| 7 | 40 | 16 | 1028 | 1165 | 711 | 698 |
| 8 | 40 | 16 | 1082 | 1458 | 565 | 561 |
| 9 | 40 | 16 | 1041 | 1215 | 558 | 545 |
| 10 | 40 | 16 | 1019 | 985 | 468 | 459 |
| 11 | 40 | 16 | 1010 | 1025 | 381 | 372 |
| 12 | 40 | 16 | 988 | 939 | 319 | 314 |
| 13 | 40 | 16 | 1041 | 964 | 306 | 318 |
| 14 | 40 | 16 | 1138 | 1077 | 184 | 215 |

were set to 77763. The EBE-LU algorithm was found to be 2× faster (in execution time) than the EBE algorithm on 16 processors. In terms of iterations required for convergence the EBE required 1.953 more iterations to converge than the EBE-LU algorithm. The metrics presented also provide a measure of the compute times for the back-solve operations in the EBE-LU algorithm. The EBE-LU back solve operations were found to contribute a negligible amount of time to the total execution time to reach converged results.

To check the speedup in parallel for the EBE-LU preconditioned system we increased the number of processors from 16 through 120. Table 2 provides the parallel performance measurements for the Kovasznay flow problem. Here $p_{level}$ denotes the order of the spectral/*hp* polynomial used for analysis, and $N_{dof}$ denotes the number of degrees of freedom in the problem and speedup is

$$Speedup = t_{N_{p1}}/t_{N_{pi}} \tag{26}$$

where, $t_{N_{p1}}$ denotes the execution time on $N_{p1}$ processors, and $t_{N_{pi}}$ is the execution time on $N_{pi}$ processors, where $N_{p1} \leq N_{pi}$ (e.g. $N_{p1}$=16 and $N_{pi}$=60, 120).

Increasing the number of processors, there is a slight increase in the number of iterations to obtain converged results. This is because of the slight differences in numerics while performing back solve and factorization processes per block in each processor. Thus we measured the execution time (in seconds) for one iteration per processor for the problem. On 16 processors this value was found to be 0.0619. On higher processor counts of 60 and 120 the values ranged from 0.002768 through 0.00136. Varying the number of processors from 16 to 60 provided superlinear increase in performance due to

Table 2. Parallel LU performance

| Index | Re | $p_{level}$ | Ndof | Np | EBE-LU | Time(s) | Speedup |
|---|---|---|---|---|---|---|---|
| 1 | 40 | 5 | 103684 | 16 | 337 | 334 | - |
| 2 | 40 | 5 | 103684 | 60 | 271 | 45 | 7.4 |
| 3 | 40 | 5 | 103684 | 120 | 337 | 55 | 6 |

effective cache memory utilization. However varying the number of processors from 60 through 120 gave a speedup of 2.08 which is slightly higher than linear speedup. Verification of the algorithm was achieved using the $L_\infty$ norm of the difference of the numerical results from the analytical solutions for the problem. The $L_\infty$ norm was found to be 0.000407 and 0.000291 for the $u$, and $v$ components of the velocity and 0.0320 for the pressure, respectively for a polynomial order of 5 for the problem.

**6. Inexact Newton-Krylov Methods** Upon implementaton of the EBE-LU preconditioner for solving the incompressible Navier-Stokes equations, we further explore using non-linear solvers for both two-and three-dimensional incompressible flows. We consider both steady state and transient analysis. Let us consider the discrete non-linear problem denoted by:

$$F(x) = 0 \tag{27}$$

where $F : R^n \to R^n$ is a nonlinear mapping with the following properties:
(1) There exists an $x^* \in R^n$ with $F(x^*) = 0$.
(2) F is continuously differentiable in a neighborhood of $x^*$.
(3) $F'(x^*)$ is non singular.

In the particular case of solution of computational fluid dynamics problems, $\mathbf{x} = \{\mathbf{u}, \mathbf{p}\}$ denotes the vector of nodal variables for the velocity field and pressures over the domain of interest. For moderate to high Reynolds numbers the non-linear convective term is dominant and the solution procedures have to be able to accommodate the strong non-linearities.

We assume $F$ is continuously differentiable in $\Re^{n_{sd}}$ where $n_{sd}$ is the number of spatial dimensions. Let us denote the Jacobian matrix by $\mathbf{J} \in \Re^{n_{sd}}$. Newton's method for solving the non-linear problem is a classical algorithm and can be formulated as follows (Elias, 2004; Elias, 2006)

Newton Algorithm
for k=0 step 1 until convergence
solve $J(x_k)s_k = -F(x_k)$
set $x_{k+1} = x_k + s_k$

Newton's method was designed for solving problems where the initial guess is close to the non-linear solution. Among the disadvantages of Newton's method are the requirements that the linear system be solved exactly, to machine accuracy which can be computationally expensive. Computing an exact solution using a direct method can also be prohibitively expensive if the number of unknowns is large and may not be justified when $x_k$ is far from the solution. In spectral computations where there are high computational costs involved with quadratures, it is imperative to resort to efficient procedures for solving this discrete system of equations. In this case Inexact Newton methods are used to compute some approximate solution leading to the following algorithm

for k=0 step 1 until convergence
Find **some** $\eta_k \in [0, 1]$ and $s_k$ that satisfy
$||F(x_k) + F'(x_k)s_k|| \leq \eta_k||F(x_k)||$
set $x_{k+1} = x_k + s_k$

or a damped version of the final update to the new solution vector as: set $x_{k+1} = x_k + \alpha s_k$ where $\alpha$ is the damping parameter. For some $\eta_k$ where $|| \cdot ||$ is the norm of choice. This new formulation allows the use of an iterative linear algebra method, one first chooses $\eta_k$ and then applies the iterative solver to the algorithm until an $s_k$ is determined so the residual norm satisfies the convergence criterion. In this work $\eta_k$ is often called the forcing term, since its role is to force the residual of the above equation to be sufficiently small. In general, the non-linear forcing sequence needs to be specified to drive the solution toward the solution of the non-linear problem as

$$\frac{||r_k||}{||F(x_k)||} \leq \eta_k \tag{28}$$

In our implementation, we use the element- by-element (EBE) BiCGSTAB method to compute the $s_k$. There exists several choices for the evaluation of the forcing function $||\eta_k||$ as mentioned in Eisenstat et al. (1996). The forcing term is chosen to be a constant $\eta_k = 10^{-04}$. Other choices require an additional evaluation of the Jacobian, which is expensive in practical computations and in implementation procedures.

The derivations of the convergence behaviour for a generic non-linear problem have been outlined in Dembo et al. (1982). The basic requirement for the convergence of the iterative method is that $F$ is continuously differentiable in the neighbourhood of $x^* \in R^n$ for which $F(x^*)$ is non singular and that $F'$ is Lipschitz continuous at $x^*$ with constant $\lambda$, i.e.

$$||F'(x) - F'(x^*)|| \leq \lambda||x - x^*|| \tag{29}$$

The construction of the tangent stiffness matrix required for the algorithm follows standard procedures in finite element analysis. We follow through some of the developments for obtaining the tangent stiffness matrices for inexact Newton-Krylov methods. The non-linear tangent stiffness matrix terms are evaluated from the residual equations based on Taylor series approximations. Let us denote the residual of the linearised non-linear system by the following:

$$\{R\} = [A(x)]\{x\} - \{b\} \tag{30}$$

We expand the residual about the (known) solution at the $r^{th}$ iteration.

$$\{R\{x\}\} = \{R\{x\}^{r-1}\} + \left(\frac{\partial\{R\}}{\partial\{U\}}\right)^{r-1} \cdot \{\delta x\} + ... \tag{31}$$

Ordering the terms of second order and higher we obtain

$$\{R\{x\}^{r-1}\} = -\left(\frac{\partial\{R\}}{\partial\{U\}}\right)^{r-1} \cdot \{\delta x\} \tag{32}$$

since we seek to find the solution for which $\{R\{x\}\} = 0$, we obtain the following equation

$$\left(\frac{\partial\{R\}}{\partial\{U\}}\right)^{r-1} \cdot \{\delta x\} = -\{R\{x\}^{r-1}\} \tag{33}$$

and

$$[T(x)]^{r-1} \cdot \{\delta x\} = -\{R\{x\}^{r-1}\} \tag{34}$$

where the tangent stiffness matrix is defined as

$$[T(x)]^{r-1} = \left(\frac{\partial\{R\}}{\partial\{U\}}\right)^{r-1} \tag{35}$$

In the following sections we present the results obtained with the Inexact Newton-Krylov methods with the procedures described earlier for solving benchmark incompressible flow problems.

**Flow Past a 2-*D* Cylinder at Different Reynolds Numbers**

Table 3. Newton Krylov Performance Metrics

| Index | Re | Ndof | Newton | Inexact | Re | Ndof | Newton | Inexact |
|---|---|---|---|---|---|---|---|---|
| 1 | 400 | 121203 | 2095 | 106 | 40 | 132591 | 2993 | 1052 |
| 2 | 400 | 121203 | 7299 | 698 | 40 | 132591 | 69466 | 2541 |
| 3 | 400 | 121203 | 6191 | 1662 | 40 | 132591 | 47876 | 2729 |
| 4 | 400 | 121203 | 13497 | 2547 | 40 | 132591 | 90142 | 4495 |
| 5 | 400 | 121203 | 9462 | 4077 | 40 | 132591 | 65057 | 5032 |
| 6 | 400 | 121203 | 9948 | 2034 | 40 | 132591 | 157600 | 15337 |
| 7 | 400 | 121203 | - | 5523 | 40 | 132591 | 155933 | 32188 |
| 8 | 400 | 121203 | - | 2804 | 40 | 132591 | 156901 | 31713 |
| 9 | 400 | 121203 | - | 2436 | 40 | 132591 | - | 65505 |
| 10 | 400 | 121203 | - | 2436 | 40 | 132591 | - | 32962 |
| 11 | 400 | 121203 | - | 2436 | 40 | 132591 | - | 31214 |
| 12 | 400 | 121203 | - | 2436 | 40 | 132591 | - | 32962 |

After demonstration of the scalability and effectiveness of spectral/*hp* based implementations for solving incompressible flow problems, we extended the computational framework for the solutions of larger problems. We consider (as a second example) both steady and unsteady flow past a circular obstacle in a Newtonian flow field at a Reynolds number of 40 and 100 respectively. We consider a large computational domain of dimensions $[46 \times 41]$ for the steady state example. A circular cylinder with unit diameter obstructs the flowfield with the center located at $[15.5, 20.5]$. The presence of the cylinder was modeled as a porous domain with a low permeability. On account of the large computational domain the flow at the top, inlet and bottom are unperturbed from the presence of the cylinder and are prescribed to have a velocity of u-1.0 and v-0. A datum pressure of 0.0 was prescribed at the inlet face to anchor the pressure. The domain was discretized into $[38 \times 32]$ elements, with a $p_{level}$ of six in each element which resulted in a discrete system with a total of $132,591$ degrees of freedom. The length of the wake and drag obtained from the present results were in excellent agreement with benchmark results. The number of iterations to convergence for the flow past a cylinder problem has been presented in Table 3 with columns following the column *Re*-40. The number of processors was set to 32. The 'forcing term' $\eta_k$ for the inaccurate solve step was taken as a uniform value of $10^{-04}$ for the problem. The problem took 12 non-linear iterations to converge. However these iterations were considerably cheaper (3.4 times) compared to the exact Newton solution of the problem.

Next we consider the flow past a cylinder for unsteady state analysis at $Re = 100$. The computational domain was taken of dimensions $[28, 16]$. The center of the cylinder of diameter 1.0 was located at $[8, 8]$. Reynolds number was based on the diameter of the cylinder, viscosity of the fluid, and the inflow free stream velocity at the inlet. The contour plot of the $u$-component of the velocity field for unsteady analysis has been presented in Figure 3. Also the $v$-component of the velocity in the near vicinity of the cylinder has been presented in Figure 4.

**Three Dimensional Lid-driven Cavity Flow**

We further test the NK algorithm for solving a benchmark three-dimensional problem with spectral/*hp* element method. As the problem of interest, we choose the three-dimensional leaky lid-driven cavity flow problem. The domain comprises of a cavity of unit dimensions that is uniformly discretized into $14 \times 14 \times 14$ spectral elements with $p_{level}$ of three in each element. The side walls of the cavity are specified to have no-slip and no-penetration boundary conditions. The top wall is set to move in a distinct orthogonal direction (either $X$, $Y$ or $Z$) with a velocity of unity. The Reynolds number is based on the velocity of the moving lid, the viscosity of the fluid, and the length of an edge of the cavity. We consider the problem for two different Reynolds numbers of Re-100 and 400 respectively. The problem has been solved by Jiang et al. (1998) in three-dimensions and benchmark results have been provided. We provide the agreement between the present results and benchmark results of Jiang et al. (1998) in Figure 5. There is excellent agreement between the present results and reference results in literature for this problem for both Reynolds numbers.

Figure 6 presents the convergence in the $L_2$ norm of the residuals of the linearized problem for three-dimensional driven cavity flow for Reynolds number 400. The convergence history for two different linear steps (1 and 3) are provided. The forcing term for the linear convergence was taken as a value of 0.01 for Re-100 and $10^{-4}$ for Re-400 respectively. The linearized system took from six through 96 iterations to converge for the Re-100 problem with a total of $318,028$ degrees of freedom. For the Reynolds number of 400 it took 22 iterations to converge to the non-linear tolerance specified. Non-linear convergence was declared when the $L_2$ norm of the velocity reduced by a factor of $10^{-04}$ when compared to the $L_2$ norm of the initial residual. In comparison, the LSFEM technique utilized by Jiang et al. (1998) utilized a $50 \times 50 \times 50$

mesh with 5 degrees of freedom per element. This resulted in a total number of 625,000 degrees of freedom. The efficacy of the spectral discretization is clear from the dof count which was found to be half of that found with lower order methods.

**Fully Immersed Cylinder in 3-*D***

We consider the flow past a cylinder for three- dimensional analysis. Earlier example considered the flow past a two dimensional cylinder in a large computational domain. Such flows provide faithful representations of the flow metrics for the actual three-dimensional problem since the domain is large in size and the flow can be assumed to be symmetric in the third dimension. Further for large computational domains the end effects are negligible in flow computations. In other situations where the cylinder lies in a relatively confined domain there are considerable interactions of the perturbed flow past the cylinder with the boundary layers and ignoring the third dimension will provide inaccurate results.

To consider such three dimensional effects we examine flow past a three dimensional domain of dimensions $14 \times 12 \times 2$ at a Reynolds number of Re-200. The axial length of the cylinder is considered as 2 units. The Reynolds number was based on the diameter of the cylinder taken as unity and density and viscosity of the fluid. We placed the center of the cylinder at the location of $[5, 6, 0]$ with a domain height of 2 units. This places the cylinder symmetrically from the walls of the computational domain. A free stream velocity of $u = 1$ was specified on the inlet, front, and back face boundaries. No-penetration boundary condition was specified at the bottom, top, front, back, and left faces. The exit was assumed to be traction free $t_y = 0$. No velocity component in the third dimension was specified at all faces other than the exit face which was assumed to be traction free with $t_z = 0$. The exit face was specified to be traction free in all three components of $t_x = t_y = t_z = 0$. Pressure was specified as the datum pressure at the inlet to the computational domain of $p = 0$ at the left face.

The problem was discretized into $28 \times 28 \times 4$ mesh with a $p_{level} = 4$ in each element. This resulted in a discrete linear system with 868292 degrees of freedom. A space-time decoupled finite element formulation was utilized for solving the non-linear system in time. The inexact-newton krylov method was used to accelerate convergence of the problem to a relative linear tolerance of $10^{-05}$. The problem was simulated on 158 processors till an end time of 82. The time increment for the simulation was set at $\delta_t = 0.1$. This required solving the above linear system over 820 time steps for generating the simulation results. Figure 7 presents the iso-contour plots of the pressure at development of periodic steady state. Figure 8 presents the iso-contour plots of the $w$ velocity component at the same instant of time. The periodic fluctuating component of the lift for the three dimensional computations have been presented in Figure 9. From the fluctuating lift coefficient one derives the Strouhal number of $St = 0.169$. A two dimensional component of the simulation was obtained after prescribing the top and bottom walls of the domain to have free stream velocity of $u = 1$, and a no penetration velocity component of $v = 0$. The 2-*D* domain was considered of size $[14 \times 12]$ with the center of the cylinder located at $[5, 6]$. The value of the Strouhal number obtained from the two-dimensional simulation was found to be $St = 0.166$. The two dimensional simulation validated the development of the full three-dimensional flow field.

**Partially Immersed Cylinder in 3-*D***

Let us consider the flow past a partially immersed cylinder in $3D$ with Inexact Newton-Krylov methods. We consider a computational domain of dimensions $[14 \times 12 \times 4]$. The height of the cylinder was considered to be h=2 units from the bottom face of the domain. Reynolds number considered for analysis was taken as Re-200. Reynolds number was based on the diameter of the cylinder and the velocity of free stream of $u = 1$. Inlet to the computational domain was considered on the left face of the domain. Similar to the fully immersed cylinder analysis we placed the center of the cylinder at the location of $[5, 6, 0]$ with a domain height of 2 units. This places the cylinder symmetrically at the center from the walls of the computational domain. We define the immersion length of the cylinder in the domain as follows

$$L_{\mathfrak{I}} = \frac{h}{H} \tag{36}$$

Here, $h$ denotes the height of the cylinder, and $H$ denotes the total height of the computational domain. On the boundaries of the immersed cylinder in the free stream $\Gamma_h$ we prescribe values of the velocity components $(u.v.w) = \{0, 0, 0\}$. A free stream velocity of $u = 1$ was specified on the inlet, front, and back face boundaries. No-penetration boundary condition was specified at the bottom, top, front, back, and left faces. The exit was assumed to be traction free $t_y = 0$. No velocity component in the third dimension was specified at all faces other than the exit face which was assumed to be traction free with $t_z = 0$. The exit face was specified to be traction free in all three components of $t_x = t_y = t_z = 0$. Pressure was specified as the datum pressure at the inlet to the computational domain of $p = 0$ at the left face. The immersion length thus considered for analysis was determined as 0.5. We subject the three-dimensional problem for analysis. Fluctuations on the lift coefficient and the $z$-component of the force on the cylinder were metrics of interest.

We determine the coefficients with the help of integration of the forces on the sides of the cylinder. Let us denote the direction cosines of the inward normal with $(n_y, -n_x, n_z)$. Let $S$ denote the surface of the cylinder. The drag, lift and tangential forces were calculated with the above formulae

$$\mathbf{F} = \int_S \sigma \cdot \mathbf{n} d\mathbf{S} \tag{37}$$

The stress tensor is evaluated for Newtonian fluids as

$$\sigma = -p\mathbf{I} + 2\mu\left(\nabla\mathbf{u} + \nabla\mathbf{u}^T\right) \tag{38}$$

The force coefficients were evaluated with the help of the above formulae

$$c_d = \frac{2F_d}{\rho\bar{U}^2 D}$$
$$c_l = \frac{2F_l}{\rho\bar{U}^2 D}$$
$$c_z = \frac{2F_z}{\rho\bar{U}^2 D} \tag{39}$$

The diameter of the cylinder $D$ was taken as the characteristic dimension for determination of the forces, the velocity of the free stream was considered as $U$. Since the cylinder was considered partially immersed in the domain considerable (non-trivial) forces were found for each of the variables. The mean drag coefficient for the partially immersed cylinder was found to be 1.20. Fluctuations on the mean $c_z$ were found to be negligible. The mean $z$ component of the force coefficient was found to be 0.27. An over-lay plot of the $u$-contours have been presented in Figure 10. From the figure an envelope over the cylinder domain was clearly identified. A similar plot for the $v$-contours of the fluid flow have been presented in Figure 11. A trace of the temporal evolution of the mean lift coefficient with time has been presented in Figure 12. It is evident from the figure that the coefficient of lift with time reaches a periodic steady state by the time instant t=50. Periodic shedding from the bottom of the cylinder were found be developed and shedding was found to be periodic from the lift.

## 7. Conclusion

We provide two distinct procedures for the solutions of incompressible Navier-Stokes equations with stabilized formulations. The first method describes the development of an LU based preconditioner for solving spectral discretizations of Navier-Stokes equations. The second method involved development of an Inexact Newton-Krylov methods for solving incompressible flow problems. Spectral accuracy to known analytical solutions were verified. While inexact newton-krylov methods are expected to perform better with low order implementations of the problems solved, the element by element preconditioner developed is better suited for spectral implementations where higher order implementations provide the most suitable framework. We utilize the algorithms developed to solve an array of problems in CFD. Performance metrics for both algorithms were demonstrated with a scalability analysis performed on a range of processor counts varying from 12 through 160 processors. We provide a computational framework for solving problems which involve higher order discretizations in CFD computations.

## Acknowledgment

The authors appreciate the funding by National Science Foundation NSF/CREST HRD 0932339. Also, the authors would like to thank the Texas Advanced Computing Center (TACC) at the University of Texas at Austin for providing high performance computing facility, which has contributed to the research results reported in this paper.

## References

Barth, T., Chan, T., & Tang, W. (1998). A parallel non-overlapping domain-decomposition algorithm for compressible fluid flow problems on triangulated domains. *Contemporary Mathematics, 218,* 23-41.

Benzi, M., Golub, G. H., & Liesen, J. (2005). Numerical solution of saddle point problems. *Acta Numerica, 14,* 1-137. http://dx.doi.org/10.1017/S0962492904000212. URL http://journals.cambridge.org/articleS0962492904000212

Benzi, M., Golub, G. H., & Liesen, J. (2005). Numerical solution of saddle point problems. *Acta numerica, 14*(1), 1-137.

Benzi, M., & Tuma, M. (1999). A comparative study of sparse approximate inverse preconditioners. *Applied Numerical Mathematics, 30*(2), 305-340.

Brooks, A., & Hughes, T. (1982). Streamline upwind/petrov-galerkin formulations for convection dominated flows with particular emphasis on the incompressible navier-stokes equations. *Computer methods in applied mechanics and engineering, 32*(1), 199-259.

Brooks, A., & Hughes, T. J. (1980). Streamline upwind/petrov-galerkin methods for advection dominated flows. In: Third Internat. Conf. of Finite Element Methods in Fluid Flow, Banff, Canada.

Brooks, A. N., & Hughes, T. J. (1982). Streamline upwind/petrov-galerkin formulations for convection dominated flows with particular emphasis on the incompressible navier-stokes equations. *Computer methods in applied mechanics and engineering, 32*(1), 199-259.

Chorin, A. J. (1968). Numerical solution of the navier-stokes equations. *Mathematics of computation, 22*(104), 745-762.

Dembo, R., Eisenstat, S., & Steihaug, T. (1982). Inexact newton methods. *SIAM Journal on Numerical analysis, 19*(2), 400-408.

Eisenstat, S., & Walker, H. (1996). Choosing the forcing terms in an inexact newton method. *SIAM Journal on Scientific Computing, 17*(1), 16-32.

Elias, R., Coutinho, A., & Martins, M. (2004). Inexact newton-type methods for non-linear problems arising from the supg/pspg solution of steady incompressible navier-stokes equations. *Journal of the Brazilian Society of Mechanical Sciences and Engineering, 26*(3), 330-339.

Elias, R., Coutinho, A., & Martins, M. (2006). Inexact newton-type methods for the solution of steady incompressible viscoplastic flows with the supg/pspg finite element formulation. *Computer methods in applied mechanics and engineering, 195*(23), 3145-3167.

Elias, R., Martins, M., & Coutinho, A. (2006). Parallel edge-based solution of viscoplastic flows with the supg/pspg formulation. *Computational Mechanics, 38*(4), 365-381.

Gervasio, P., & Saleri, F. (1998). Stabilized spectral element approximation for the navierstokes equations. *Numerical Methods for Partial Differential Equations, 14*(1), 115-141.

Gustafsson, I., & Lindskog, G. (1986). A preconditioning technique based on element matrix factorizations. *Computer methods in applied mechanics and engineering, 55*(3), 201-220.

Hughes, T. J. (1995). Multiscale phenomena: Green's functions, the dirichlet-to-neumann formulation, subgrid scale models, bubbles and the origins of stabilized methods. *Computer methods in applied mechanics and engineering, 127*(1), 387-401.

Hughes, T. J., & Brooks, A. (2016). A multidimensional upwind scheme with no crosswind diffusion. *Finite element methods for convection dominated flows, AMD*, submitted.

Hughes, T. J., & Brooks, A. (1982). A theoretical framework for petrov-galerkin methods with discontinuous weighting functions: Application to the streamline-upwind procedure. *Finite elements in fluids, 4,* 47-65.

Hwang, F., & Cai, X. (2005). A parallel nonlinear additive schwarz preconditioned inexact newton algorithm for incompressible navierstokes equations. *Journal of Computational Physics, 204*(2), 666-691.

Jiang, B. (1998). The least-squares finite element method: theory and applications in computational fluid dynamics and electromagnetics. Springer

Kovasznay, L. (1948). Laminar ow behind a two-dimensional grid. In: *Proc. Camb. Philos. Soc, 44*, 58-62. Cambridge Univ Press.

Meister, A., & Vömel, C. (2001). Efficient preconditioning of linear systems arising from the discretization of hyperbolic conservation laws. *Advances in Computational Mathematics, 14*(1), 49-73.

Mittal, S., Kumar, V., & Raghuvanshi, A. (1997). Unsteady incompressible ows past two cylinders in tandem and staggered arrangements. *International Journal for Numerical Methods in Fluids, 25*(11), 1315-1344.

Naumov, M. (2011). Incomplete-lu and cholesky preconditioned iterative methods using cusparse and cublas. Tech. rep., Technical Report and White Paper.

Nour-Omid, B., & Parlett, B. (1985). Element preconditioning using splitting techniques. *SIAM journal on scientific and statistical computing, 6*(3), 761-770.

Quarteroni, A., Saleri, F., & Veneziani, A. (1999). Analysis of the yosida method for the incompressible navierstokes equations. *Journal de mathématiques pures et appliquées, 78*(5), 473-503.

Ranjan, R., & Feng, Y. (1979). Spectral/hp supg/pspg stabilized formulation for navier stokes equations. *International Journal of Computational Fluid Dynamics, 34*, 19-35.

Ranjan, R., & Reddy, J. (2012). On multigrid methods for the solution of least-squares finite element models for viscous flows. *International Journal of Computational Fluid Dynamics, 26*(1), 45-65.

Reddy, J. N., & Gartling, D. K. (2010). The finite element method in heat transfer and fluid dynamics. CRC press

Sampath, R., & Zabaras, N. (2001). Numerical study of convection in the directional solidification of a binary alloy driven by the combined action of buoyancy, surface tension, and electromagnetic forces. *Journal of Computational Physics, 168*(2), 384-411.

Shadid, J., Tuminaro, R., & Walker, H. (1997). An inexact newton method for fully coupled solution of the navier stokes equations with heat and mass transport. *Journal of Computational Physics, 137*(1), 155-185.

de Sturler, E., & Liesen, J. (2005). Block-diagonal and constraint preconditioners for nonsymmetric indefinite linear systems. part i: Theory. *SIAM Journal on Scientific Computing, 26*(5), 1598-1619.

Tezduyar, T. (1992). Stabilized finite element formulations for incompressible ow computations. *Advances in applied mechanics, 28*(1), 1-44.

Vannieuwenhoven, N., & Meerbergen, K. (2010). Imf: An incomplete multifrontal lu-factorization for elementstructured sparse linear systems. TW Reports.

Wang, K., Lawlor, O., & Kale, L. (2005). The nonsingularity of sparse approximate inverse preconditioning and its performance based on processor virtualization. Tech. rep., Technical Report, Department of Computer Science, University of Illinois at Urbana-Champaign.



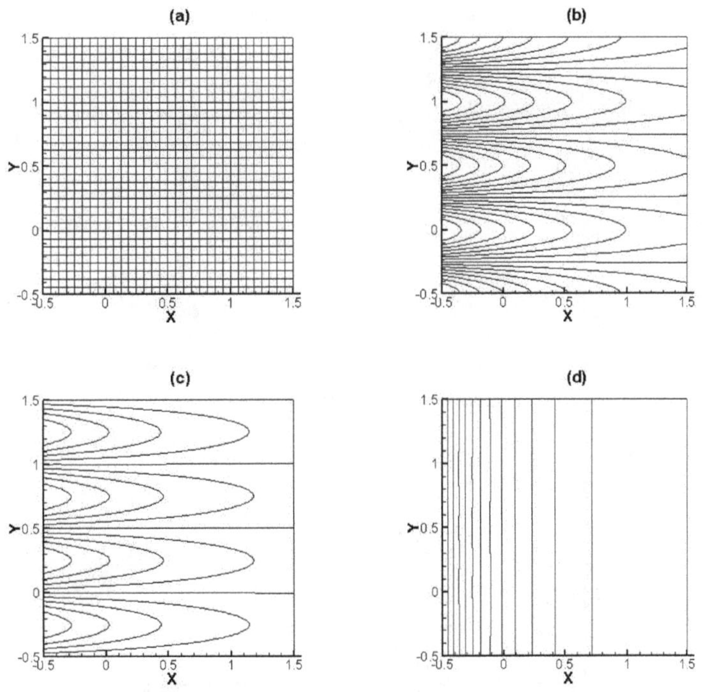


Figure 1. Kovaszany Flow Re-40

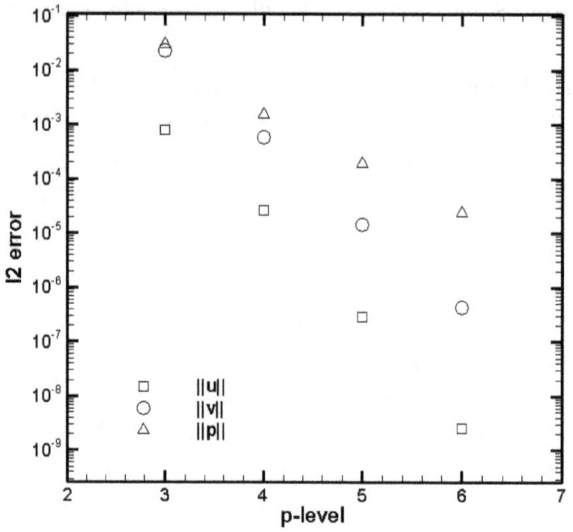


Figure 2. $L_2$ error Kovaszany Flow Re-40

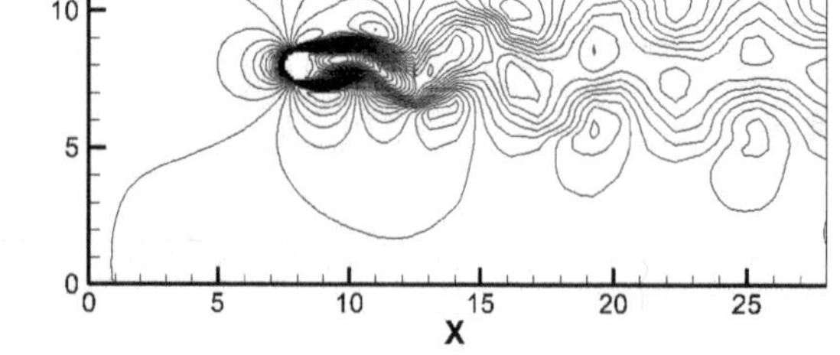

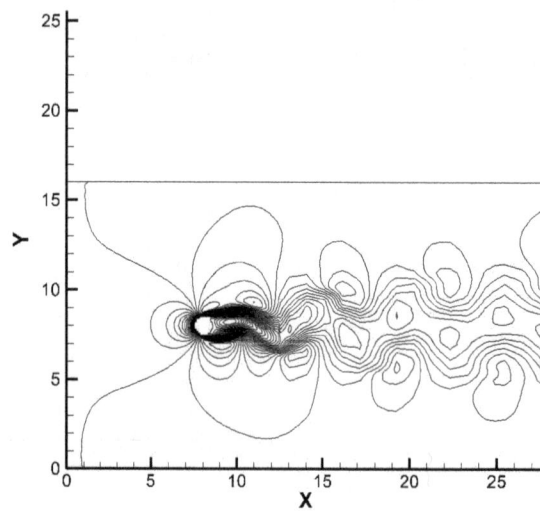

Figure 3. $U$ contour plot flow past a cylinder Re-100

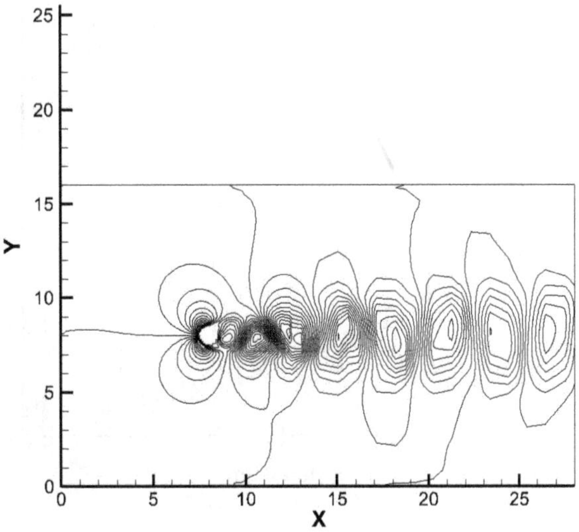


Figure 4. *V* contour plot flow past a cylinder Re-100

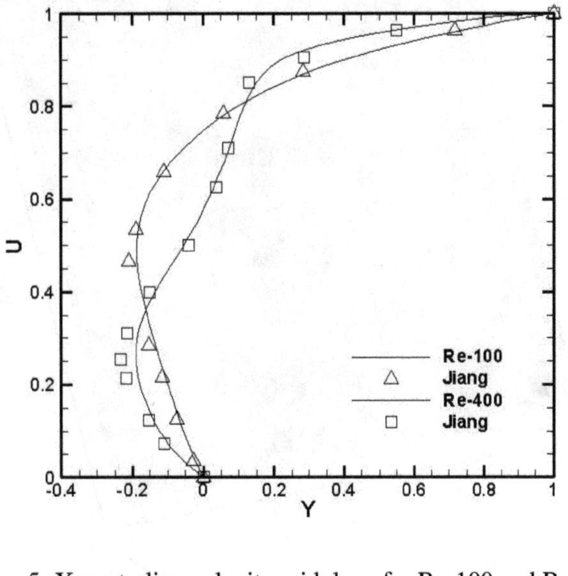


Figure 5. X-centerline velocity midplane for Re-100 and Re-400

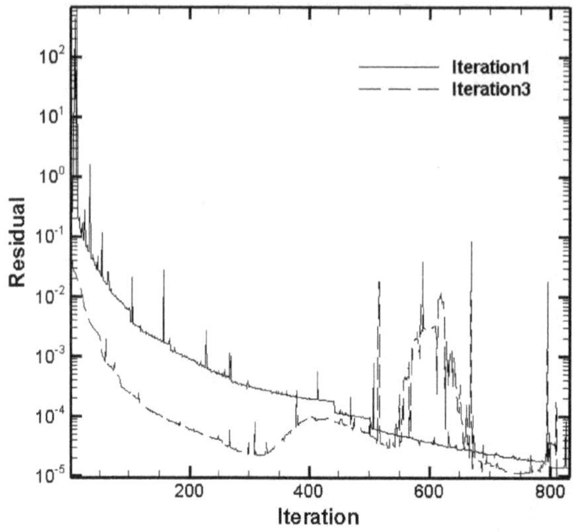


Figure 6. Linear Iterations for convergence for 3-d cavity Re 400

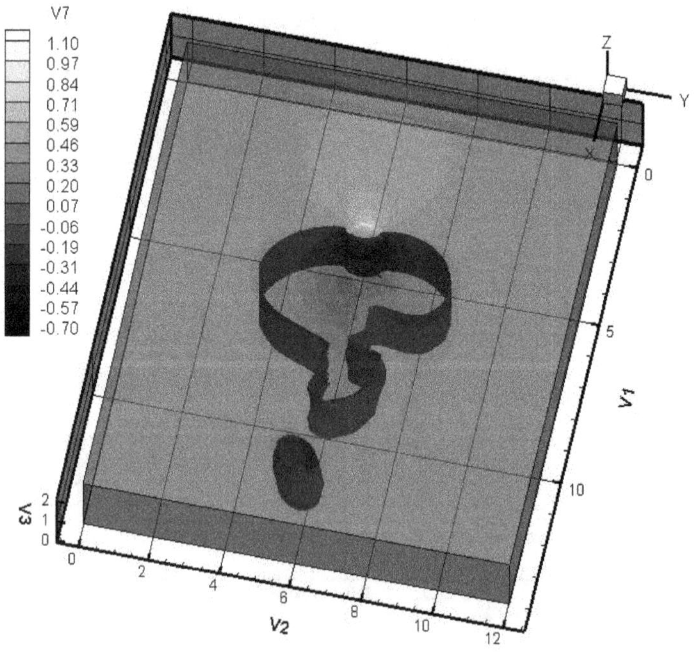


Figure 7. Pressure iso-contours Re-200

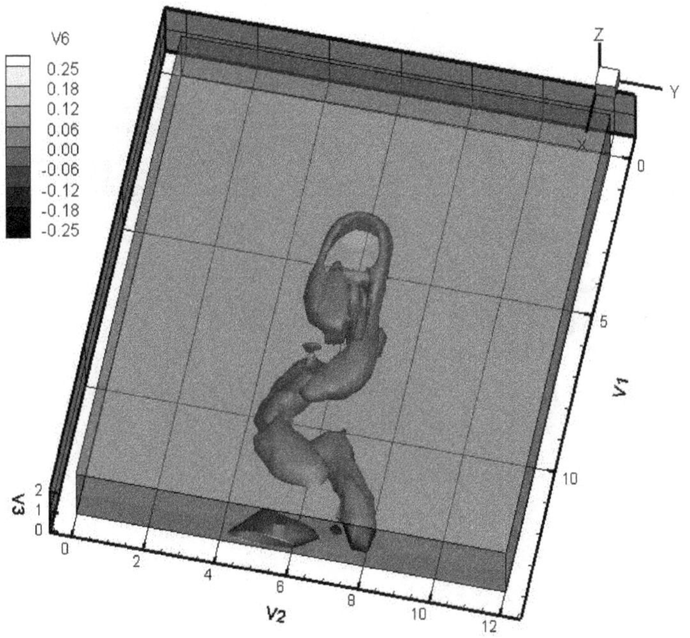


Figure 8. w-component of velocity Re-200

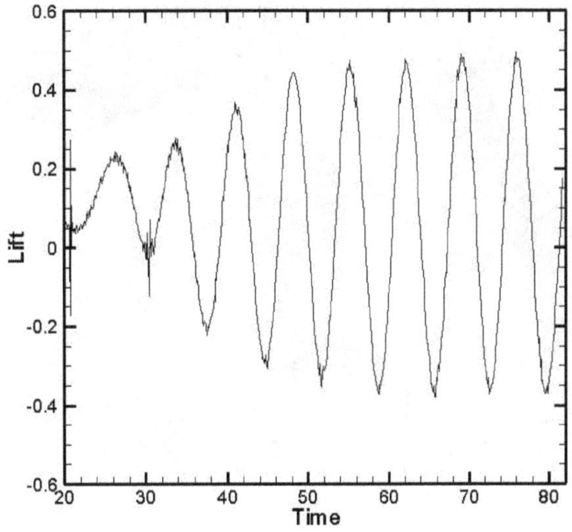


Figure 9. Fluctuating component of lift Re-200

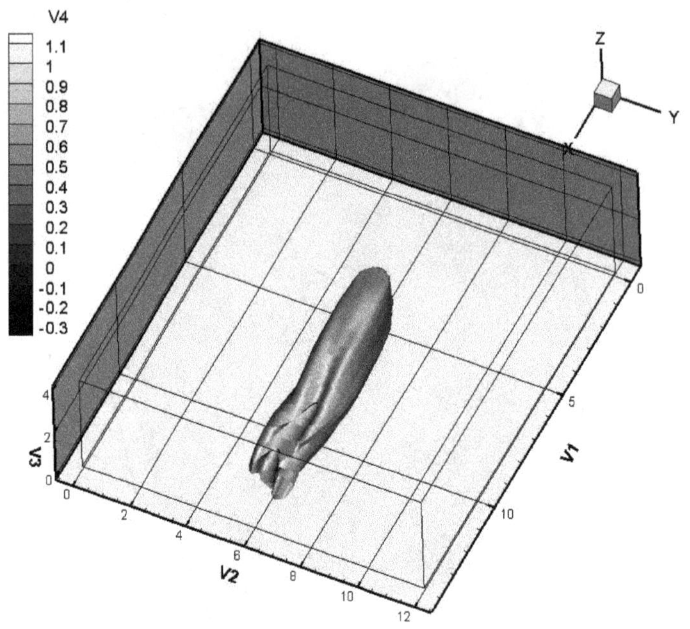


Figure 10. *u*-component of velocity Re-200

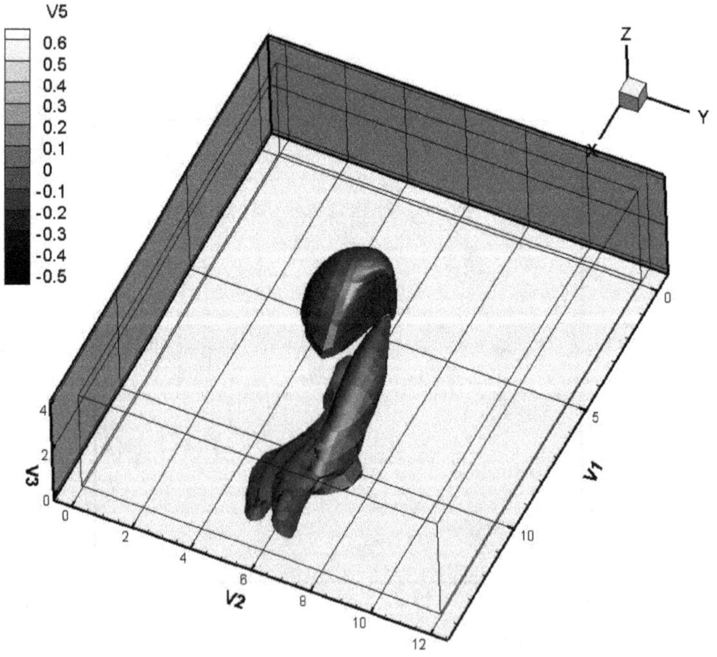


Figure 11. *v*-component of velocity Re-200

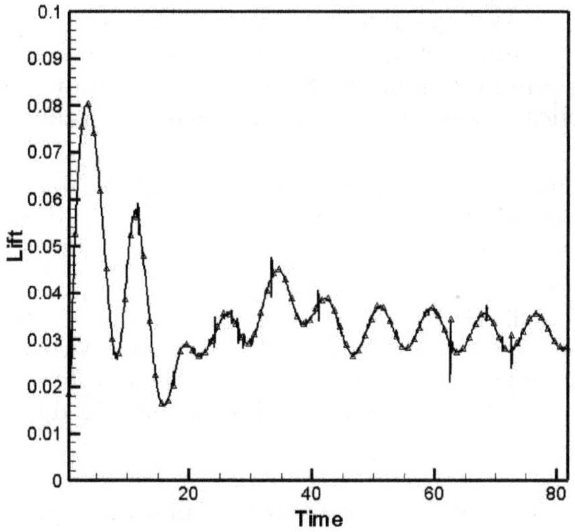


Figure 12. Lift coefficient half immersed cylinder Re-200

# Permissions

All chapters in this book were first published in JMR, by Canadian Center of Science and Education; hereby published with permission under the Creative Commons Attribution License or equivalent. Every chapter published in this book has been scrutinized by our experts. Their significance has been extensively debated. The topics covered herein carry significant findings which will fuel the growth of the discipline. They may even be implemented as practical applications or may be referred to as a beginning point for another development.

The contributors of this book come from diverse backgrounds, making this book a truly international effort. This book will bring forth new frontiers with its revolutionizing research information and detailed analysis of the nascent developments around the world.

We would like to thank all the contributing authors for lending their expertise to make the book truly unique. They have played a crucial role in the development of this book. Without their invaluable contributions this book wouldn't have been possible. They have made vital efforts to compile up to date information on the varied aspects of this subject to make this book a valuable addition to the collection of many professionals and students.

This book was conceptualized with the vision of imparting up-to-date information and advanced data in this field. To ensure the same, a matchless editorial board was set up. Every individual on the board went through rigorous rounds of assessment to prove their worth. After which they invested a large part of their time researching and compiling the most relevant data for our readers.

The editorial board has been involved in producing this book since its inception. They have spent rigorous hours researching and exploring the diverse topics which have resulted in the successful publishing of this book. They have passed on their knowledge of decades through this book. To expedite this challenging task, the publisher supported the team at every step. A small team of assistant editors was also appointed to further simplify the editing procedure and attain best results for the readers.

Apart from the editorial board, the designing team has also invested a significant amount of their time in understanding the subject and creating the most relevant covers. They scrutinized every image to scout for the most suitable representation of the subject and create an appropriate cover for the book.

The publishing team has been an ardent support to the editorial, designing and production team. Their endless efforts to recruit the best for this project, has resulted in the accomplishment of this book. They are a veteran in the field of academics and their pool of knowledge is as vast as their experience in printing. Their expertise and guidance has proved useful at every step. Their uncompromising quality standards have made this book an exceptional effort. Their encouragement from time to time has been an inspiration for everyone.

The publisher and the editorial board hope that this book will prove to be a valuable piece of knowledge for researchers, students, practitioners and scholars across the globe.

# List of Contributors

**Joseph Dongho**
Department of Mathematics and Computer Science, University of Maroua, Cameroon

**Alexander P. Buslaev**
Moscow Automobile and Road State Technical University, Moscow, Russia

**Alexander G. Tatashev**
Moscow Technical University of Communications and Informatics, Moscow, Russia

**Rakesh Ranjan and Yusheng Feng**
Center for Simulation, Visualization and Real-Time Prediction (SiViRT)
Department of Mechanical Engineering, University of Texas, San Antonio, USA

**Anthony Theodore Chronopoulos**
Center for Simulation, Visualization and Real-Time Prediction (SiViRT)
Department of Computer Science, University of Texas, San Antonio, USA

**S.S. Sandhya**
Department of Mathematics, Sree Ayyappa College for Women, Chunkankadai, India

**E.Ebin Raja Merly**
Departmentof Mathematics, Nesamony Memorial Christian College, Marthandam, India

**S.D.Deepa**
Research Scholar, Nesamony Memorial Christian College, Marthandam, India

**Alexander P. Buslaev and Alexander G. Tatashev**
Moscow Automobile and Road State Technical University, Moscow, Russia
Moscow Technical University of Communications and Informatics, Moscow, Russia

**Luis Teia**
Berlin, Germany

**Bienvenu Ondami**
Univesrité Marien Ngouabi, Brazzaville, Congo

**Diakarya Barro**
Université Ouaga II BP: 417 Ouagadougou 12, Burkina Faso

**Alexander P. Buslaev and Alexander G. Tatashev**
Moscow Automobile and Road State Technical University, Moscow, Russia
Moscow Technical University of Communications and Informatics, Moscow, Russia

**Mouhamadou A.M.T. Baldé and Babacar M. Ndiaye**
Laboratory of Mathematics of Decision and Numerical Analysis, LMDAN-FASEG, University of Cheikh Anta Diop, BP 45087 Dakar-Fann, 10700 Dakar, Senegal

**Labelling S.S. Sandhya**
Department of Mathematics, Sree Ayyappa College for Women, Chunkankadai-629 003,Tamilnadu, India

**E.Ebin Raja Merly**
Department of Mathematics, Nesamony Memorial Christian College, Marthandam-629 165,Tamilnadu, India

**G.D.Jemi**
Department of Mathematics, Narayana guru College of Engineering, Manjalumoodu-629 151,Tamilnadu, India

**Modou Ngom**
LERSTAD, Universit´e Gaston Berger de Saint-Louis, SENEGAL

**Gane Samb Lo**
LERSTAD, Universit´e Gaston Berger de Saint-Louis, SENEGAL and LSTA, Universit´e Pierre et Marie Curie, France (affiliated), gane-samb

**Lina Zhou**
College of Mathematics and Information Science, Hebei Normal University, Shijiazhuang 050024 Hebei, China

**Weihua Jiang**
College of Science,Hebei University of Science and Technology, Shijiazhuang 050018, Hebei, China

**Liang Kong**
Department of Mathematical Sciences, University of Illinois at Springfield, Springfield, Illinois

**Ali Musaddak Delphi**
Department of Mathematics, College of Basic education, University of Misan, Iraq

**S. T. Aleskerova**
Institute of Mathematics and Mechanics of ANAS, Baku, Azerbaijan

**Li Yang**
College of Mathematics and Information, China West Normal University, China

# Index

www.ingramcontent.com/pod-product-compliance
Lightning Source LLC
Chambersburg PA
CBHW070153240326
41458CB00126B/4507
* 9 7 8 1 6 8 2 8 5 4 4 9 5 *